New Horizons in Mobile and Wireless Communications

Volume 4
Ad Hoc Networks and PANs

For a list of recent related titles in the *Artech House Universal Personal Communications Series*, please turn to the back of this book.

New Horizons in Mobile and Wireless Communications

Volume 4
Ad Hoc Networks and PANs

Ramjee Prasad
Albena Mihovska

Editors

ARTECH HOUSE

BOSTON | LONDON

artechhouse.com

Library of Congress Cataloging-in-Publication Data
A catalog record for this book is available from the U.S. Library of Congress.

British Library Cataloguing in Publication Data
A catalog record for this book is available from the British Library.

ISBN-13: 978-1-60783-973-6

Cover design by Igor Valdman

© 2009 ARTECH HOUSE
685 Canton Street
Norwood, MA 02062

10 9 8 7 6 5 4 3 2 1

Contents

Acknowledgments

First of all, the editors would like to acknowledge Ms. Dua Idris from CTIF at Aalborg University for her big effort towards the completion of this book. Dua was involved in the FP6 IST project SIDEMIRROR as a research engineer. Her consistent hard work and passionate efforts toward improving the quality of the content are very commendable.

Further, this book would not have been possible without the strong support of the European Union project officers from the Directorate General Information Society (DG INFSO) of the European Commission in Brussels, namely, Dr. Jorge M. Pereira, Dr. Manuel Monteiro, and Dr. Francisco Guirao Moya, who guided the project work and ensured strong cooperation with the rest of the FP6 IST projects in the area. Their effort was an essential prerequisite for the successful editing of the available research results. The material in this book was collected and structured with the support of the European Commission within the framework of the IST supporting project SIDEMIRROR, in an effort that spanned a period of five years. It originated from the technical reports and documentation of the projects involved in the FP6 IST R&D initiatives *Mobile and Wireless Systems Beyond 3G* and *Broadband for All.*

The project SIDEMIRROR consisted of two member organizations, Aalborg University and Artech House Publishers. The editors would like to acknowledge the support of their colleagues from Artech House toward the completion of this manuscript. Finally, the editors would like to thank the administrative and IT staff from Aalborg University for providing the required administrative and IT project support.

Ramjee Prasad
Albena Mihovska
Aalborg, Denmark
September 2009

CHAPTER 1

Introduction

Ad hoc networks are important enablers for next generation communications. Generally, such networks can be formed and deconfigured on-the fly and do not require a particular underlying infrastructure (i.e., there is no need for a fixed base station, routers, or cables). Ad hoc networks can be mobile, stand-alone or internetworked with other networks.

A *wireless* ad hoc network is a collection of autonomous nodes or terminals that communicate with each other by forming a multihop radio network and maintaining connectivity in a decentralized manner. Just as any radio network, wireless ad hoc networks, suffer from issues such as noise, interference, fading, insufficient bandwidth, and so forth.

Ad hoc networks are at the basis of concepts such as reconfigurability, self-organization, pervasiveness, ubiquity, dynamics, and user-centricity, which have been recognized as the main characteristics of next generation communication systems. Therefore, research in the area of ad hoc networking is receiving much attention from academia, industry, and government. These networks pose many complex issues and many open problems.

The European-funded research under the umbrella of the Framework Program Six (FP6) provided significant contributions to the advancement of the ad hoc network concepts and standards [1, 2]. In particular, it contributed to the developments in the area of sensor networks, personal networks and federations, wireless mesh networks, multihop communications and their integration with cellular systems and the backbone infrastructure, reconfigurability, routing, and some others.

This book provides a comprehensive overview of the main contributions realized by the FP6 EU-funded projects in the area of Information Society Technology (IST) and under the themes *Mobile and Wireless Systems Beyond 3G* and *Broadband for All*. This chapter introduces into the topic of ad hoc networks. In particular, Section 1.1 gives an overview of the current state of the art in the area, and describes the main types of ad hoc networks and their characteristics. Section 1.2 gives an overview of the standardization and regulation activities in the area. Section 1.3 provides and overview of the book.

1.1 Wireless Ad Hoc Networks

A wireless ad hoc network can be understood as a computer network with wireless communication links where each node has the capacity to forward the data to other nodes. The decision for determining what nodes are to forward the data and to whom are made dynamically based on the connectivity in the concerned network. This feature makes it different from the wired network technologies where designated nodes with custom hardware are forwarding the data or from the managed wireless networks where a special node is assigned to manage the communication between other nodes. The main features of the ad hoc networks are their minimum requirements for configuration and quick deployment.

Each node in a wireless ad hoc network functions as both a host and a router, and the control of the network is distributed among the nodes. The network topology is in general dynamic, because the connectivity among the nodes may vary with time due to node departures, new node arrivals, and the possibility of having mobile nodes. Hence, there is a need for efficient routing protocols to allow the nodes to communicate over multihop paths consisting of possibly several links in a way that does not use an excess of network resources.

There are two main types of ad hoc networks, *mobile ad hoc networks* (MANETs) and *wireless sensor networks* (WSNs).

1.1.1 Mobile Ad Hoc Networks (MANETs)

A mobile ad hoc network (MANET) is a self-configuring network consisting of mobile routers that are interconnected by (usually bandwidth-constrained) wireless links. The vision of MANETs is to support robust and efficient operation in mobile wireless networks by incorporating a routing functionality into the mobile nodes. Such networks are envisioned to have dynamic, sometimes rapidly-changing, random, multihop topologies, which are likely composed of relatively bandwidth-constrained wireless links. The network is decentralized and all network activity including the discovery of the topology and the delivery of messages is executed by the mobile nodes themselves.

MANETs can range from small, static networks that are constrained by power sources, to large-scale, mobile, highly dynamic networks. The design of network protocols for these networks is a complex issue [3]. Regardless of the application, MANETs need efficient distributed algorithms to determine network organization, link scheduling, and routing. The determination of viable routing paths and delivering messages in a decentralized environment where network topology fluctuates is not a well-defined problem. While the shortest path (based on a given cost function) from a source to a destination in a static network is usually the optimal route, this idea is not easily extended to MANETs. Factors such as variable wireless link quality, propagation path loss, fading, multiuser interference, power expended, and topological changes, become relevant issues. The network should be able to adaptively alter the routing paths to alleviate any of these effects. Preservation of security, latency, reliability, intentional jamming, and recovery from failure are significant concerns, especially when MANETs are used in the context of defense or dual-use applications [4]. A lapse in any of these requirements may degrade the performance and dependability of the network.

Mobile ad hoc devices are able to detect the presence of other ad hoc devices, establish communication links among each other, and communicate information such as packetized digital data. The traditional work around MANETs had largely focused on homogeneous nodes and single parameter optimization [5]. Within the FP6 IST program, the projects MAGNET and MAGNET Beyond [5] extended these concepts in support of PN technologies.

1.1.1.1   Self-Organization Capabilities

Current radio technologies offer, up to a certain extent, self-organizational capabilities at the link layer: 802.11 provides link-level self-organization, Bluetooth networks organize themselves by forming piconets or even scatternets. Self-organization at the network layer is also receiving a lot of attention in the context of MANETs, in which nodes need to cooperate to organize themselves and to provide network functionality, due to the absence of any fixed infrastructure.

In the context of PNs [5], the problem has a completely different dimension, because self-organization spans over multiple network technologies and strongly builds on the concept of trust between the *personal nodes* and *devices*. Security, privacy, and trust solutions were another research challenge undertaken within the frames of the FP6 IST projects.

A PN should be self-organizing and self-maintaining, handling mobility and thereby providing its own addressing and routing mechanisms for its internal communication (i.e., support of dynamic behavior). Therefore, developing PN networking solutions can build to a certain extent on ad hoc networking techniques and concepts. Due to the specific nature and context of PNs, existing solutions for mobile ad hoc networks cannot be adopted directly. A PN has a specific wireless/wired geographically dispersed network topology, which, to a certain extent, can rely on the fixed infrastructure (e.g., edge routers, for providing networking solutions). Also, PNs are built around a specific trust relation concept, on which higher layer protocols can rely, which is absent in traditional ad hoc networks. Therefore, the overall PN architecture proposed in the IST project MAGNET was quite novel [5].

1.1.1.2   Mobility

The field of mobile ad hoc networks has seen a rapid expansion due to the proliferation of wireless devices, witnessed by the efforts in the IETF MANET working group [6]. A lot of attention was given to the development of routing protocols, with the MANET group working on the standardization of a general reactive and proactive routing protocol, and, in a lesser extent, to addressing and Internet connectivity [7]. Mobility of individual devices, mobility of complete clusters and splitting and merging of clusters (especially in the context of PNs), requires efficient mobility solutions. Worth mentioning in this context are the activities on mobile networks within the Mobile IP (MIP) Working Group [8] of the IETF, the work on the extensions of MIP for mobile ad hoc networks interconnection [9], and the work within the Network Mobility (NEMO) working group that is concerned with the mobility of an entire network. People are beginning to carry multiple Internet-enabled devices such as cell phones, PDAs, laptop computers, and music players. Instead of each device connecting to the Internet separate, all of the devices could connect to the Internet through a PAN. Using NEMO, one device, such as a cell phone, would act as the mobile router providing continuous access to the rest of the devices. Mobility solutions for PNs can borrow from this work, but should be adapted to fit the underlying PN architecture and addressing schemes.

The NEMO architecture was adopted in the FP6 IST project DAIDALOS to allow a network to move as a whole [10]. A single point, the *mobile router* (MR), manages the mobility of the network.

Initially, the DAIDALOS architecture was designed and developed as a stand-alone subsystem, supporting the following three main functionalities:

- Network mobility basic support, by implementing partially the NEMO basic support protocol [11];
- Route optimization support for local fixed nodes and nested configurations [12];
- Multicast support for mobile networks [13].

Integration efforts are required to develop new solutions that make possible the provision of QoS and security to mobile networks, as well as the efficient support of mobile nodes roaming between moving networks and infrastructure-based access networks [14]. In addition to these features, the NEMO architecture was enhanced by means of improved route optimization solutions (supporting also VisitingMobileNodes), by enriched multicast mechanisms, and the support for localized mobility solutions (based on an extension of the *network-based localized mobility management*—NetLMM protocol) [15].

The MANETs in DAIDALOS II can be seen as multihop networks connected to the core network by means of one or more gateways. Because the access clouds could be considered as local mobility domains, the integration of the MANET within the overall architecture required the analysis of the interaction between these networks with the local mobility management protocol. Such interactions depend on the number of gateways supported and their location, in the same or different local domains. This, in turn, has impact on the ad hoc nodes address configuration and on the mobility management.

The concept of multihoming can also be applied to MANETs where multiple egress/ingress points (gateways) are considered. In the case of multiple gateways inside the same local domain, or in different local domains, the implementation and integration of MANETs (e.g., the DAIDALOS implementation) also provides QoS and security support. Inside the MANET the solutions can be specifically built according to the MANET unstable and dynamic characteristics; however, once these are integrated with the infrastructure, they are not visible to the outside of the MANET. The integration of the MANET concept in the overall DAIDALOS architecture is shown in Figure 1.1.

Figure 1.1 includes the basic network components and protocols, which build-up a mobility and QoS integrated architecture, including the MANET and NEMO networks, and the support of multicast services [15]. The various network devices exchange differing control information depending on the service that is being delivered.

The requirements taken into account for the design of the architecture are the following:

- Access network operators can implement their own mobility solution (within their domains). The solution must be independent of external mobility operators (including home);
- Minimize complexity in the terminal;
- Efficient use of wireless resources;

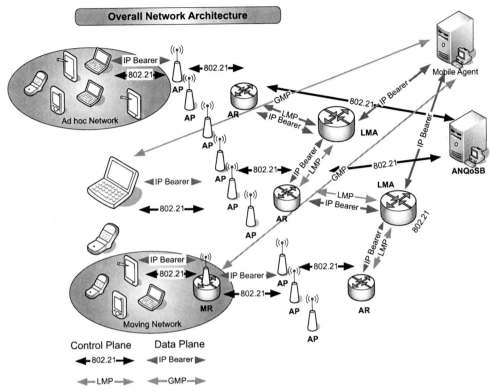

Figure 1.1   Integration of the MANET and NEMO concepts in the DAIDALOS architecture [15].

- Reduced signaling overhead in the network;
- The solution must be security friendly;
- Seamless handover support;
- Multihoming support;
- Scalability for routing;
- Minimize network side nodes modifications;
- Support for heterogeneous networking;
- QoS provision.

1.1.1.3   Security, Privacy, and Trust for Ad Hoc Networks

Easy, secure, and simple establishment of connections between devices and networks is the key to support data exchange in various scenarios. Current solutions and protocols are very heavy and not user friendly for secure access to network, services, and devices, which makes these unsuitable for use in MANET0 scenarios. Routing is one of the main processes on the networking abstraction level, which is responsible for the finding and establishment of the routes among the communicating nodes. Current ad hoc routing protocols inherently trust all participants. Most ad hoc routing protocols are cooperative by nature and depend on neighboring nodes to route packets. Such a naive trust model allows malicious nodes to paralyze an ad hoc network by inserting erroneous routing updates, replaying old messages, changing routing updates or advertising incorrect routing information. None of the protocols such as the *Ad Hoc On Demand Distance Vector Routing* (AODV), *Dynamic Source Routing* (DSR), Ariadne, the *Authenticated Routing for Ad Hoc Networks* (ARAN), the *Security Aware Routing* (SAR), or the *Server Routing Protocol* (SRP), provide a solution to the requirements of certain discovery, isolation, or robustness. The routing process must be shielded by solutions that will grant the integrity and the availability of the networking procedures [16].

The capability to provide secure context transfer is essential in achieving fast performance in a wireless environment. Secure fast context transfer in handovers between heterogeneous access technologies/network types is needed. Furthermore, providing context-aware, adaptive and personalized services to the user, poses many opportunities, challenges, and risks. The ability to offer secure, intuitive, and easy-to-use solutions for accessing contextual services that have to be location-aware and time-sensitive; personal preference and network bandwidth aware, and finally, device-capability aware, is a big challenge. In addition to routing and context transfer, key management is a fundamental prerequisite for any cryptography-based security solution. For example, in personal area networks (PANs) most personal nodes are low-capacity devices, therefore, the security protocols in both lower and upper layers, must rely on optimized cryptographic algorithms, which consume less resources but hence provide the necessary security features.

In the future ad hoc network, trust, identity management, and privacy will need considerable effort if end-to-end security is required. Thus, a mechanism of enabling extension of the trust between individual nodes needs to be defined. Also, protection of user location, identity, and privacy need to be considered. The user's location, identity and privacy requirements must be taken into account by the mobility procedures. The precise nature of these requirements may have a considerable impact on the mobility procedures.

1.1.1.4 Vehicular Ad Hoc Networks (VANETs)

The *vehicular ad hoc network* (VANET) is a form of MANET, which is used for communication among vehicles and between vehicles and roadside equipment. The VANET formed by a number of nearby peers is organized in a fully decentralized fashion, in order to cope with continuous arrival and disappearance of peers.

The 802.11p [17] amendment of the IEEE 802.11 family of standards is currently drafted specifically for such vehicular ad hoc communications. However, IEEE 802.11p is predicted to be limited in situations with a high peer density, compromising reliable transmission of time-critical safety related messages. Safety improving messages may also arrive from sensors installed along roads through a low data rate air interface based on, for example, IEEE 802.15.4. Access via roadside units that are part of some VANETs are a further possibility. Additionally, some peers may be part of a *wireless personal area network* (WPAN) based on e.g., Bluetooth. Besides, some peers might have access to the infrastructure. Possible access networks are UMTS, WLANs available in hot spots, or WiMAX.

The *intelligent vehicular ad hoc network* (InVANET) makes use of WiFi IEEE 802.11b/802.11g/802.11p and WiMAX IEEE 802.16 for providing easy, accurate, effective communication between multiple vehicles on dynamic mobility.

For efficient support of the VANET applications, new communications protocols have to be developed and standardized. These protocols concern all layers from the physical to the application layer and they are expected to provide both vehicle-to-vehicle and vehicle-to infrastructure communications. For the efficient development and deployment of the vehicular communication systems, many aspects of the radio channel in the 5-GHz band are yet to be understood. Especially, the time-selective fading statistics of vehicle-to-vehicle communication links differ substantially from those of cellular communication networks. Such statistics must be investigated in the 5-GHz band. Furthermore, the characteristics of propagation channels for *multiple input multiple output* (MIMO) systems are to be defined, even though MIMO is gaining more and more importance in the IEEE 802.11 standard family. Use MIMO or smart antenna systems can provide better radio spectrum use, especially in dense urban environments, in which the density of traffic peers is enormous. The beamforming capabilities of smart antennas systems could help to transmit the information in determined directions. For example, at a high-speed motorway, the car could transmit the information only in cars in the same direction of the road. An example is shown in Figure 1.2.

Security is an indispensable prerequisite for advanced vehicle applications. The system must ensure that the transmission comes from a trusted source and that the transmission has not been altered. As far as security is concerned, it is essential to make sure that critical information cannot be inserted or modified by an attacker. However, most security mechanisms end up consuming too many system resources.

Figure 1.2   Use of smart antenna systems in VANETs.

This might seriously degrade the system capabilities in terms of latency and channel capacity. Privacy is another major challenge. The vehicle safety communication applications might broadcast messages about a vehicle's current location, speed, and direction. It is important that the privacy is maintained and the identities of the users are not revealed. On the other hand, to ensure accountability, messages need to be uniquely signed. However, the unique signatures will allow the signer to be tracked and eventually reveal his true identity.

1.1.2 Wireless Sensor Networks (WSNs)

A *wireless sensor network* (WSN) is a wireless network consisting of spatially distributed autonomous devices using sensors to cooperatively monitor physical or environmental conditions, such as temperature, sound, vibration, pressure, motion, or pollutants, at a variety of locations [19]. In WSNs, sensors with wireless communication capability and some level of intelligence for signal processing and networking of the data are spread across a geographical area. WSNs can have different applications, from military sensor networks that detect and gain as much information as possible about enemy movements, explosions, and other phenomena of interest; to sensor networks that are used to detect and characterize *chemical, biological, radiological, nuclear, and explosive* (CBRNE) attacks and materials; to sensor networks that are used to detect and monitor environmental changes in plains, forests, oceans. As already mentioned, wireless traffic sensor networks can be used to monitor vehicle traffic on highways or in congested parts of the city; wireless surveillance sensor networks can be used to provide security in shopping malls, parking garages, and other facilities; wireless parking lot sensor networks that help to determine the degree of occupancy, and so forth.

1.1.2.1 Requirements for WSNs

The following requirements can be summarized for WSNs:

- *A large number of (mostly stationary) sensors*: Aside from the deployment of sensors on the ocean surface or the use of mobile, unmanned, robotic sensors in military operations, most nodes in a smart sensor network are stationary. Scalability could be a major issue.
- *Low energy use*: Since in many applications the sensor nodes will be placed in a remote area, service of a node may not be possible. In such a case, the lifetime of a node may be determined by the battery life, thereby requiring the minimization of energy expenditure.
- *Self-organization*: Given the large number of nodes and their potential placement in hostile locations, it is essential that the network should be able to self-organize itself because the manual configuration is not feasible. Moreover, nodes may fail (either from lack of energy or from physical destruction), and new nodes may join the network. Therefore, the network must be able to periodically reconfigure itself so that it can continue to function. Individual nodes may become disconnected from the rest of the network, but a high degree of connectivity must be maintained.

- *Collaborative signal processing*: Another factor that distinguishes WSNs networks from MANETs is that WSNs are focused on the detection/estimation of some events of interest, and not just communications. To improve the detection/estimation performance, it may be required to fuse data from multiple sensors. This may put constraints on the network architecture as data fusion requires the transmission of data and control messages.
- *Querying ability*: To obtain information collected in the region, a user may have to query an individual node or a group of nodes and depending on the amount of data fusion required to be performed, it may not be feasible to transmit a large amount of the data across the network. Instead, various local sink nodes can collect the data from a given area and create summary messages. A query could be directed to the sink node nearest to the desired location.

1.1.2.2   Protocols and Architectures

Sensor network research has gained significant attention of the research community in recent years. Consequently, a variety of protocols has been developed by researchers across the world, addressing different issues in this area. Examples are the node and data centric communication protocols (ranging from energy aware medium access control to networking, routing and transport), support protocols such as time synchronization and localization or middleware frameworks to abstract the complexity of low-level protocols in order to ease application development [20]. While some of the solutions only concentrate on providing a particular service function within a sensor node, (e.g., MAC), others are aimed at realizing a variety of system functions cutting across multiple layers of the protocol stack in order to provide more efficient solution, mostly tailored to a particular application environment.

Although these protocols represent efficient solutions that operate well in their intended environments, the extent, to which those solutions are able to adapt to different application situations is quite limited. Furthermore, researchers and system designers are now confronted with a plethora of mostly incompatible protocol solutions.

Like the traditional communication networks, sensor networks can also be analyzed in terms of seven OSI layers. For tiny, low power sensors the most important issue is the power consumption. This means that all protocols and applications for sensor networks must consider the power consumption issue and try to the best to minimize it [21].

Recent research by academia has addressed the adaptability and compatibility of protocolsolutions by proposing frameworks and architectures to unify the wireless sensor network stacks in the nodes [22, 23] or make them more adaptive to address a variety of application requirements [24].

Apart from academic efforts, work has been ongoing within standardization bodies for defining the specifications for WSNs. One supporting standard for wireless sensing andcontrol systems is the ZigBee specification [25]. Based on the IEEE 802.15.4 standard [18], the ZigBee standard specifies a network layer, an application layer, and security functions. The ZigBee specification represents an important first step in order to ensure compatibility and allow the unified application development for an industry standard for WSNs. While it was designed for mainly static sensing and control applications, the existing ZigBee protocol stack architecture is constrained by limited protocol options and a lack of service functions and support mechanisms.

In order to support the requirements of new applications that may emerge with the integration of WSNs to next generation (e.g., Beyond 3G) systems, a more flexible architecture would be required. Such an architecture was proposed and developed by the FP6 IST project e-SENSE [26]. The e-SENSE architecture is a flexible framework with the following characteristics:

- Ability to take advantage of new communication paradigms and protocols, (e.g., data centricity);
- Easy configuration capabilities for various application environments based on the selection of appropriate protocols and service functions;
- Ability to provide a relevant infrastructure within the system to allow more efficient operation of protocols through cross-layer interaction;
- Adaptation capabilities of the WSN system and its behavior according to environmental conditions;
- Enhanced support functionality required for many applications, (e.g., localization);
- A gateway functionality in support of Beyond 3G network access.

Figure 1.3 shows the logical architecture of the e-SENSE protocol stack.

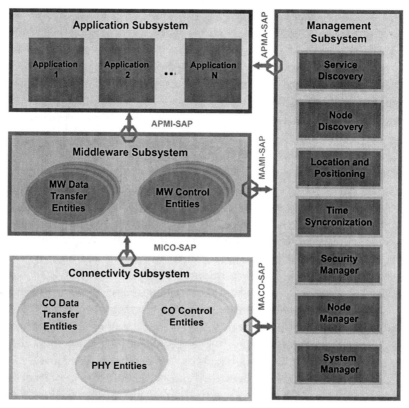

Figure 1.3 Logical architecture of the e-SENSE protocol stack [20].

The architecture is divided into four logical subsystems, namely the application (AP), management (MA), middleware (MI), and connectivity (CO) subsystems. Each subsystem comprises various protocol and control entities, which offer a wide range of services and functions at *service access points* (SAP) to other subsystems. This entire stack was implemented in a full-function sensor node and in a gateway node; a reduced-function sensor node has fewer functions (e.g., a reduced-function node is not able to perform routing or create a network); a relay node has a simpler structure with only the connectivity subsystem in its protocol stack.

The protocol stack as proposed by the project e-SENSE is not as strictly layered as specified in the classic OSI seven-layer model in order to facilitate cross-layer optimization. Each subsystem contains some entities, which can be combined in many ways to create several protocol stacks according to the node role in the network and application requirements. The protocols that are supported by the protocol stack in Figure 1.3 are implemented as modules, namely protocol elements. Although these elements can be organized in any order, usually only a pre-set configuration of protocols is allowed.

The connectivity subsystem of the e-SENSE protocol stack consists of functions that are required for operating the physical layer (PHY), the medium access control (MAC), and the network layer. The PHY functions comprise the radio transceiver. The MAC functions control the access of nodes to the shared radio channel and provide means for synchronization, beaconing, and reliable message transmission and reception. It is initially assumed that the e-SENSE PHY and MAC functions are very similar to those defined in the IEEE 802.15.4 low-rate wireless PAN standard. The network functions include mechanisms to create and release a network, to let devices join and leave the network, to discover and maintain routes between nodes, to route messages from a source to the destination, to perform neighbor discovery, and to apply security and reliability to the message transfer.

The service offered by the network functions mainly defines the service of the connectivity subsystem because most of the PHY/MAC functions are an integral part of the connectivity subsystem and are thus invisible to other subsystems.

The management subsystem is responsible for the configuration and initialization of the connectivity and middleware subsystems. Based on the information provided in the application profiles, it defines the role of the node within the sensor network. The management subsystem interfaces to the middleware subsystem to configure the middleware and to exchange management protocol messages with peer nodes and interacts with the connectivity subsystem to manage the network, the MAC, and the PHY functions of the sensor node.

In the management subsystem several functions are implemented: security manager, service discovery, node manager, network manager, location, and positioning services and node discovery. The security manager controls all issues related to authentication, privacy and trust of data at different levels of e-SENSE protocol stack. Service discovery and node discovery enable the subsystems to find a service or a node in the network. The node manager handles all tasks related to correct operation of the nodes. Location and positioning services supply information to subsystems on absolute or relative position of the node in space. The management subsystem provides the means to program and manage the lifecycle of the e-SENSE system by updating software code according to application requirements. An information database in the management subsystem includes permanent parameters and attributes related to all subsystems. Each subsystem can access the database and by using it get and set primitives; it can also write and read parameters in the information base.

The middleware subsystem has the general purpose to develop and handle an infrastructure where information sensed by nodes is processed in a distributed fashion and, if necessary, the result is transmitted to an actuating node and/or to the fixed infrastructure by means of a gateway. The middleware subsystem provides two types of data transfer services for the transport of application data packets: a node centric service and a data centric service. The former enables the sensor nodes to transfer data to particular nodes or group of nodes by using an explicit address. The latter is a data transfer service based on publish–subscribe paradigm. This is a messaging system where senders and receivers do not communicate directly with each other but via a broker. The role of the middleware subsystem is to provide a publish–subscribe abstraction in order to facilitate information exchange among nodes and from nodes to back-end applications. In addition to the transport mechanisms for sensor information, the middleware subsystem also provides an infrastructure where information sensed by nodes is processed in a totally distributed fashion.

The application subsystem hosts several sensor applications. Each application can send and receive sensor data by using the data transfer services provided by the middleware subsystem. The application subsystem configures the management subsystem according to node functionalities and role in the sensor network.

The connectivity subsystem of the e-SENSE protocol stack can generate and maintain WSNs, which support a *star*, *tree*, and *mesh* topology. A WSN consists of various network nodes, namely coordinators, clusterheads, and end devices. In a *star* topology, a single coordinator controls a network. It is responsible for forming and releasing the network and communicates directly with all devices in the network. In a tree and mesh network, clusterheads take part in controlling a subset of end devices of the WSN.

Unlike end devices, clusterheads never sleep, generally remain on as long as the network is active, and provide a routing service to other network nodes. In a *tree* network, clusterheads forward data and control messages using a hierarchical routing protocol, while in a *mesh* network full peer-to-peer communication is required to implement an appropriate routing protocol. A clusterhead can act as an end device. End devices, which are grouped in a cluster, can only directly communicate with the clusterhead or coordinator they cannot relay messages for other nodes, and can sleep to save energy.

1.1.2.3 Security, Privacy, and Trust in WSNs

Research on WSN security until recently was mainly focused on aspects related to security architectures, privacy, and trust frameworks. A great deal of research has been carried out on energy-efficient and light-weight security mechanisms but mainly for the so called *first generation* of WSN architectures—namely networks consisting in numerous unattended sensor nodes, applicable for example in environmental monitoring, military, and surveillance applications. Solutions, which have been proposed for privacy, focused on the aspect of *location privacy* (e.g., VANET), data anonymity or delayed disclosure. Solutions for trust establishments were proposed for ad hoc networks, but very few can be applied to sensor networks. A few published works on security, trust and privacy frameworks for WSNs aim at providing complete solutions with a set of suitable mechanisms, and they do not offer flexibility, adaptability and context-awareness for all security, privacy, and trust services.

In ubiquitous sensorized environments diverse type of data for an individual, the user location and environment, the devices, health status (e.g., in the case of body area networks-BANs), will be generated and processed, containing in most of the cases very confidential and sensitive information related to the private life of the persons. For use of WSNs in medical scenarios, for example, information for the current health status and patient's medical history is communicated. All these diverse contexts require different types of information related to an individual user to be *securely* disclosed to parties with different levels of trust.

WSNs are very sensitive to failures. As sensor nodes die, the topology of the sensor networks changes very frequently. Therefore, the algorithms for sensor networks should be robust and stable. The algorithms should work in case of node failure. When mobility is introduced in the sensor nodes, maintaining the robustness and consistent topology discovery become much difficult. Besides, there are huge amount of sensors in a small area. Most sensor networks use broadcasting for communication, while traditional and ad hoc networks, use point-to-point communication. Hence, the routing protocol should be designed considering these issues as well.

Sensors typically have limited resources and, therefore, can benefit from tailored air interfaces (e.g., IEEE 802.15.4). In [27], IEEE802.15.4a standard compliant PHY-layer models were provided, both for preamble detection and bit error rate. This is also a good basis for further PHY-MAC optimization. A specific innovation is to take interference into account using realistic channel models and two kinds of receiver architectures (coherent and noncoherent). An IR-UWB PHY model was also provided in [27] based on an actual existing implementation. A particular attention was paid to the synchronization scheme, which performance is a key issue in an energy-efficient implementations of IR-UWB systems. The optimization was done in the channel context of BANs.

Cross-optimization of lightweight PHY/MAC and MAC/transport combinations can be beneficial to achieve efficiency in WSNs [27]. The purpose of this activity is, therefore, to optimize the air interface in terms of, e.g., power consumption.

1.2 Wireless Mesh Networks

Wireless Mesh Networks (WMNs) provide a cost-effective approach to extend the range of the air interface. A WMN is a type of network that is dynamically self-organized and self-configured comprising of nodes that automatically establish and maintain mesh connectivity among themselves. WMNs provide a fault-tolerant architecture. Unlike the traditional MANETs, WMNs are a combination of infrastructure (mesh routers) and infrastructureless (*mesh clients*) networks to offer a broader and more flexible array of services and applications. WMNs can have devices, which can operate on multiple radio interfaces to increase capacity and throughput and decrease interference. WMNs are formed by two types of nodes: *mesh routers* and *mesh clients*. *Mesh routers* have minimal mobility and form the backbone of the WMNs, whereas the *mesh clients* can be either stationary or mobile and form a client mesh network among themselves and with mesh routers.

A WMN can be deployed incrementally, (i.e., one node at a time as per the requirement. To increase reliability and connectivity for the users, more nodes can be installed. The WMN technology could be applied for broadband home networking, community, and neighborhood networks, enterprise networking, building automation, intelligent transportation systems, and so forth. For example, the infrastructure part for the ITS can be provided by the wireless mesh routers. The mesh routers can be mounted at appropriate areas to provide ubiquitous coverage to mesh clients. The mesh clients can communicate with the mesh routers on the appropriate air interface to get the required information (e.g., congestion warning, accidents notification, weather updates, and so forth). The mesh clients should be able to communicate with each other in an ad hoc fashion also through a suitable air interface, which can handle high mobility (e.g., 802.16e, 802.20). Vehicular mesh networks also impose new challenges when it comes to transferring multimedia content at high speeds. The available bandwidth becomes an important issue along with the interference (in congested urban areas).

WMNs are a promising way for *Internet service providers* (ISPs), carriers, and others to roll out robust and reliable wireless broadband service access that requires minimal up-front investments.

The following are the characteristics of mesh networks [28]:

- Multihopping is used to achieve higher throughput without sacrificing the effective radio range via shorter link distances. There is less interference between nodes and more efficient reuse of frequency.
- Flexible network architecture, easy deployment and configuration, fault tolerance, and mesh connectivity due to which the upfront investment requirement is low.
- Mesh routers have minimum mobility, whereas mesh clients can be stationary or mobile.
- Both backhaul access to Internet and *peer-to-peer* (P2P) communications are supported. WMNs also enable the integration of multiple wireless networks.
- Mesh routers usually do not have strict constraints on consumption of power.
- WMNs should be compatible and interoperable with existing networks.

1.2.1 Wireless Mesh Backhaul Design

Wireless mesh networks have been investigated intensively recently in relation to their ability to substitute the wired network infrastructure functionality by a cheap, quick, and efficient solution for wireless data networking in urban, suburban, and even rural environments [29]. The growth of wireless broadband data services demanding higher bit rate than *Asynchronous Digital Subscriber Lines* (ADSL) requires wider capacity solutions not only in the access and core networks, but also on the side of the backhaul [30]. The demand for these services will be distributed among many different environments, from offices zones to rural areas. In some of these scenarios, the traditional wired high-speed fiber implementation in the backhaul network is not attractive for the operators anymore, both from an economical and technical point of view.

For wide area access, the access points (or base stations) are typically located at high towers or at the rooftop of buildings. However, as the capacity demands increase, the access point is moving closer to the user and it could be placed at below-the-rooftop heights. In this way, it can provide better signal reception and higher spatial frequency reuse factor [30]. For a cellular network point of view this means that the cell size has to shrink in order to satisfy the increased capacity demands, which in turn demands the placement of more base stations.

[29] proposed an approach that provides higher speed backhaul, capitalizing on emerging trends, such as the evolving IEEE 802.16m standard, and allowing affordable availability of broadband access in rural areas or in countries with limited existing wireline infrastructure. The approach is based on the combined use of MIMO technology and multiple hops to overcome the disadvantages of traditional wireless backhaul solutions, which require line of sight, high towers, and fine tuning to operate properly and can easily fail if any of these conditions are not met. In order to meet distance, aggregate rate, and physical mounting requirements (e.g., nonline of sight) at the backhaul, multiple antenna techniques are proposed to be deployed, both for the rural (macrocell) and urban (micro/picocell) scenarios. For a given network layout, the approach of [29] optimizes the design of the topology of the backhaul network in terms of load, delay, and resiliency by taking into account channel and interference conditions in order to provide adaptive scheduling, routing and power control. The joint design of intelligent/opportunistic scheduling and routing enhanced via the spatial dimension through intelligent antennas is a novel approach that can lead to unprecedented network throughput gains and QoS performance [30].

It must be mentioned, however, that the unlicensed band in the 60-GHz range provides new opportunities for low-cost deployment of WMNs with much increased bandwidth (5000 MHz) than lower frequency bands. The propagation characteristics at 60 GHz help to overcome the limited range of lower frequency bands. The much lower wavelength allows the design of very small transceivers. The narrow beam forming of antennas at 60 GHz can allow directional communication with the intended receivers with little or no interference. Thus, WMNs can form a high frequency links from the mesh routers to provide a high-capacity backbone for cellular networks; connect buildings with Ethernet connectivity, and so forth. The same very high-frequency radio links can be used by the mesh routers, which can be mounted on traffic lights or buildings to send directional traffic consisting of time-critical information (e.g., weather updates, routes experiencing congestion, or multimedia advertisements).

1.2.2 Multihop Concepts in Next Generation Wide Area Systems

Multihop concepts can be beneficial in the context of the successful deployment of ubiquitous radio system concepts. The FP6 IST projects WINNER and WINNER II [33] proposed a ubiquitous radio system with multihop capabilities.

The very high data rates envisioned for next generation communication systems, such as IMT-Advanced [34, 35] in reasonably large areas do not appear to be feasible with the conventional cellular architecture due to the following two basic reasons [36]. First, the transmission rates envisioned for next generation communication systems are two orders of magnitude higher than those of third generation systems. This demand creates serious power concerns because for a given transmit power level, the symbol (and thus bit) energy decreases linearly with the increasing transmission rate [36].

Second, the spectrum that is considered for IMT-Advanced systems (e.g., WINNER) will be located above the 2-GHz band used by third generation systems. The radio propagation in these bands is significantly more vulnerable to non line-of-sightconditions, which is the typical mode of operation in today's urban cellular communications.

One solution to these two problems is to significantly increase the density of base stations, resulting in considerably higher deployment costs that would only be feasible if the number of subscribers also increased at the same rate. This is cost-demanding and environmentally nonfriendly. On the other hand, the same number of subscribers will have a much higher demand in transmission rates, making the aggregate throughput rate the bottleneck in future wireless systems. Subscribers might not even be willing to pay the same amount per data bit as for voice bits, another factor that is of economical disadvantage of a drastic increase in the number of base stations.

As more fundamental enhancements, advanced transmission techniques and co-located antenna technologies, together with some major modifications in the wireless network architecture itself can enable effective distribution and collection of signals to and from wireless users [36]. The integration of multihop capability into conventional wireless networks is perhaps the most promising architectural upgrade. It allows that the rate can be traded for range of coverage and vice versa.

[47] investigated the achievable decrease of deployment costs, together with the achievable network performance compared to traditional cellular concepts. The proposed relay-based concepts were divided in homogenous multihop concepts based on one radio interface mode only, heterogeneous deployment concepts based on multihop links that interconnect on more than one radio interface mode or two different radio interface technologies. The benefits of relays are that they improve link budgets and may hence ease covering remote (or shadowed) areas. There also exist issues that need to be considered, like in the case of rapidly moving users, which may, due to inherent delay issues, not be able to take advantage of the relay nodes.

Furthermore, scenarios may exist where the deployment of relay nodes may be undesirable due to resulting complicated interference management issues, similar to a problem reported for hierarchical cellular systems where the deployment of overlapping macrocells and microcells is always a source for unpredictable interference and near-far effects.

Making the relays mobile also introduces new challenges beyond those that are seen when dealing only with fixed relays. For example, dealing with interference between mobile relays may be challenging especially in situations where the mobile relay population is dense [36]. Mobile relays, when acting as moving gateways, may also offer benefits by reducing the number of simultaneous handovers in the network (only the moving gateway is involved in the handover) and by allowing terminals to operate under more favorable link conditions.

The deployment and operation of next generation ubiquitous networks needs to be easier in order to prove economically viable and stimulate easy adoption. Thus, the operational aspects need to be taken into account in the design, together with requirements such as the support of autoconfiguration and autotuning. These requirements are especially important in relation to relay nodes, as the amount of relay nodes may exceed significantly the number of existing cellular base station sites.

1.2.2.1 Requirements and Functionalities for Multihop Networks

One critical function related to radio resource management (RRM) of multihop networks is *routing* [36]. It is intended to find the best routes to forward packets on.

Routing schemes can be classified into the two basic groups: *proactive* routing, and *reactive* routing.

In the first type of routing, network nodes continuously evaluate the routes within the network, so that when a packet needs to be forwarded, the route is already known and can be immediately used. Reactive protocols, on the other hand, invoke a route determination procedure on demand only. Thus, when a route is needed, some sort of global search procedure is employed.

Two main scenarios may be envisioned for the routing functionality. In the first scenario, the main control functionality of the routing resides in the access points only (i.e., *centralized* routing). The routing functionality can reside in all network nodes on the route between some access point and some user terminal (for one connection) that also participates in the route determination (i.e., *distributed* routing). In the case of centralized routing, the terminals and relay nodes should conduct the necessary measurements to assist the access point. The time scale of when the routing protocol should be invoked is highly dependent on the employed routing protocol and is, therefore, only applicable when discussing one specific routing protocol. The load status of nearby relay nodes, the available resources at relay nodes, the signal strength of nearby relay nodes (e.g., the averaged downlink pilot power, and the distance or path loss between the terminal of interest and nearby relay nodes are the main decision variables to be considered [36].

Multimode macro-diversity relaying (i.e., multiple hops over multiple different transmission modes of the same system) may also be envisioned [37]. An example of this is the case were the access point has the possibility to either forward some packets directly (single-hop) to some UTs using one mode or to forward the packet over a two-hop path using the same mode over the first hop and another mode over the second hop. To enable this possibility the access point needs to be aware of the possible paths to reach the UT and the quality over each link in this path.

It is envisioned that a flexible logical architecture where a single physical node is foreseen with multiple logical functionalities, can have a number of benefits in terms of improved network performance [38]. It must be noted, that such an approach might have impact in the context of multihop networks on layer 2 procedures, such as *buffer management* and *security procedures* (e.g., ciphering), which are handled on a per-hop basis. If the route through the multihop radio network is changed, (e.g., a relay node is added or deleted), the layer 2 context has to be transferred to the new node to avoid loss of data and guarantee a ciphered connection [36]. In addition, the user has to trust the intermediate relay nodes.

The concept of the *mode convergence protocol* (MCP) was proposed to enable a well-structured design of B3G protocols of layer 2 and 3 according to the various operating modes based on the same split of generic and specific user and control plane functionalities [36]. This MCP allows by means of cross-layer optimization and cross-plane management, for a context transfer for an efficient handover between the two radio modes. The MCP allows, for example, the transfer of status information of the ARQ mechanism.

1.2.2.2   Flexible Architecture Design

An example architecture comprising several different physical network nodes is shown in Figure 1.4. In general, two different modes of network operation can be distinguished.

The first operational mode relies on a network infrastructure similar to the one of existing cellular mobile communications systems. It can be called the infrastructure-based mode. Elements of this infrastructure are the *fixed routing relay nodes* (FR-RNs). Transmissions between FRRNs and FRRN and *access points* (APs) may take place in different transmission resources. Because visibility (i.e., line-of-sight) between FRRNs can be guaranteed, it is possible to use a higher frequency band than the one used for the other transmissions. While the transmission in the same frequency band may be useful during the initial network start, the multihop transmission may be shifted to a different band if a greater network capacity is required.

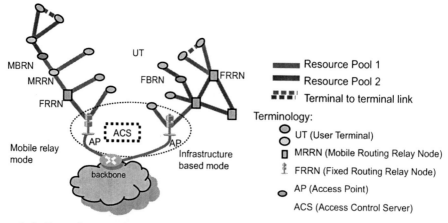

Figure 1.4   Example network architecture in support of multihop communications [36].

In this concept, it is assumed that a homogeneous air interface is used for both type of communications (fixed-to-fixed, mobile-to-mobile).

The second mode is characterized by ad hoc behavior. This mode does not rely on any network infrastructure and is shown on the left hand side of Figure 1.4. The role of forwarding data packets is taken over by mobile devices that are called *mobile routing relay nodes* (MRRN), which could be subscriber devices, such as the user terminal (UT). In addition, UTs are able to directly communicate with one another without intervention of any intermediate node. Because the fixed network elements are not mandatory in this mode, connectivity cannot be guaranteed and will be a problem if no fixed elements are available or cannot be reached because of the lack of routing capable terminals. Therefore, a combination of infrastructure and infrastructure less mode is likely, which is shown in Figure 1.4. In this architecture, the infrastructure is dynamically extended via MRRNs. This flexible mode of operation allows to extend the coverage and even to increase the spectral efficiency on demand. With increasing user densities it becomes possible that this architecture scales towards higher overall system throughput and larger coverage by means of mobile multihop extension. The MRRNs become part of the extended infrastructure-based architecture.

A number of problems must be mentioned that need to be addressed if the pure infrastructureless mode is used. Because the user equipment is used for forwarding data and valuable battery power is consumed by other users, a decent beneficial system for users allowing their equipment to act as relay needs to be introduced.

Furthermore security and authorization issues need to be addressed.

1.2.3 Classification of Ad Hoc Networks

Ad hoc networks can be broadly classified into the following types [39]:

- Isolated ad hoc networks;
- Integrated ad hoc networks;
- Cellular ad hoc networks.

Isolated ad hoc networks are networks in which the nodes communicate with each other within the same ad hoc network and have no connection with any infrastructure-based communication network. Isolated ad hoc networks can be categorized further depending on their size into *large-scale* and *small-size* isolated ad hoc networks.

Large-scale isolated ad hoc networks are made of a large number (in the order of thousands to hundred thousands of nodes). These type of networks maybe used to transport some very important messages with a small amount of data (e.g., in defense communications). It has been proposed to apply these networks to form a radio *metropolitan area network* (MAN) or maybe even a *wide area network* (WAN) to compete with the wired telecommunication networks. The problems associated with the large-scale isolated ad hoc networks is related to the severe security issues, extremely high network organization and very low traffic performance, mainly due to the lack of centralized control server, signaling costs associated with routing and the flooding of route requests, and the large delays due to the many hops [39].

The small size isolated ad hoc networks are suitable for connecting home appliance, exchanging information during meetings, instant digital presentations, playing a game among a group of people, and so fourth.

The *cellular ad hoc* network can be understood as a self-organized multi-hop network with multiple access points that are connected to the broadband core network. Here, most of the traffic is to/from access points unlike in anisolated network. In the early deployment of second generation networks, operation and maintenance (O&M) was based on in site operation. For existing third generation (3G) systems, O&M relies on software applications that manage the wireless system in a centralized way [40]. The expected rise in complexity of next generation wireless systems has attracted research activities towards the design of beyond 3G systems infrastructures by exploiting *cognitive networking* capabilities. Leaving all management responsibilities to the O&M prevents from exploiting the full potential of wireless networks. Conversely, a large degree of self-awareness and self-governance is to be considered in the development of future wireless networks and systems. In response, there has been a major push for self-managing networks and systems in recent years.

Self-organization allows the network to detect changes, make intelligent decisions based upon these inputs, and then implement the appropriate action. The systems must be location and situation-aware, and must take advantage of this information to dynamically configure themselves in a distributed fashion. Applied to RRM, self-organizing networks (SON), for example, allow for dynamic management and coordination of radio resources at the cell border, in order to avoid performance loss or degradation of service. Self-organization of access points and mobile nodes can be useful in situation like frequency planning in the range above 3 GHz and are an important step towards pervasive communication environments.

1.3   Pervasive Communications and Systems

Pervasive environments present a number of challenges to conventional approaches in networking and service provisioning: scalability, heterogeneity, and complexity. Such environments are characterized by huge information and communication technologies (ICT) systems foreseen by a high degree of dynamism, where networks, which are subject to frequent disconnections will be used to provide a large variety of services. These complex systems will greatly exceed the ability of humans to design them and predict their behavior, especially the interaction patterns among their individual components. The research activities, therefore, focus on investigating and providing new methodologies and tools for the design and management of pervasive ICT systems.

Pervasiveness requires a step beyond the traditional engineering design approaches, (e.g., predictability) towards autonomous behavior and based on the concepts of uncertainty. This requires a fundamental shift in the focus of the design process itself, moving from performance-oriented design to design for extreme robustness and resilience. At the same time, to cope with the dynamism intrinsically present in such systems, an ability to dynamically adapt and evolve in an unsupervised manner) has to be understood as a constituent system property.

In this perspective, paradigms based on decentralized, bottom-up approaches (like bio-inspired ones) are of particular interest for overcoming the limitations of more conventional top-down design approaches.

Ubiquitous access to services means that a service a user wants to use should be available when needed regardless of network technology, device architecture, the service type or user location. Such a concept was realized in a personalized way by the FP6 IST project DAIDALOS [10] for an almost unlimited type of services. A number of other FP6 IST projects achieved significant results in the same area, among those, the FP6 IST projects MAGNET and MAGNET Beyond [5], COMET [41], SPICE [42], OPUCE [43], E2R [44], and others [1, 2].

Ubiquitous access to services can be achieved by imposing close interactions across multiple subsystems, such as device and user mobility, authorization, and accounting (AAA), QoS, and security. The development of the pervasive platform concept, and associated cross-device concepts is an essential block for triggering these coordinated interactions.

The pervasive support architecture proposed in [45] consists of a service management infrastructure and user devices infrastructure. The service management infrastructure provides ubiquitous access to services. The functional entities run on top of a *pervasive service provisioning platform* (PSPP), which discovers services available to a user at any time, composes atomic services into composed pervasive services, and provides runtime mechanisms. A *service management* infrastructure includes the basic infrastructure for using multiple virtual identities (VIDs) for protecting user privacy. A *user device* infrastructure would support the context-aware interaction with pervasive services. This infrastructure should include functional entities for context management, personalization, and context-aware VID negotiation. The DAIDALOS pervasive service platform was described in detail in [46].

Context-aware access to services implies that the services should not only be accessible everywhere, but also that they should be adaptable to the user's preferences and networking context. Innovations within FP6 focus on mechanisms for collecting and processing of the relevant information from the network infrastructure and the services themselves and its use. The information can be used for the provisioning of services as well as to assist the user in the selection or customization the service. Due to the continuous changes of the user's context, which may be rapid and unexpected, a service may have to reconfigure itself frequently. The service adaptation can be a source of information for personalization and adaptation to user behavioral patterns. Context information can affect not only services in use, but also can influence the process of discovery, selection and composition of services. A pervasive architecture should support discovery and (re)composition of services and session management at a level at least two degrees above the current solutions in telecommunication environments.

Personalized and context-aware access to services and applications will be a driving force and one important factor for differentiating services in future. Today, only some applications can interact with a other sources of information in the environment to provide the basis for context awareness and extended personalization by means of user-centered configuration and control. Context engines will acquire context information from the user environment and will feed the information to pervasive applications.

Ambient intelligence (AmI) is a new research field aiming at building digital environments that are aware of the humans' presence, their behaviors and needs. Among the key features, *context awareness* plays the most important role in any ambient intelligence system. Ambient intelligence builds on the development of various ad hoc networking capabilities that exploit highly portable and numerous, very-low-cost computing devices.

1.4   Preview of the Book

This book is organized as follows. Chapter 2 addresses the basic concepts related to ad hoc networks, like topology formation, capacity bounds, and problems of self-organization in cooperative ad hoc networking. This chapter also gives a classification of routing protocols briefing their main characteristics. Cross-layering, which is proven as a successful design tool in ad hoc networking is discussed in the later section of the chapter. The IST project ADHOCSYS [21], e-SENSE [19], MAGNET [20], and MAGNET Beyond [20] have made major contributions in this chapter.

Chapter 3 focuses on the general concepts in achieving quality of service (QoS) guarantees in wireless ad hoc networks. It covers issues such as definitions of novel QoS frameworks, applications, and parameters; quality of service provisioning; classification of QoS solutions; and QoS architecture proposals for ad hoc networks. The chapter also provides an overview of the security aspects in wireless ad hoc networking. An introduction to security peculiarities, challenges, and needs in a wireless ad hoc domain is given. Possible threats and attacks, important security mechanisms, and several security architectures are explained.

Chapter 4 provides an overview of Personal Area Networks. A comparison between WLAN and PAN is established for better understanding of the concept of PAN. Relevant radio technologies are discussed along with the topics of interoperability and personalization and individualization of devices and services. The challenges, open issues, and security of PAN are also discussed in the last sections of the chapter.

Chapter 5 gives the complete details of the research and outcomes of the IST projects MAGNET and MAGNET Beyond. The concepts of PAN, private PAN, and personal networks is presented along with the associated topics of networking, services, and security.

References

[1]   FP6 IST Projects at http://cordis.europa.eu/ist/ct/proclu/p/mob-wireless.htm.

[2]   FP6 IST projects in *Broadband for All* at http://cordis.europa.eu/ist/ct/proclu/p/broadband.htm.

[3]   Gharavi, H., "Multichannel Mobile Ad Hoc Links for Multimedia Communications," *Proceedings of the IEEE*, Vol. 96, No. 1, January 2008.

[4]   Finocchio, P., R. Prasad, and M. Ruggieri, ed. *Aerospace Technologies and Applications for Dual Use* (Subtitle: A New World of Defense and Commercial in 21st Century Security), River Publishers: 2008, ISBN: 978-87-92329-04-2.

[5]   FP6 IST Projects MAGNET and MAGNET Beyond, at www.ist-magnet.org.

[6]   IETF MANET Working Group, http://www.ietf.org/html.charters/manet-charter.html.

[7]   Hoebeke, J., et al., "An Overview of Mobile Ad Hoc Networks: Applications and Challenges," in *Journal of the Communications Network*, Part 3, July–September 2004, pp. 60–66.

[8]   IP Routing for Wireless/Mobile Hosts, http://www.ietf.org/html.charters/mobileip-charter.html.

[9]   Jönsson, U., et al., "MIPMANET—Mobile IP for Mobile Ad Hoc Networks," in *Proceedings of the IEEE/ACM Workshop on Mobile and Ad Hoc Networking and Computing*, August 2000, Boston, MA.

[10]   FP6 IST Project DAIDALOS and DAIDALOS II, at www.ist-daidalos.org.

[11]   Devarapalli, V., et al., "Network Mobility (NEMO) Basic Support Protocol," IETF Internet RFC, January 2005.

[12]   Bernardos, C., J. Bagnulo, and M., Calderón, "MIRON: MIPv6 Route Optimization for NEMO," in *Proceedings of the Fourth Workshop on Applications and Services in Wireless Networks*, August 2004, pp. 189–197.

[13]   Hugo, D. v., et al., "Efficient Multicast Support within Moving IP Sub-Networks," in *Proceedings of the IST Mobile and Wireless Communications Summit 2006*, June 2006, Myconos, Greece.

[14]   FP6 IST Project DAIDALOS II, Deliverable DII-241, "Architecture and Design: Routing for Ad-hoc and Moving Networks," October 2007, at http://cordis.europa.eu/ist/ct/proclu/p/mob-wireless.htm.

[15]   FP6 IST Project DAIDALOS II, Deliverable DII-211, "Concepts for Networks with Relation to Key Concepts, especially Virtual Identities," September 2006, http://cordis.europa.eu/ist/ct/proclu/p/mob-wireless.htm.

[16]   Prasad, N., R., et al., "Secure Service Discovery Management in MAGNET," in *Proceedings of the IST-MAGNET Workshop*, *"My Personal Adaptive Global Net: Visions and Beyond,"* November 2004, Shanghai, China.

[17]   IEEE 802.11p, at http://standards.ieee.org/board/nes/projects/802-11p.pdf.

[18]   IEEE 802.15.4 "IEEE Std 802.15.4TM-2003: Wireless Medium Access Control (MAC) and Physical Layer (PHY) Specifications for Low- Rate Wireless Personal Area Networks (LRWPANs)," October 2003, at www.ieee.org.

[19]   Römer, K., and F. Mattern, "The Design Space of Wireless Sensor Networks," in *IEEE Wireless Communications*, Vol. 11, No. 6, December 2004, pp. 54–61.

[20]   FP6 IST Project E-SENSE, "System Concept and Architecture," November 2006, at http://cordis.europa.eu/ist/ct/proclu/p/mob-wireless.htm.

[21]   Ahmed, S., "TRLabs Report on Current Researches on Sensor Networks," May 2004, at www.geocities.com/certificatessayed/publications/trlabs_sensor_survey.pdf.

[22]   Culler, D., et al., "Towards a Sensor Network Architecture: Lowering the Waistline," in *Proceedings of the Tenth Workshop on Hot Topics in Operation Systems*, 2005, Santa Fe, NM, USA.

[23]   Polastre, J., et al., "A Unifying Link Abstraction for Wireless Sensor Networks," in *Proceedings of the Third International ACM Conference on Embedded Networked Sensor Systems (SenSys)*, 2005, San Diego, CA.

[24]   Marron, P. J., et al., "Tinycubus: A Flexible and Adaptive Framework for Sensor Networks," in *Proceedings of the Second European Workshop on Wireless Sensor Networks*, January 2005, Istanbul, Turkey, pp. 278–289.

[25]   ZigBee

[26]   FP6 IST Project E-SENSE, at http://cordis.europa.eu/ist/ct/proclu/p/mob-wireless.htm.

[27]   FP6 IST Project e-SENSE, Deliverable D3.3.1, "Efficient and Light Weight Wireless Sensor Communication Systems," May 2007, at http://cordis.europa.eu/ist/ct/proclu/p/mob-wireless.htm.

[28] Akyildiz, I. F., X. Wang, and W. Wang, "Wireless Mesh Networks: A Survey," in *Computer Networks and ISDN Systems*, Vol., 47, No. 4, March 2005, pp. 445–487.

[29] FP6 IST Project MEMBRANE, at http://cordis.europa.eu/ist/ct/proclu/p/mob-wireless. htm.

[30] FP6 IST Project MEMBRANE, Deliverable 2.1, "Most Relevant Community/Market Needs and Target Deployment Scenarios for Wireless Backhaul Multi-Element Multihop Networks," July 2006, at http://cordis.europa.eu/ist/ct/proclu/p/mob-wireless.htm.

[31] Gkelias, A., and K. K., Leung, "Multiple Antenna Techniques for Wireless Mesh Networks," Book chapter in *Wireless Mesh Networks: Architectures, Protocols, and Applications*, Springer Science, 2007.

[32] FP6 IST Project MEMBRANE, Deliverable 6.1, "Final Report," September 2008, at http:// cordis.europa.eu/ist/ct/proclu/p/mob-wireless.htm.

[33] FP6 IST project WINNER and WINNER II, at http://cordis.europa.eu/ist/ct/proclu/p/mob-wireless.htm.

[34] Next Generation Mobile Network (NGMN) Alliance, "Next Generation Mobile Networks Beyond HSPA and EVDO: A White Paper," December 2006, at www.ngmn.org.

[35] RECOMMENDATION ITU-R M.1645, "Framework and Overall Objectives of the Future Development of IMT 2000 and Systems Beyond IMT 2000," at www.itu.int.

[36] FP6 IST Project WINNER, Deliverable 3.1, "Description of Identified New Relay Based Radio Network Deployment Concepts and First Assessment by Comparison against Benchmarks of Well Known Deployment Concepts Using Enhanced Radio Interface Technologies," November 2004, at http://cordis.europa.eu/ist/ct/proclu/p/mob-wireless.htm.

[37] FP6 IST Project WINNER, Deliverable 7.1, "System Requirements," August 2006, at http://cordis.europa.eu/ist/ct/proclu/p/mob-wireless.htm.

[38] FP6 IST Project WINNER II, Deliverable D6.13.14, "WINNER II System Concept Description," November 2007, at http://cordis.europa.eu/ist/ct/proclu/p/mob-wireless.htm.

[39] Xu, B., S. Hischke, and B. Walke, "The Role of Ad hoc Networking in Future Wireless Communications," in *Proceedings of ICCT2003*, pp. 1353–1358.

[40] CELTIC EU Project WINNER+, Deliverable 1.1, "Advanced Radio Resource Management Concepts," January 2009, at www.celtic-initiative.org/Projects/WINNER+/default.asp.

[41] FP6 IST Project COMET, at https://www.comet-consortium.org/.

[42] FP6 IST Project SPICE, at www.ist-spice.org.

[43] FP6 IST Project OPUCE, at http://cordis.europa.eu/ist/ct/proclu/p/mob-wireless.htm.

[44] FP6 IST Project E2R, at http://cordis.europa.eu/ist/ct/proclu/p/mob-wireless.htm.

[45] FP6 IST Project DAIDALOS II, Deliverable D122, "Updated Daidalos II Global Architecture," November 2007, at http://cordis.europa.eu/ist/ct/proclu/p/mob-wireless.htm.

[46] Prasad, R. and A. Mihovska, Ed., *New Horizons in Mobile Communications Series:* Networks, Services, and Applications, Artech House, 2009.

[47] ICT Program, FP7, "Working Program, 2008-2010," available at www.cordis.eu/fp7.

CHAPTER 2

Protocols and Algorithms for Ad Hoc Networks

Ad hoc networks are composed of a large number of nodes, which communicate by means of (wireless) local network protocols and short-range technologies. Algorithms are the tools and procedures that are applied in order to optimize the performance, capacity, or reliability of the ad hoc system.

Ad hoc networks are complex environments, where a mobile user might not have the chance to retrieve all the necessary information about the neighboring access points or other types of nodes, therefore, the network is required to implement intelligent functions to manage the information exchange, the mobility, the resources, and the quality of service (QoS). Other main issues connected to ad hoc networks are trustworthiness, security, and flexibility.

The Sixth framework program (FP6) cofunded a number of projects that investigated core issues in the development of trustworthy wireless communications protocols and devices, and delivered a methodology based on rigorous tools and techniques for scenarios and types of ad hoc networks. In an ad hoc network environment, mobility, resource management, and QoS cannot be regarded anymore as independent issues [1]. Interoperability between devices is another core concept.

This chapter describes different network topologies and management strategies for ad hoc networks, including aspects related to routing, quality of service (QoS), and mobility.

A number of FP6 IST projects [2, 3] investigated issues related to the management of networks and information in ad hoc networks. This was done for stand-alone ad hoc networks, or in the context of their integration with other networks. The FP6 IST projects MAGNET and MAGNET beyond investigated and proposed network layer protocol solutions and architectures in the context of personal area networks (PANs) [4]. The goal was to connect securely personal devices, nodes, services, and applications located around the person and in distant personal clusters (e.g., car, home, or office). PNs and federations of PNs are examples of a new class of dynamic and autonomous networked systems, for which optimized solutions are needed, striking a balance between ease of use, trustworthiness and flexibility. The FP6 IST projects DAIDALOS and DAIDALOS II validated an IPv6-based based architecture supporting heterogeneous access and integrating advanced ad hoc network concepts [5]. The FP6 IST project ADHOCSYS [6] developed a wireless ad hoc network with improved system reliability and availability through the automatic configuration of network parameters and self-healing mechanisms and the development of supporting frameworks for network security and user authentication, and for QoS and multimedia services to end-users. The FP6 IST project CRUISE [7] focused on the communication and application aspects of wireless sensor networking.

This chapter is organized as follows. Section 2.1 introduces into the most essential challenges related to ad hoc network management. Section 2.2 gives a thorough overview and performance evaluation of the routing aspects related to the different types of ad hoc networks. Section 2.3 describes similar aspects, specifically related to self-organizing networks (e.g., sensor networks). Section 2.4 concludes the chapter.

2.1   Introduction

The extension of the Internet protocols from fixed to mobile and wireless environments for access purposes created a scenario where mobility, resource management, and QoS require joint solutions. In communication systems, the common optimization problem generally concerns of increasing the capacity, QoS of the network and reducing the overall cost. These issues require the optimization of nonlinear and discrete functions [8]. Moreover, when multiobjective goals are involved, the problem becomes even more complex.

One of the challenges in ad hoc networks relates to the number of different network topologies that a user device can encounter while on the move, which requires that some degree of awareness (or intelligence), is introduced into the system. Other advantages of making an ad hoc network intelligent are increased reliability and flexibility (e.g., self-healing and self-organization capabilities), and increased capacity (e.g., self-optimization). An ad hoc network can be a large-scale one, and in such case, it is important to reduce the number of parameters that need to be set and tuned during deployment during optimization of end-to-end performance parameters (e.g., capacity). A number of new research challenges, relate to how nodes of large-scale networks (e.g., wireless sensor networks-WSN) are interconnected and in a number of scenarios how the nodes are self-configured [9].

Research has concentrated on the convergent integration of ad hoc wireless networks, such as *wireless local area networks* (WLANs) and *wireless personal area networks* (WPANs) with the infrastructure. Research has also been carried out on hierarchical ad hoc networks with different classes of nodes. However, very little has been done on the integration of ad hoc networks, which use proprietary wireless technologies and protocols (e.g., sensor networks), with other wireless networks. One challenging area is the study of the *wireless hybrid networks* (WHNs), which represent the merging of WSN and cellular systems and are characterized as multitier, heterogeneous networks supporting low-power sensing and control. Academic research on sensor networks has mainly employed architectures, which fit theoretical and research needs: homogeneous, isolated, very large, multihop. In practice, wireless sensing and control needs to be connected to the existing infrastructure through *wireless body area networks* (WBANs), WPANs, WLANs, beyond 3G cellular or directly wired IP networks. Integration of MANETs with the existing cellular architectures was a focus of the FP6 IST projects DAIDALOS and DAIDALOS II with the objective to ensure location independent and optimized personal services. For example, by successfully integrating a MANET with the IETF *network mobility* (NEMO) architecture, a user terminal can roam, among infrastructure and infrastructureless access networks (e.g., a vehicular mobility scenario)

[5]. Some main architectural issues related to such integration are how to cope with Layer 3 local mobility, addressing, support of handover in the ad hoc and between the infrastructure and the ad hoc network, and multihoming, to mention a few [10]. In this context, integration of the *Ad Hoc On-Demand Distance Vector* (AODV) and Mobile IP integration in the scope of the development of a suitable routing strategy is one approach.

Figure 2.1 shows the MANET local mobility management solution (LMM) in a local mobility domain (LMD).

In infrastructure networks, the movement detection (i.e., attachment/detachment) can easily be done recurring to the standard mechanisms present in the IPv6 protocol stack (such as the ICMPv6 neighbor solicitation and neighbor advertisement mechanism). However, in the multihop environment provided by the MANETs this standard mechanism cannot be used, and thus has to be replaced for either an explicit signaling performed by the mobile node, or by the help of explicit signaling done by the terminal's neighbors in the MANET. Because the terminal has to perform explicit signaling upon handover to maintain QoS reservation and to get the needed authorization to perform handover, the most reasonable solution is that terminal also notifies the MANET access router (AR) or gateway (GW) every time it performs an attachment. In this case, the local mobility protocol will be responsible for the detachment at the old gateway.

Topology is an important model of the ad hoc network state as it implicitly gives a lot of information about the active nodes and the connectivity/reachability map of the system. Topology characteristics are very important for support of flow

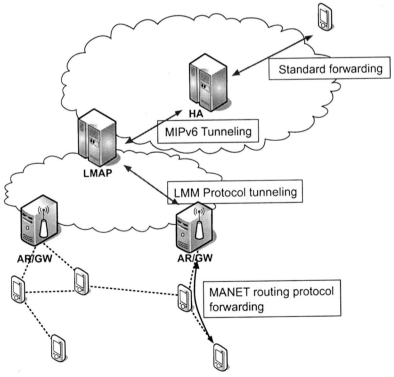

Figure 2.1 Integration of an ad hoc network in a local mobility domain [10].

distribution in the case of multihoming. The concept of multihoming in MANET is shown in Figure 2.2.

In the network/system configuration of Figure 2.2, multihoming is deployed at all levels. The multihoming extensions within the LMD (i.e., MAP and AR/GW) and the *home agent* (HA) are transparent to the MANETs, (related to the execution of multiple registration and load balancing). The specifics of the ad hoc environment are handled only at the mobile terminal level. These can be summarized as follows:

- *Multihop registration*: when a node registers with a network, the *protocol for carrying authentication for network access* (PANA) is used. Such a protocol can be used when there is a multihop Layer 3 path between the supplicant and the *nonaccess stratum* (NAS), therefore it can be used in ad hoc networks, but it has to be triggered every time a gateway is discovered.
- *Load balancing*: the length of the multihop paths must be considered when assigning a flow to a specific gateway.
- *Load balancing policies communication*: the engine taking the decision to load balance the flows among available access routers within an LMD is the MT. For downlink traffic such decision must be communicated to the AR/GW and then to the MAP. The LMM copes with the latter, while the first task is achieved by media-independent handover (MIH) signalling [10, 11]. Such signalling must be defined in order to be usable in ad hoc networks too (i.e., it has to enable multihop Layer 3 communication).
- *Load balancing uplink execution*: the data packets of a flow assigned to a gateway must be routed to the gateway itself. This can be achieved by adding routing header extensions to the packets: such extensions are statically bound with the gateway, i.e. all traffic destined to a gateway will contain a routing header with the address of that gateway.

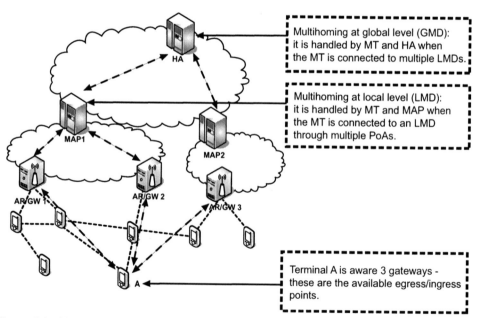

Figure 2.2 MANET and multihoming concept [10].

The ad hoc mobile node has a double role, acting as a host, which produces and consumes application traffic and acting as a router that forwards the traffic of other nodes. The mobile node needs to be able to retrieve the QoS parameters from the application characteristics, trigger the check for QoS resources along the ad hoc path and check the available resources in its wireless medium. It can also classify and mark the packets according to its class, and ensure the QoS differentiation.

Topology discovery algorithms have to retrieve the network state information crucial for network management and to enable network administrators to monitor topology control decisions, such as the need for power and resources regulation.

At the same time, the topology determines sensor gaps and coverage and could be used as a routing framework and for calculating certain network characteristics: coverage, connectivity, usage patterns, cost model, and so fourth. It provides an adequate model of the network, which can be used for management functions such as the future node deployment, and setting duty cycles. Topology is also a key attribute because it aids in network performance analysis.

Figure 2.3 to Figure 2.6 show the different possible topologies for the operation of a radio access network, namely, single-hop, multihop, and mesh. In the single-hop mode, the information is transmitted between radio access points (e.g., BS) and mobile stations MS directly in a single hop (see Figure 2.3).

In a multihop mode, the information is transmitted between radio access points to the mobile stations in more than one hop. The intermediate points between the access points (i.e., BS) and the destination (i.e., relay stations) regenerate and re-transmit radio signals. The topology of the multihop mode is shown in Figure 2.4.

In a mesh mode, the relay stations are supposed to have connections between each of them, if physically possible. This puts a requirement for routing algorithms between the relay nodes. An example of a mesh network topology is shown in Figure 2.5.

In a peer-to-peer topology, the MSs are connected directly or through relay nodes, but no radio access points are explicit in their connections. Peer-to-peer communication has been studied in 3GPP [12] as an ad hoc multihop relaying protocol based on [13]. It requires the MS to probe the neighborhood in regular intervals for potential relays, which in turn results in an overhead [14]. Therefore, the integration of a peer-to-peer mode in a radio access network should take this implication into account.

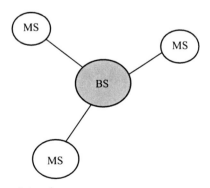

Figure 2.3   Single-hop network topology.

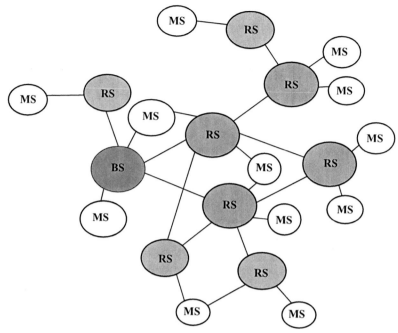

Figure 2.4 Multihop network topology.

An example of a peer-to-peer network topology is shown in Figure 2.6.

In general, the different types of ad hoc networks have not only different topologies but also different node classifications and capabilities, which impose different constraints. For example, in terms of mobility of the devices, the nodes can be classified as fixed and portable [15]. This is a typical scenario for multihop or mesh networks. Depending on the role of a node and the type of network it belongs to, it would implement different functionalities (e.g., routing optimization, service management), and as a result will need different processing capabilities, power supply requirements, and so forth.

QoS is usually required in all parts of the communication system because there are different applications in the various scenarios possible, which in turn have stringent constraints either in terms of delays, or data rates, or both [16].

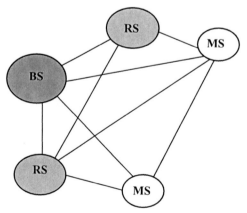

Figure 2.5 Mesh network topology.

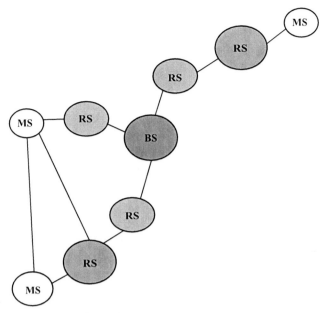

Figure 2.6   Peer-to-peer network topology.

The guarantee of providing end-to-end QoS is challenged by the following four factors:

1. The selection of the version of the standard and the underlying algorithms;
2. The interface at the level of the node providing the routing (mapping);
3. The routing, relaying, and other (e.g., cooperative) algorithms;
4. The other functionalities and protocols that are present in the nodes and that must be coordinated (e.g., ARQ).

2.2   Routing

In any distributed network an efficient and reliable protocol is needed to pass pack-ets between nodes. A routing algorithm should ensure that, provided that the source and destination nodes are connected, each packet sent should eventually reach its destination. If the source and destination nodes are not directly connected then this will take multiple hops. The problem is more complicated if the network is dynamic and links may be made or broken autonomously [17].

Various routing algorithms have been proposed for packet forwarding. A rout-ing algorithm, in general, decides, which output link an incoming packet should be transmitted on. A routing "table" contains the mappings from the networks and host addresses to output ports on the router. The routing algorithm builds this "table" and regularly updates it.

The goal of all routing algorithms is to discover and use the *sink trees* for all routers. A sink tree is formed by the optimal routes, determined without regarding the network, its topology, and traffic.

There are two main types of routing algorithms, namely the following: *adaptive* and *nonadaptive*. A *nonadaptive* or *static* routing algorithm follows a predetermined path, on which a packet is sent, and is unaffected by changes to the topology of the network or the number of packets that are sent along particular routes. An example of such an algorithm is the *shortest-path* algorithm, proposed by Dijkstra [18]. The nodes may or may not know the topology of the whole network but they must know the neighboring nodes, to which they are directly connected. Another example is the *flooding* algorithm. Every incoming packet is sent out on every outgoing line except the one it arrived on. Although this methodology avoids the creation of routing tables, it can lead to a large number of duplicated packets.

An *adaptive* algorithm determines the path, on which a packet is sent based on the current perception of the state of the system, the topology, or the number of packets passing along each link. It is required that this information is regularly updated. This is done by propagating messages about changes in the state throughout the system. The choice of path, along which to send these messages and packets is also regularly updated. An example of an adaptive algorithm is the *distance vector routing* algorithm [19]. The idea behind this routing algorithm is to come up with a forwarding database, which helps to determine the neighbor to which to send to for each possible destination. This is achieved by exchanging distance vectors, which tell the transmitters distance to each destination. A disadvantage of the *distance vector routing* algorithm is that there can be a considerable time lag between a link breaking and all nodes in the network learning about that break.

The distance vector routing algorithm can be replaced by the *link state routing algorithm* [20]. In this algorithm each router maintains a map recording its current belief about the state of the network; whenever the state of a link changes it comes up or goes down, this information is flooded throughout the network.

Hierarchical routing would help to make the above algorithms scale to large networks.

The MANETs topology may change frequently as nodes move, thus turning multihop routing into a particularly difficult problem. Two approaches have been proposed to combat the above problem, namely, *topology-based*, and *position-based* routing schemes. In *topology-based* approaches, the nodes exchange connectivity information with their neighbors to create routing tables. The first *topology-based* routing protocols were directly derived from the *distance vector IP* protocols. In wireless environments, however, these protocols waste resources when collecting topological information also for unnecessary destinations. To address this problem, newer versions of these protocols have been proposed that postpone the collection of routes up to the moment when nodes need them. One shortcoming of *topology-based* routing is that nodes have to collect routing information that might be arbitrarily distant. This might be difficult in a wireless network, therefore, *position-based* routing has been proposed as an alternative. By using location of the destination, of the node itself and of its neighbors, it is possible to reach the destination with only a very limited view of the network, often restricted to the immediate one-hop neighborhood. Position-based routing schemes can be, in turn, divided into *routing* algorithms and *pre-processing* algorithms. The *routing* algorithm runs at each node and determines which neighbor should be the next hop for a packet.

The *preprocessing* algorithm serves to create a graph for the routing algorithm. This usually means to create in each node a local view of a global planar graph. Position-based routing schemes are attractive because of the simple structure but are costly in terms of the technology required to determine the position and in routing failures, even in extremely simple scenarios.

2.2.1   Models and Algorithms for Routing and Forwarding in MANETs

Routing is a crucial radio resource management (RRM) function allowing for finding the best routes to forward packets on [21].

Routing schemes could be basically classified into two groups: *proactive* routing, and *reactive* routing. In the first type of routing, network nodes continuously evaluate the routes within the network, so that when a packet needs to be forwarded, the route is already known and can be immediately used. Reactive protocols, on the other hand, invoke a route determination procedure on demand only. Thus, when a route is needed, some sort of global search procedure is employed.

Two main scenarios may be envisioned for the routing functionality; either the main control functionality of the routing resides in the access points only (*centralized* routing) or it resides in all network nodes on the route between some access point and some user terminal (for one connection) that also participates in the route determination (*distributed* routing). For centralized routing, the terminals and relay nodes should conduct the necessary measurements to assist the access point.

The time scale of when the routing protocol should be invoked is highly dependent on the employed routing protocol and is, therefore, only applicable when discussing this specific routing protocol [21].

2.2.1.1   Dynamic Source Routing

The *Dynamic Source Routing* (DSR) protocol is an on-demand (reactive) routing protocol that is based on the concept of *source* routing [22, 23]. Mobile nodes are required to maintain route caches that contain the source routes of which the mobile is aware. Entries in the route cache are continually updated as new routes are learned. In DSR, different packet types are used for route discovery, forwarding, and route maintenance.

DSR uses two types of packets for route discovery, *Route Request* (RREQ) and *Route Replay* (RREP) packet. When a mobile node has a packet to send to some destination, it first consults its route cache to determine whether it already has a route to the destination. If the node does not have an unexpired route to the destination, it initiates route discovery by broadcasting RREQ packet. This route request contains the address of the destination, along with the address of the source node and a unique identification number. Each node upon receiving the packet checks whether it knows of a route to the destination. If it does not, it adds its own address to the route record of the packet and then forwards the packet along its outgoing links.

A RREP is generated when either the route request reaches the destination itself, or when it reaches an intermediate node, which contains in its route cache an unexpired route to the destination. By the time the packet reaches either the destination or such an intermediate node, it contains a route record yielding the sequence of hops taken and the respective addresses of the intermediate nodes.

If the node generating the route reply is the destination, it places the route record contained in the route request into the route reply. If the responding node is an intermediate node, it will append its cached route to the route record and then it will generate the route reply. To return the route reply, the responding node must have a route to the initiator. If it has a route to the initiator in its route cache, it may use that route. Otherwise, if symmetric links are supported, the node may reverse the route in the route record. If symmetric links are not supported, the node may initiate its own route discovery and piggyback the route reply on the new route request.

Collisions between route requests propagated by the neighboring nodes (insertion of random delays before forwarding RREQ should be avoided as much as possible. To avoid multiple hosts replying simultaneously from their cache, each host would delay its reply slightly. The time of delay depends on the distance measured in hops from the RREQ source. This approach can serve to prevent a node from sending an RREP if it hears another RREP with a shorter route.

Route maintenance is accomplished through the use of *route error* (RERR) packets. RERR packets are generated at a node, when the data link layer encounters a fatal transmission problem. When a route error packet is received, the hop in error is removed from the node route cache and all routes containing the hop are truncated at that point.

Improvements of the protocol are possible, when a relay node that forwards a data packet observes the entire route in the packet and updates its route cache. Closely related to this approach, is the update of the routing table by nodes that work in promiscuous mode and overhear unicast packets being sent to other nodes, comprising route information.

The main advantage of DSR is that routes are maintained only between nodes, which need to communicate. This reduces the overhead of the route maintenance. Furthermore, route caching can further reduce the route discovery overhead. A single route discovery may yield many routes to the destination, due to the intermediate nodes replying from local caches.

The main disadvantages are that the packet header size grows with the route length due to source routing, which leads to a major scalability problem [21].

2.2.1.2   Ad Hoc On-Demand Distance Vector (AODV)

The *Ad Hoc On-Demand Distance Vector* (AODV) routing [24] is a reactive routing protocol intended for use by MS and, which determines the routes to destinations in ad hoc networks. It offers quick adaptation to dynamic link conditions, low processing, and memory overhead, and low network resource use. The AODV is based on the *Destination Sequence Distance Vector* (DSDV) routing algorithm and uses destination sequence numbers to ensure loop freedom. The destination sequence number is created by the destination to be included along with any route information it sends to the requesting station that allows the source to select the most recent and fresh route [24].

In AODV, different packet types are used for the route discovery, forwarding, and route maintenance. AODV uses the two basic packet types for route setup, the

RREQ and RREP. During the route discovery, when a route to a new destination is needed the station broadcasts an RREQ packet. This is shown in Figure 2.7.

Each station receiving the request caches a route back to the originator of the request, so that the RREP packet can be transmitted via unicast from the destination along a path to that originator. The packet is rebroadcast by every station as far as this packet has not been received before, which can be checked by means of its unique packet ID.

To decrease the overhead during flooding, a hop-count is introduced that limits the forwarding. With an expanded ring search the hop-count is increased when no responds to a request have been received after a predefined time-out, resulting finally in the flooding of the whole network with RREQ packets.

A route is determined when the RREQ reaches a station that knows a valid route to the destination (e.g., the destination itself). This route is made available by transmitting a RREP back to the originator of the RREQ.

Due to reverse-path setup based on caching the RREP packet, every station on the route knows the next-hop neighbor they should address to forward a packet to the destination. AODV provides a routing table with information for all known routes. An entry in the routing table contains the destination, the next neighbor to forward a packet, and a metric (e.g., the hop-count) for that route. In Figure 2.7, a packet that contains the source and destination address can be forwarded along the path that the RREP packet traversed before with the help of the routing table without further information.

When a link break in an active route is detected, the broken link is invalidated and a RERR message is typically transmitted back to the source to notify other nodes that the loss of that link has occurred. To detect a link break each station periodically transmits *HELLO* packets with their ID [21].

The station will setup a new route by means of the mechanisms described above. This results in large delays and overhead, especially when a long route breaks near to the destination. Hence, a local route repair is defined in AODV. After a station has recognized a link break it tries to recover the route by means of transmitting an RREQ packet like it is defined for initial route discovery. If this attempt fails it will send a RERR back to the source, otherwise it receives a RREP with a fresh route to the destination.

It is not guaranteed that the station that locally recovers a broken link succeeds, resulting in undesirable delays and wasted bandwidth, because the source and in-

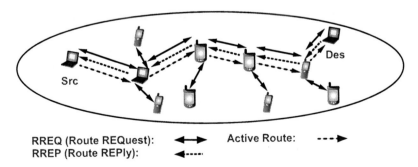

RREQ (Route REQuest):   ←→   Active Route: - - -►
RREP (Route REPly):   ←-----·

Figure 2.7 Principle of AODV [21].

termediate stations continue transmitting packets to that station. Therefore, local route-repair should be favored the closer the link break happens to the destination, whereby closer is defined by the routing metric, (e.g., the number of hops or the geographical distance). Typically, local route-repair is initiated if the link break happens closer to the destination than to the source.

The main advantages of AODV are that routes are maintained only between nodes that have to communicate, similar to DSR, and that the route information is locally stored in each intermediate relay-based on reverse path setup procedure, and the whole route information has not to be sent in the packet header, making AODV more scalable than DSR. Thus, AODV will perform better when there is a small set of source and destinations on each node. For example, in an architecture where the greatest part of flows takes place between the nodes and a gateway, AODV will lead to a smaller overhead because the routing table only features the route entries to active destinations [5].

In the presence of very dynamic and very dense network topologies, however, AODV has some deficiencies.

If routes break during an active communication, a route-repair process should be initiated, that imposes additional overhead and needs considerable time. For that purpose respective methods for local repair are proposed for AODV. But the proposed solutions seem to be far from optimum, because these need considerable time and transmit resources.

2.2.1.3   Optimized Link State Routing Protocol (OLSR)

The *Optimized Link State Routing Protocol* (OLSR) is a proactive table-driven protocol developed for mobile ad hoc networks. As a table-driven link-state protocol, OLSR has all existing routes in an ad hoc network, to all the nodes. In order to keep the routing information up to date, OLSR exchanges its topology information with other nodes in the network regularly.

In order to reduce the number of link-state updates, OLSR uses a special mechanism of *multipoint relays* (MPR). Only the MPRs selected from the surrounding nodes are allowed to forward control traffic. The MPRs provide an efficient mechanism for flooding the control traffic by reducing the number of transmissions required. OLSR also supports a way to advertise networks that are reachable through a node, using the *host network association* message [10].

In comparison to the OLSR protocol, AODV uses significantly less network resources when no communications are taking place, at the expense of a larger latency in finding new routes. Although, AODV sends small periodic HELLO messages to update the state on the adjacent nodes, these messages are smaller than the ones used by OLSR. If the MANET topology is mainly static, some routes can be reused, leading to a more efficient routing process [10]. AODV quickly reacts to topology changes by using the route repair mechanism. AODV, however, does not allow a node to know beforehand if a destination is available, without flooding a route-request message.

OLSR relies on periodic HELLO messages and TC messages that proactively build a routing table on each node, to each node inside the MANET. These messages lead to a systematic overhead, even if there are no flows inside the network, but they allow a node to have all the routing information needed to start a flow as soon as possible, without further delay [10]. It also enables a node to know if a destination is available or not, just by looking at the local routing table. OLSR is also slower than AODV to react to topology changes, but this issue can be minimized by carefully setting smaller intervals for the periodic messages.

AODV will perform better when there is a small set of source and destinations on each node. It will lead to a smaller overhead than OLSR, and the routing table will only feature route entries to active destinations.

Support of AODV and OLSR in an Integrated MANET Architecture

There are three options possible for the support of both protocols into an architecture that integrates MANET and an infrastructure-based network [10]:

- The gateways and the nodes support and run both protocols;
- The gateways and the nodes support both protocols, but each node only runs on one of them;
- The gateways and the nodes support both protocols, but only one is active at a time on all the nodes.

In the first approach, all the nodes are capable of running AODV and OLSR. OLSR would be always active, and AODV would be used each time when an OLSR route is invalid. If both protocols are supported, there will always be the systematic overhead from OLSR. If AODV were used as well, in order to handle fast topology changes, there would be an AODV overhead as well. The daemons from both routing protocols would consume more CPU time and memory, as well as battery power.

The second option can be described by a scenario, in which the gateways are running both routing protocols and let each node to decide what routing protocol to use. If a node only has one routing protocol, it automatically chooses to use that one. This approach would allow the terminals for only having a single routing protocol, saving storage space, and a dynamic selection of routing protocols based on some metric. The problem about this approach is that the routing information from one node running one protocol is meaningless to nodes running the other protocol. To allow all the nodes to interoperate, they would have to at least support one of the protocols. The choice would be AODV [10]. Some nodes would support AODV as a fallback mechanism for situations where an OLSR route was not available.

The problems with this approach are that the nodes running OLSR would be running two daemons and sending messages for both protocols, experiencing the problems connected to the simultaneous use of both protocols. In some situations, a node running OLSR would not have neighbors running OLSR, making it useless. Even if all the nodes were running OLSR at a given moment, AODV would still be required to be running, because the nodes would not be aware that there were no nodes running AODV only. Nodes would not be able to know if a destination is present in the MANET by consulting the local routing table, that one of the desirable features from OLSR. Finally, there is the problem about how a node would select one protocol instead of another. The routing protocol decision must take into account the state of the whole MANET, but it is not feasible to make that information available locally on each node.

The third option is to make all the nodes and the gateways to support both OLSR and AODV, but to force them to use the same at any given time. The gateway can decide to switch to a different routing protocol, and then advertise the chosen one to all the nodes inside the MANET. Every time a node receives a message with a different routing protocol, it enables it and disables the previously active one. A problem of this solution is the fact that all the terminals must have both daemons installed (but not running) in the memory. It is also difficult to define a metric that allows the gateway to know the topology of the MANET, without increasing the signaling overhead, and to make all the gateways of the same MANET to use the same routing protocols.

This is the easiest and the more efficient solution to really enhance an architecture, with two routing protocols [10].

Choice of Routing Protocol in an Integrated MANET Architecture

There are two options available for choosing the right routing protocol [10]:

- The gateways dynamically choose the routing protocol based on real-time information of the MANET;
- The gateways advertise a routing protocol that is statically configured by the operator, or based on policies.

The first option is not always feasible to provide the needed information to the gateway, so that the gateway can take a decision in real-time. To choose the right routing protocol, the gateway would have to know the topology of the MANET to be able to infer the density and the mobility characteristics of it. To do so, a new protocol would have to be introduced, that would consume network resources. In addition, the metrics could change dramatically in few seconds, especially in the situation where two (or more) MANETs merge.

In the second option, each gateway advertises the preferred routing protocol that is chosen based on policies and/or previous planning. As an example, an ad hoc gateway operator can configure AODV as the preferred routing protocol on locations, where it expects that the MANET is not very dense, and the topology is relatively dynamic. Because switching routing protocols can be an expensive operation, especially when the switch is from a reactive protocol to a proactive one, that operation must be infrequent [10].

Running Different Routing Protocols on a Node with Multiple MANETs

It is not always desirable to run both protocols on all the nodes, and all the nodes must run the same routing protocol to have an efficient operation [10].

There are two possible approaches to support this scenario:

- Make all the nodes from the MANETs run the same routing protocol;
- Only support the merging of MANETs running the same routing protocol.

In the first option, all the nodes would fall back to AODV when a MANET with AODV was reachable. A switch to OLSR would be more expensive in terms of overhead, because all the nodes would have to learn the routes to all the other nodes that were switching from AODV to OLSR. Gateways would be required to run both AODV protocols every time, to enable the nodes to attach to all of them, but running two routing protocols on the gateway would not be a problem because gateways are generally not battery operated, and are provided with some CPU power and memory. When a gateway detects the merging of two MANETs running different routing protocols, it would have to change the preferred routing protocol to AODV.

A problem might occur though when the MANET splits. The gateways that were running OLSR after the merge would start advertising it as the preferred routing protocol as before. The nodes that only reach those gateways would require the switch from AODV to OLSR, leading to a delay in obtaining the routes that were available after the merging. The problem would be more severe if merges and splits are frequent. This is shown in Figure 2.8.

Supporting the merge of MANETs running the same routing protocol does not have any major drawback, except for the fact that a node can be stopped from using all the reachable gateways, because some of them are running a different protocol (Figure 2.9). To minimize this, it desirable for each operator to run the same routing protocol on their own gateways in order to allow a node for connecting to every available gateway on each geographical site. Even thus, the operator can still decide the best routing protocol for each location, trading it of for the support for multiple gateways.

It has been proposed that a unique prefix is assigned to each node, and that the routing protocols are aware of this [10]. In this way, the OLSR *host and network association* (HNA) message supports the advertisement of the network and the prefixes. The OLSR announces the node address, and then advertises its prefix,

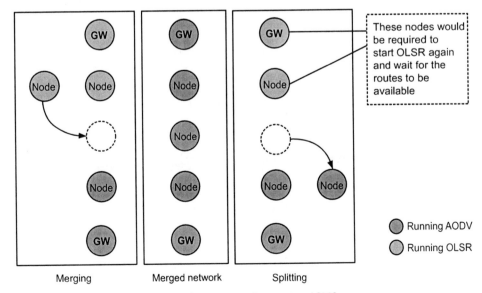

Figure 2.8   Split of MANET running the same routing protocol [10].

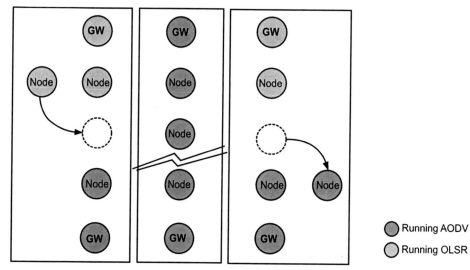

Figure 2.9 *Separation of nodes running different protocols [10].*

using the HNA message. AODV can easily be extended to support this functionality
as well. The easiest way is not to alter the AODV messages. All the nodes would
send a *route request* message for the destination host address. The destination node
would inspect that address and send a *route reply* message if the address has its prefix. The routes would be host-based. A second option would allow AODV to add
routes to networks and not to hosts. To support this, AODV either would have to
be modified, or would have to ensure that the nodes inside an LMD have the same
prefix length.

Each routing protocol must be aware if the address that each node is using is
valid inside the MANET and also the virtual interfaces where the addresses have
been configured. Each routing protocol must also know what routing table it will
manage.

To support both routing protocols, and to enable the usage of more routing protocols without changes to the architecture, the ad hoc routing can be managed by a
routing manager entity [10]. The *routing manager* works in a way that it enables all
the routing protocols to stay transparent for the rest of the architecture. This allows
the architecture to be extended with new routing protocols in the future. The *routing manager* would be responsible for starting and stopping the routing protocols,
and to inform them about changes in the prefixes and/or virtual interfaces.

An ID is given to each routing protocol, allowing for the routing protocol dependent operation to take place using the same messages for all of them. An example of routing IDs is shown in Table 2.1.

Table 2.1 Example of Routing IDs [10]

RP	ID
AODV	0x0
OLSR	0x1

2.2.1.4 Greedy Perimeter Stateless Routing (GPSR)

The *Greedy Perimeter Stateless Routing Protocol* (GPSR) is a location-based routing protocol. Such protocols use information about the geographical position of nodes in the network.

Distinct from topology-based approaches, location-based routing comprises only the forwarding step. There is no root establishment, nor a root maintenance procedure. Every node chooses the next-hop neighbor or a set of next-hop neighbors according to their mutual position with respect to destination.

The only additional information that is required is the position of the destination. Acquiring this information is comparable to the route establishment in *topology-based routing* schemes. This process is termed *location service*. The responsibility of a location service is to make a position of every node available to every other node in the network. The simplest approach to acquire the location of the final destination is to flood the network with a respective request. Thus, every node receives position of every other node in the network. However, pure flooding results in large overhead.

2.2.2 Multicast Routing Protocols

There are three fundamental methods for transmitting data over a network: *unicast, broadcast,* and *multicast* [25]. Unicast traffic is sent to a single specific destination such as a host computer, Web server, or a particular end user. Broadcast traffic is forwarded to all users of a network. Multicast traffic is delivered to a specific subset of the network's users. For example, an entire department or location may share information by using multicast data transmissions.

Both unicast and broadcast traffic are easy for networks to implement; data packets will either be delivered to a single unique destination, or they will be propagated throughout the network for all end users. Supporting multicast traffic is considerably more complex because participants must be identified, and traffic must be sent to their specific locations.

The network should also refrain from sending traffic to unnecessary destinations to avoid wasting valuable bandwidth.

A multicast routing protocol should deliver packets to all intended recipients with high probability to avoid that delay-sensitive applications get impacted.

There are different multicast routing protocols, and each one has its own unique technological solution. The *Distance Vector Multicast Routing Protocol* (DVMRP) is the earliest protocol for multicast routing. A key concept introduced by DVMRP is the use of separate forwarding trees for each multicast group; this fundamental principle continues to be used in the newer multicast routing protocols.

Another multicast routing protocol was an extension of the OSPF protocol called multicast OPSF (MOPSF). OSPF is designed explicitly to be an interior gateway protocol, meaning that it resides within a single autonomous system. Hence, any extensions to OSPF, such as MOSPF, would also reside within the confines of one autonomous system. MOSPF provides an effective means for a single corporation, university, or other organization to support multicast routing, but it cannot support wide-scale applications that require the use of the Internet [25].

An integrated multicast support solution was developed in the FP6 IST project DAIDALOS for the support of a MANET integrated with a NEMO architecture (see Section 2.2.1) [10]. A multicast mechanism for the dynamic selection of multicast agents (DMA) was added at the ARs of the NEMO architecture with the goal to reduce the frequent reconstruction of multicast tree and to optimize the multicast routes simultaneously through selecting a new *multicast access router* (MAR) dynamically. Contrary to original proposals of selecting static *local multicast agents* (LMA) [26] and the dynamic multicast agent approach [27], the decision for the AR serving as DMA is done at the mobile router (MR).

The intelligent selection algorithm based on the movement and distance maintains a recorded NEMO attachment history and checks the path increment of recently joined subnets (ARs). After start-up and in the case of a high handover rates, since the reconstruction frequency attempts to increase again, a decision mechanism is applied as a solution before the dynamic selection process whether to choose a *bi-directional tunneling* (BT) between the MR_HA and the MR (default DMA) or the dynamic approach.

A decision for the multiple paths (DMA tunnels) for the different multicast groups (flows/sessions) is possible.

This functionality was implemented in a NEMO multicast module, located in both the MR and the HA [10].

A *Multicast MANET Routing Protocol* (MMARP) was chosen to provide the MANET multicast routing. The MMARP provides efficient multicast routing inside the MANET using a mesh-based distribution structure, which can offer a good protection against the mobility of the nodes. The nodes involved in the multicast communication can be located in the same MANET, in different MANETs, or in the infrastructure network. The nodes could be static or mobile nodes, and there would be handover of one node belonging to one MANET to another during mobility. When several paths exist between two nodes participating in a multicast communication, or between a node and its gateway, the path with best *expected transmission count* (ETX) metric [28] should be selected, which is likely to select the path with the best quality.

The MMARP described here [10] was especially designed to interoperate with the fixed multicast-enabled routers, placed in the access network. The protocol uses the *Multicast Listener Discovery* (MLD) protocol to interoperate with the multicast-enabled routers, so it is independent of the multicast routing protocol used in the fixed network.

The MMARP was further enhanced with routing metrics taking into account the quality of the wireless links. MMARP, as most of the ad hoc protocols, finds the path by using the minimum hop count metric. Minimizing the hop-count maximizes the distance traveled by each hop, which is likely to minimize the signal strength and maximize the loss ratio. In addition, in a dense network there may be many routes of the same minimum hop-count with different qualities; the arbitrary choice is not likely to select the best. A wireless routing algorithm can select better paths by explicitly taking the quality of the wireless links into account [10].

2.2.3 Routing/Forwarding in Infrastructure-Based Multihop Networks

The basic routing/forwarding operation is relatively straightforward in infrastructure-based ad hoc networks in comparison to the pure ad hoc networks (i.e., MA-

NETs) [21]. Establishing and maintaining connectivity is a relatively easy task in infrastructure-based networks due to the presence of a common source or sink, which may have considerable complexity and intelligence (e.g., BS or AP). Therefore, other more involved goals can be targeted in choosing routes in infrastructure-based networks, such as QoS optimization (including throughput maximization, delay minimization, etc.) and traffic diversion due to load-balancing [21]. Further, routing among fixed relay stations (RSs) and routing among mobile/movable RSs have different characteristics, which should be taken into account when developing the routing algorithms.

MANET networks are characterized by a varying topology and require a completely distributed operation of the routing protocols. The routing mechanisms for MANETs usually target to minimize the protocol overhead needed to manage frequent topology changes by means of *on-demand route* formation [22–24] or hierarchical schemes [29]. Therefore, route selection is based on inaccurate network state information, and the metrics adopted try to optimize the energy consumption rather then the use of radio resources [21], which creates a challenge of providing QoS guarantees to traffic flows [30, 31] and several algorithms have been proposed [32–36]. The majority of the QoS routing algorithms proposed for MANETs are tailored to specific Medium Access Control (MAC) layers able to provide information on resource availability and to control resources assigned to traffic flows. The more common approach is to consider a TDMA-based ad hoc network [35]. Each connection specifies its QoS requirement in terms of time slots needed on its route from a source to a destination. For each connection, the QoS routing protocol selects a feasible path based on the bandwidth availability of all links and then modifies the slot scheduling of all nodes on the paths.

This task is not easy and even the calculation of the residual bandwidth along a path is NP-complete [21] and it has been shown that the slot assignment problem is equivalent to the graph coloring problem [37].

Infrastructure-based wireless networks can be extended in their coverage by mobile nodes with relay capabilities spontaneously acting as forwarder for other nodes outside the range of components belonging to the infrastructure, specifically, the APs [21]. In particular, self-organizing networks are promising candidates to wirelessly extend the fixed network infrastructure. Though those networks cannot provide any guarantees with respect to QoS and grade-of-service, they extend the coverage for best-effort services in a spontaneous, flexible, and cost-efficient way. In such networks nodes can communicate without the need of an infrastructure with each other by using intermediate nodes as relays, resulting in multihop communication. Thus, data from/to the Internet have to traverse several hops between the AP and those stations over a respective path. One basic challenge in such kind of networks is the routing [21].

In general (see Section 2.2.1), existing routing approaches can be classified in three categories: *proactive, reactive*, and *position-based* algorithms. In proactive schemes, all nodes maintain routing information about the available paths in the whole network even if these paths are not currently used. Hence, proactive schemes do not scale well with network size, and frequent network changes will result in high traffic load caused by signaling messages used for route maintenance, making this approach less suitable for very flexible and dynamic network topologies.

Reactive routing schemes, also called *on-demand* routing, establish and maintain only paths that are currently in use, thereby reducing the signaling traffic. Nevertheless, they have two inherent limitations when the topology changes frequently: first, even though less routes have to be maintained than in proactive approaches, route maintenance may still generate a significant amount of signaling traffic, and second, packets on their way to the destination are likely to be lost if the route to the destination breaks.

The *position-based* routing algorithms, eliminate some of the mentioned deficiencies of the proactive and the reactive algorithms. They only require information on the physical position of the participating nodes. A comprehensive overview and investigation can be found in [38]. It can be expected that in the near future all devices can be located (e.g., for the purpose of location-based services), which increases the options for use of this type of routing. The forwarding decision in position-based routing is based on the destination's position contained in the packet and the position of the forwarding node's neighbors. Position-based routing, thus, does not require the establishment or maintenance of routes. The nodes neither store routing tables nor do they need to transmit messages to keep routing tables up-to date.

The FP6 IST project WINNER [21] established strategies and methods for the efficient deployment of fixed relay stations, in a way that the overall cost of the network is minimized. Efficient radio resource allocation to network elements is a critical part of the overall network cost optimization effort. By identifying a criterion to determine, in which conditions a single-hop fixed radio link can be efficiently (from a spectral efficiency perspective) replaced with a chain of (multihop) links was found also helpful in determining the advantages and disadvantages of multihop routing.

One of the open questions regarding the deployment of wireless networks using fixed relays is the optimal number of hops between the source and destination radio stations. It is important to be able to decide with reasonable accuracy in what conditions it is more advantageous to send a signal directly to a destination (may be by increasing the allocated transmit power, bandwidth, or time) or route the same signal over a number of relay stations, each using much less resources compared to the replaced link. The ITU requirements for next generation systems have identified that use of solutions with at least two hops (whenever necessary), one of the hops being between the fixed relay and the mobile terminal, can be quite beneficial for meeting the high-level performance requirements of the systems [39]. The remaining radio link going back to the base BS, referred to as the "feeder system" in [21], is comprised of a set of fixed relays interconnected through radio links, and can have one or more hops. In order to determine the optimal number of hops of the feeder portion of the B3G wireless system, [21] proposed a novel approach based

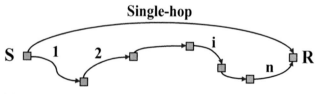

Figure 2.10   Multihop link topology with fixed relays [21].

on analyzing the aggregate spectral efficiency of the multihop communication system. The additional radio power introduced in the network by relays comes from the "wall plug" and should not be added to the radio resource costs. More significant is the effect of the relaying schemes over the aggregate end-to-end spectral efficiency. Higher overall spectral efficiency allows for better use of the available spectrum license.

Figure 2.10 shows the multihop link *S-R* in the feeder part of a fixed relay network with an *n*-hop link, where *S* and *R* are the source and the recipient stations, respectively.

The message can be either sent directly from *S* to *R* (single-hop operation), or the message can be sent via *n*–1 intermediate fixed relays over *n* hops. The relays are of the digital regenerative type. Because the relays are fixed, the topology of the system is considered known and nondynamic, and no routing algorithm is required. The overhead messaging related to multihop functionality was found insignificant, and is not considered as a factor in the discussion here. The processing delays in the relays are considered much smaller compared to the transmission time of relayed data packets.

The system uses a time-slotted resource allocation scheme, where each radio link is assigned a channel in the frequency–time domain. Each intermediate radio link adopts an appropriate modulation scheme resulting in the best possible spectral efficiency based on the given SNR conditions. The foolwoing assumptions were made in [21]:

- The amounts of data received and transmitted by each relay are equal.
- For simplicity, all *n* hops of the multihop link, and as well the single hop *S-R* are allocated the same amount of bandwidth *B*.

The individual time required to pass a message over the hop *i* is calculated by

$$t_i = \frac{M}{Bx_i} \tag{2.1}$$

where *M* is the message size in bits and x_i is the spectral efficiency (in bits/sec/Hz) of the radio link over hop *i*.

- The timeslots t_i allocated to each hop in the multihop link are considered orthogonal to each other. For simplicity, but without loss of generality, it can be assumed that all links operate on the same carrier frequency. Although the orthogonality condition seems conservative [21], in most cases the number of hops envisioned for the multihop link is three or less; with such a low number of intermediate links and given the high SNR required by the desired data rates, it is unlikely that any frequency or time slot reuse would be possible, unless advanced processing techniques or antenna architectures are employed. In summary, the total time, *T*, required to pass a message of size *M* from *S* to *R* is the sum of all intermediate timeslots:

$$T = \sum_{i=1}^{n} t_i = \frac{M}{B} \sum_{i=1}^{n} \frac{1}{x_i} \tag{2.2}$$

Because the channels used by each hop in the multihop link are orthogonal, no cochannel interference is present.

- All radio links in the system have similar radio propagation parameters.

In the case of a n-hop link with "short" intermediate links, that is, all intermediate hops lengths d_i respect the condition:

$$\frac{d_i}{d_0} < \frac{K_2^{\frac{1}{p}}}{e^2} \tag{2.3}$$

As a consequence all intermediate hop SNRs γ_i are given by:

$$\gamma_i > e^{2p} \tag{2.4}$$

On the other side, if $f(d_i)$ is strictly concave, (2.5) holds:

$$\sum_{i=1}^{n} \frac{1}{\log_{10}\left(K_2\left(\frac{d_i}{d_0}\right)^{-p}\right)} \le \frac{n}{\log_{10}\left(K_2\left(\frac{\sum_{i=1}^{n} d_i}{nd_0}\right)^{-p}\right)} \tag{2.5}$$

Then, such an n-hop link would have a better aggregate spectral efficiency (smaller message transfer time) compared to an n-hop link having the length of all intermediate hops the same and equal with the mean hop length [21].

Further, in the particular case when all relays are placed on the straight line S-R, any configuration has a better aggregate spectral efficiency compared with the situation when the n-1 relays are distributed along equal intervals on the line S-R. In other words, if all relays are placed on the straight line S-R, evenly spaced relays achieve the worst performance (lower bound) in terms of spectral efficiency.

Considering the opposite case of an n-hop link with "long" intermediate links, it can be assumed that all hops lengths d_i respect the condition:

$$\frac{K_2^{\frac{1}{p}}}{e^2} < \frac{d_i}{d_0} < K_2^{\frac{1}{p}} \tag{2.6}$$

then the intermediate SNRs γ_i are as in (2.7):

$$1 < \gamma_i < e^{2p} \tag{2.7}$$

and $f(d_i)$ is strictly convex. In this case (2.8) holds:

$$\sum_{i=1}^{n} \frac{1}{\log_{10}\left(K_2\left(\frac{d_i}{d_0}\right)^{-p}\right)} \ge \frac{n}{\log_{10}\left(K_2\left(\frac{\sum_{i=1}^{n} d_i}{nd_0}\right)^{-p}\right)} \tag{2.8}$$

Geometrically, the smallest value of the sum of all n individual hop lengths is reached when all relays are placed exactly on the straight line S-R. In that case:

$$\sum_{i=1}^{n} d_i = D \tag{2.9}$$

As a result, the best possible geographical locations for the n-1 intermediate fixed relays, which would minimize the sum term in Equation 2-10 giving the total time required to pass the message over the multihop link:

$$T = \frac{M}{BK_1} \sum_{i=1}^{n} \frac{1}{\log_{10} \gamma_i}$$

$$= \frac{M}{BK_1} \sum_{i=1}^{n} \frac{1}{\log_{10}\left(K_2 \left(\frac{d_i}{d_0}\right)^{-p} \right)} \tag{2.10}$$

and the total message transfer time T, are along equal intervals on the straight line S-R. Then:

$$d_i = \frac{D}{n} \tag{2.11}$$

for $i = 1, 2, ..., n$.

This result shows that there is no unique optimal configuration for the locations of the n-1 relays. For "long" intermediate hops, the configuration with evenly distributed relays can be considered optimal, while for "short" intermediate hops, it is the worst. Figure 2-11 shows that the geometric possible locations for d_1, $d_2 \leq D$ are within the intersection of two circles of radius D and centers S and R, respectively.

Assuming the system parameters P_T = 30 dBm, G_T = 10 dB, G_R = 10 dB, P_n = –95 dBm, d_0 = 2m, λ = 0.06m, an intermediate hop is "short" if:

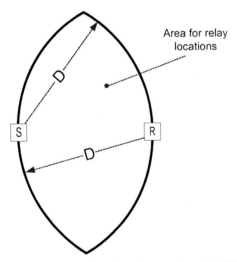

Figure 2.11   Geometric possible locations for relays in a two-hop link [21].

$$d_i < d_0 \frac{K_2^{\frac{1}{\alpha p}}}{e^2} \approx 119.6\text{m} \qquad (2.12)$$

Figure 2.12 (a) plots the time difference against the changed and normalized variables $(d_1 + d_2) / D - 1$, $(d_1 - d_2) / D$. The various surfaces are plotted for different values of the distance S-R D, which determines the 2-hop link S-R to have "short" intermediate hops or not.

The results in Figure 2.12 show that the best relay location is as close as possible to either S or R, which seems to suggest that for "short" intermediate hops the single-hop link outperforms the 2-hop link (within the framework of the initial assumptions, the 2-hop link with the relay placed at one end is equivalent with a single-hop link). For larger distances S-R, for example $D = 220$m, if the relay is located close to the straight line S-R (where $d_1 + d_2 = D$), the two hop lengths are "short," and the best relay locations are close to S or R; however, for relays located farther of the S-R line, the best relay location is at equal distances from S and R. For even larger distances S-R, the middle of the segment S-R is the optimal location for the relay (with the exceptions of locations very close to either S or R), and the use of the relay improves the overall performance compared with the single-hop.

To find out the threshold value when the 2-hop link becomes more efficient than the single-hop S-R, the condition that the message transfer time is the same for the single-hop and for the 2-hop link with the relay is placed in the middle of the segment SR [21].

2.2.3.1 Multihop Criterion

It might be more advantageous to use the n-hop link if the end-to-end spectral efficiency is improved. With the message size M and bandwidth B being the same, a single-hop link should be replaced by a multihop link if (2.13) holds:

$$\sum_{i=1}^{n} \frac{1}{x_i} < \frac{1}{a} \qquad (2.13)$$

where a represents the spectral efficiency for the single-hop link S-R. Equation (2.13) simply states that in order for the multihop link to be more efficient, the time required to pass a message of a given size M from S to R over the single-hop link must be larger that the time required for the same operation over the multihop link.

Using (2.9), (2.8), and (2.14):

$$d_1 = d_2 = \ldots = d_n \qquad (2.14)$$

the lower bound on the message transfer time using n-hops can be defined from:

$$T \geq \frac{M}{BK_1} \frac{n}{\log_{10}\left(K_2\left[\frac{D}{nd_0}\right]^{-p}\right)}$$

$$= \frac{M}{BK_1} \frac{n}{\log_{10}(n^p \gamma)} \qquad (2.15)$$

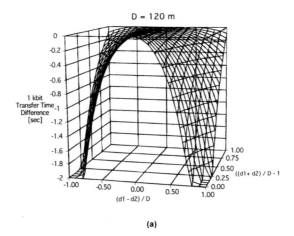

(a)

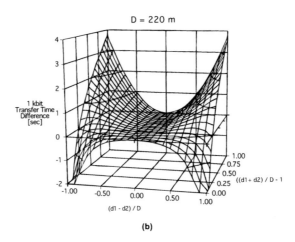

(b)

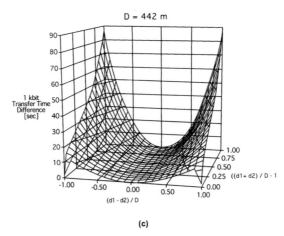

(c)

Figure 2.12 Difference between message transfer times of a two-hop link and the link with the two hops of different length: (a) D = 120m, (b) D = 220m, (c) D = 442m [21].

Equation (2.15) shows the smallest possible message transfer time using n-hops, given that (2.7) is true. If the single-hop link can transfer the data in a shorter or equal time, then there would be no point in using relays. Using (2.15), the inequality in (2.14) can be rewritten as:

$$\frac{n}{\log_{10}(n^{p}\gamma)} < \frac{1}{\log_{10}\gamma} \qquad (2.16)$$

which can be further simplified to obtain the criterion:

$$\gamma < n^{\frac{p}{n-1}} \qquad (2.17)$$

If for a given single-hop link the condition in (2.17) is not met, an n-hop link replacement with a better overall spectral efficiency does not exist, no matter where the relays are located. Equation (2.17) represents *a quantitative criterion,* which can be used to decide in which situation a multihop link could be used [21].

The SNR values in (2.17) are plotted in Figure 2.13 for various values of the path loss exponent p.

Figure 2.13 shows the variation of the threshold SNR value, at which a more efficient multihop alternative to the given single-hop link becomes feasible. For example, if the path loss exponent of all intermediate hops happens to be $p = 3.6$, then, according to (2.17), there exists a possible 3-hop configuration (the 2 relays evenly distributed along the line SR) with a better aggregate spectral efficiency, as long as the single-hop link SNR is less than 8.6 dB.

The criterion in (2.17) uses as a performance metric the bandwidth needed for transferring the messages from the source to the destination. The energy required

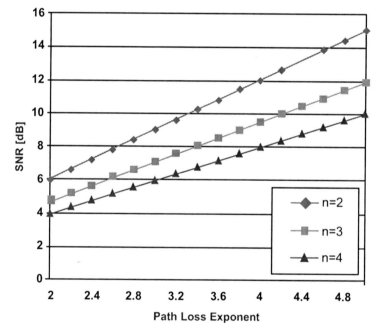

Figure 2.13 SNR values, above which the single-hop has better spectral efficiency compared to any n-hop [21].

for the transfer is not considered as a metric here; that is, the additional power inserted in the system by the intermediate relays is considered "free" [21].

The discussion above is based on the mean SNR values not including shadowing, therefore, these results are only applicable for a statistical average over a set of multihop links, and the results are not binding on a given particular realization of a multihop link.

The multihop criterion is in general applicable to multihop links with all individual hops having similar radio propagation characteristics (i.e., equal path loss exponents). In the special case when the mobile access link (the last hop of a B3G cellular link using fixed relays) has the same path loss exponent as the feeder system, the multihop criterion can be extended to cover the entire link between the BS and the mobile terminal.

2.2.3.2 Estimation of the Gains of Multihop Based on Shannon Capacity

The Shannon capacity can be used as upper bound for the achievable throughput on a single link for a given SNR and for a given bandwidth, W [40]. The Shannon capacity, C_S, over an *additive white Gaussian noise* (AWGN) channel is:

$$C_S = W \log_2\left(1 + \frac{P_{rx}}{N}\right) \tag{2.18}$$

with the received signal power, P_{rx}, and average noise power, N at the receiver. On each link of a multihop connection with n hops, the data rate has to be n times higher than the data rate of the corresponding one-hop connection to end up in the same end-to-end throughput. The required SNR for the multihop connection becomes:

$$\left(\frac{P_{rx}}{N}\right)_{MH} = \left(1 + \left(\frac{P_{rx}}{N}\right)_{one}\right)^n - 1 \tag{2.19}$$

On one side, the required transmit power has to be increased to achieve the same data rate, but on the other hand, the distance on the individual links decreases with the increasing number of hops. To incorporate this dependency in the capacity calculation, the receive power is substituted by the transmit power, P_{tx}, and the distance between source and destination, d, based on a simple one-slope pathloss model with a distance-power gradient (attenuation exponent), γ, and a constant c_0 taking into account the attenuation at 1m or:

$$P_{rx} = P_{tx} + c_0 - 10\gamma \log(d) \tag{2.20}$$

Substituting (2.20) in (2.19), the transmit-power-to-noise ratio (TNR) for the multihop connection as a function of the TNR of the corresponding one-hop connection becomes:

$$\left(\frac{P_{tx}}{N}c_0\right)_{MH} = \left[\left(1 + \left(\frac{P_{tx}}{N}c_0\right)_{one}\left(\frac{1}{d}\right)^\gamma\right)^n - 1\right] \cdot \left(\frac{d}{n}\right)^\gamma \tag{2.21}$$

This derivation assumes that transmissions cannot take place at the same time to preserve the orthogonality of the received signals in the time domain. However, concurrent transmissions increase the interference but can be acceptable if the required *signal-to-interference-and-noise ratio* (SINR) is sufficiently high. Furthermore, different interference suppression or cancellation techniques can be adopted to decrease the interference, respectively increase the SINR. In addition intelligent combining of all transmitted signals at the relays and final destination can increase the received signal power, exploiting the diversity of the radio channel. To consider the frequency reuse to come up with a higher spectral efficiency, the Shannon formula can be updated [21] to take into account the interference from simultaneous transmissions. The new parameter that is introduced is the reuse distance, d_{reuse}, measured in meters. If an equidistant spacing of relays is assumed, the reuse distance can take values equal to $i \cdot d/n$, with $i \in [1; n - 1]$. Considering the same end-to-end throughput as for a one-hop connection, the average SINR for multihop communication in the case of a 4-hop connection with frequency reuse becomes:

$$\left(\frac{P_{tx}}{N}c_0\right)_{MH}^{reuse} = \frac{\left[\left(1+\left(\frac{P_{tx}}{N}c_0\right)_{one}\left(\frac{1}{d}\right)^{\gamma}\right)^{\frac{n}{2}} - 1\right]\cdot\left(\frac{d}{n}\right)^{\gamma}}{1-\left[\left(1+\left(\frac{P_{tx}}{N}c_0\right)_{one}\left(\frac{1}{d}\right)^{\gamma}\right)^{\frac{n}{2}} - 1\right]\cdot\left(\frac{1}{k}\right)^{\gamma}} \qquad (2.22)$$

If it is assumed that two stations simultaneously transmit at a reuse distance of two hops (i.e., $d_{reuse} = d/2$), this would result in half of the number of transmission cycles compared to multihop communication without frequency reuse. In (2.22), the separation of simultaneous transmission is incorporated in the variable k. If no interference cancellation is introduced, then $k = 2$. The more interference is suppressed, the larger k becomes, unless the whole denominator in (2.22) becomes 1. In this case, the power needed for the 4-hop connection becomes equal to the power needed for two hops only. Hence, for an n-hop connection with even values n the capacity is doubled because two transmissions can take part at the same time, which is reflected in the exponent n in (2.22) that is divided by 2 compared to (2.21).

Figure 2.14 shows the resulting TNR for the multihop connection with and without frequency reuse over the respective TNR for a one-hop connection.

The capacity break-even is indicated by the line through the origin with gradient one. At the respective points both approaches, multihop and one-hop, require the same transmit power. The curves above this line indicate that one-hop connections do need less power than multihop connections, whereas curves below indicate scenarios, in which the multihop connections should be favored.

The gain for a multihop connection strongly depends on the transmit power, P_{tx}, and the attenuation exponent, γ [21, 40]. The smaller the transmit power, and the stronger the attenuation, the more attractive becomes the introduction of intermediate hops. It can be seen that with increasing the interference reduction (i.e.,

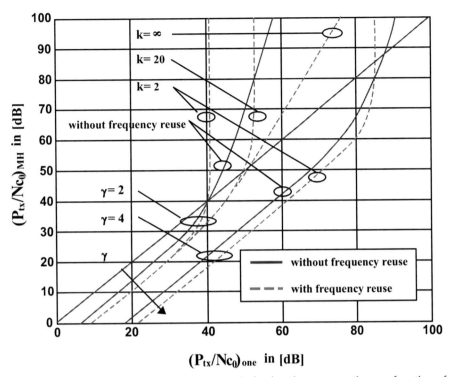

Figure 2.14 Required transmit-power-to-noise ratio for four-hop connections as function of the TNR for an one-hop connection with and without frequency reuse at 5 GHz for d = 100, N = –80 dBm [21].

with increasing the value k) the breakeven point is shifted to larger TNRs when introducing frequency reuse. Comparing the gain with and without frequency reuse, there is a constant offset of about 3 dB as long as the TNR is low enough. Beyond some points the required transmit power rapidly increases and there cannot be achieved any gains in increasing the transmit power even more because of self-interference. Nevertheless, high gains can be obtained under high attenuation, (e.g., approx. 10 dB (20 dB) for $\gamma = 2$ ($\gamma = 4$) and $n = 4$ hops).

For realistic and fair capacity estimation the protocol, respectively, physical overhead, should be taken into account [40]. The latter impact can be assessed by the worse-scenario assumption that each transmission requires some fixed overhead, ovh, and n transmissions need n-times that overhead. In this case the exponent n in (2.21) becomes

$$n \cdot (n \cdot ovh + 1)/(ovh + 1) \tag{2.23}$$

The relation of the TNR for the mutlihop connection to the TNR for the one-hop connection as a function of the introduced number of hops without considering frequency reuse is shown in Figure 2.15.

For values below 0 dB, the multihop connection needs less power than the one-hop connection, whereas values above indicate scenarios, in which a one-hop connection should be favored.

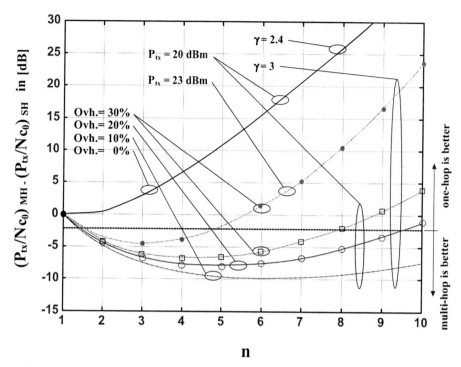

Figure 2.15 Relation of transmit-power-to-noise ratio for mutlihop and one-hop connections as a function of the number of hops, d = 100m [21].

The gain for a multihop connection strongly depends on the transmit power, P_{tx}, the distance between source and destination, d, the attenuation exponent, γ, and the number of hops, denoted by n. The smaller the transmit power, the larger the distance, and the stronger the attenuation, the more attractive becomes the introduction of intermediate hops. Typical transmit powers of P_{tx} = 20 dBm (23 dBm), a distance of 100m, attenuation exponents of γ = 2.4 (3), and different values for the overhead between 0% and 30% have been chosen. It can be seen that with increasing number of hops the gain for relaying is increasing up to a point where the overhead is dominating. For example, for P_{tx} = 20 dBm and γ = 3, the highest gains can be obtained for a number of 5 and 4 hops for overhead values of 20% and 30%, respectively. For better link conditions, respectively higher transmit power (P_{tx} = 23 dBm), the highest gain can be achieved for 3 hops only taking into account an overhead of 30%. However, the most dominant factor is the attenuation exponent, which is expected to decrease to the free-space value with decreasing distance on the one-hop connection. In this case, even for no overhead no benefits are expected from relaying for small transmit power of P_{tx} = 20 dBm and γ = 2.4.

All the aforementioned facts and results show that a number of 4 hops, which is called *oligohop*, should not be exceeded to benefit from relaying [21].

The developed quantitative criterion in (2.17) offers threshold mean SNR values below which an n-hop replacement should be considered over a single-hop link. Additional research into the statistical distribution of spectral efficiencies for multihop links is expected to give further clarifications on the properties of cellular systems using fixed relays for multihop communications.

2.2.3.3   Hybrid Routing Schemes

Location-aided AODV was proposed in [21] as a routing scheme that combines the characteristics of *on-demand* and *position-based* routing algorithms to come to a more flexible and efficient scheme avoiding the drawbacks of both at the same time.

The description of AODV in Section 2.2.1.2 showed that route breaks can have severe impacts on the system performance, especially, in dynamic network topology where route breaks happen frequently. Furthermore, the local route-repair in AODV is not optimal, since it introduces delays and overhead owing to the repair process.

Motivated by the good performance of geographical routing schemes in highly mobile environments, [21] proposed to combine both routing schemes (i.e., *on-demand topology-based* routing with *location-based stateless* routing). *Location-based* routing, also referred to as *geographical* routing, does not require route maintenance, since it determines the next hop towards the destination *on-the-fly* based on the location of the destination and the neighbors. If a station discovers a neighbor closer to the final destination, it forwards the packet to it, leaving the further delivery up to that station. To select the next neighbor towards the destination, different methods can be used, (e.g., greedy forwarding, restricted directional flooding, or hierarchical schemes). The described here hybrid routing scheme is based on greedy forwarding, which is a simple and efficient method with knowledge of only the location of the neighbors and the final destination. The combination of AODV and location-based forwarding for efficient route maintenance is called *location-aware AODV* (L-AODV).

Different to the protocol described in [41], which is used for the route discovery process, focus here is on the route maintenance. In addition, greedy-forwarding is used instead of directional flooding. [41] proposed to enhance AODV with geographical information in order to realize geocasting.

The basic operation of L-AODV is shown in Figure 2.16.

The main steps of L-AODV are the same as in conventional AODV (i.e., after route discovery and setup the packets are transferred to the destination). But when a link breaks close to the destination, L-AODV automatically switches to the greedy-forwarding mode. In this mode, it first checks whether there is a neighbor with a location closer to the destination. If this is true it continues to send the packets to that neighbor and updates its routing table, respectively. If the selected neighbor has no route to the destination in its cache, it also uses greedy forwarding mode. If the link breaks in greedy-forwarding mode, (i.e., the next-hop neighbor previously selected via greedy-forwarding is not in the communication range), the station sends an RERR packet back to the source and drops all packets for the respective destination in its queue. The source starts then again in AODV mode with a route discovery.

The two main questions are how to ensure that a station knows the location of the destination, and how to know the location of the neighbors [21].

The location of the destination is simply added in the RREP packet that is sent back to the source and all intermediate stations cache this information. The absolute location is encoded in latitude and longitude, which are each four byte in length. The location of a station is derived by means of positioning technology [e.g., from the Global Position System (GPS)].

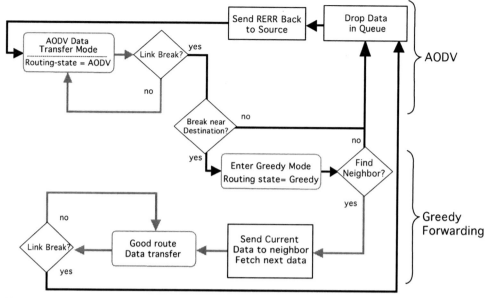

Figure 2.16 Basic description of L-AODV blocks [21].

The next challenge is to derive the locations of the neighbors in order to determine the closest neighbor to the destination, (i.e., the station to forward the packet to in case of a link break in AODV mode). Two approaches are possible. In the first approach the location of a station is included in the HELLO packets that are used to check the availability of the next-hop neighbor in AODV maintenance. The second approach is based on reactive beacon packets that are sent out by the station that has recognized a route break. As a response to this beacon each neighbor reports its current location back to that station by means of the so-called position-beacons (pos-beacon). It has been found out via simulations [41] that the second approach based on the on-demand beacon packets is more efficient than adding the eight-byte location information in every HELLO packet. Hence, only the on-demand beacon approach is used to gather the location information of the neighbors and to determine the candidate station for greedy-forwarding.

Figure 2.17 shows the required routing overhead for a scenario with five stations and use of AODV and L-AODV with beacons.

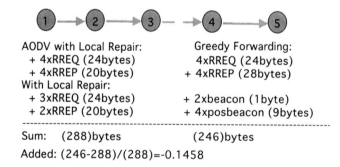

Figure 2.17 Routing overhead for AODV and L-AODV [21].

The overhead needed to discover the route is almost identical for both AODV and L-AODV, except that the RREP packets are 8-byte longer in the case of L-AODV because of the additional location information.

In case of a route break between station 3 and 4, AODV needs three RREQ packets, one by station 3 and a retransmission by station 2. If assumed that another station does not know an alternative route to station 5, say station 7 (not shown), this station also sends out an RREQ packet. Then station 5 and station 7 send back an RREP to station 3. In case of L-AODV, station 3 sends one beacon packet, to which all neighbors respond. The new station 7, which is assumed to be closer to the destination 5, responds with a pos-beacon. When station 7 receives the next data packet and if it does not have an active route to the destination, it also sends out a beacon, which is responded with two pos-beacon from station 3 and station 5, ending up in a total of two beacons and four pos-beacons for this example. Comparing the routing overhead for this example, L-AODV needs approximately 14% less overhead than AODV. In combination with *location-aided routing* (LAR) [42], to improve the route discovery, L-AODV can be a promising candidate. In addition, the integration of infrastructure components with fixed positions will further improve the performance of this hybrid routing scheme.

2.2.3.4 QoS Routing Algorithm for Fixed Relays

In a *fixed relay network* (FRN), support of multihop is mandatory and efficient routing algorithms are required to exploit at best the network resources and to efficiently manage the internal connections and load from and to the core network resources. Even if FRNs are multihop wireless networks, similar to MANETs, their characteristics make the routing problem quite different [43]. FRNs have a defined topology and the topology changes are mainly due to node failures, which is not very frequent. Therefore, the distribution of the network state information is similar in costs to that in wired networks, and even a centralized control of route selection can be adopted [44]. Finally, energy consumption is not a problem for the network nodes.

With such considerations, [21] proposed a novel QoS routing algorithm based on an ad hoc heuristic solution and on a new model for the QoS routing problem in multihop wireless networks.

The model of the wireless multihop routing problem as proposed by [21] is an extension of the multicommodity flow problem [45] where link capacity constraints are replaced with new ones that take into account interference constraints among different radio links. The model guarantees that the rates of routed flows are compatible with radio channel capacity, but does not require the explicit solving of the scheduling problem.

The proposed approach results in the complete separation between the construction of a route that satisfies the required QoS and the definition of proper scheduling at MAC level.

This means that this approach could be even applied to MAC that does not support strict resource control, (e.g., IEEE 802.11), taking a proper margin on the radio link capacity in order to take into accounts all overheads due to the protocol and the contention on the channel. In this case if strict control on the flows entering the network is enforced, the new approach assures that no persistent congestion occurs in the network.

In wireless multihop networks it is not possible to associate a capacity to each link in the graph, because the parallel transmissions on the different links may be prevented due to interference and the inability of the stations to transmit and receive at the same time. Therefore, it is assumed that when a transmission occurs on a link between nodes, the transmissions of nodes directly connected to the transmitting or the receiving nodes are prevented. Such an assumption assures that the hidden terminal problem does not occur, both for the considered transmission and that of the acknowledgement packet, (e.g., for IEEE 802.11 systems).

Let consider a graph $G = (V, E)$, whose N vertexes, $V = \{1, 2, ..., N\}$ represent the wireless stations and the M edges (i, j) connect stations within a transmission range.

If a set of K pairs of vertexes

$$(s^k, t^k) \text{ for } k = 1, 2 ... K$$

representing the source and destination node associated with K commodities, for which a path should be found on the graph G. For each commodities k can be introduced the following variables:

$f_{i,j}^k$ = Units of flow of commodity k routed on link from i to j;

F^k = Total units of flow to be sent from S^k to t^k;

$f_{i,j} = \sum_{k=1}^{K} f_{i,j}^k$ = Total units of flow on arc from i to j.

Then for each node n, represented in G by a vertex n, the following sets are defined:

$$A(n) = \{j \in V \mid (n, j) \in E\}$$
$$B(n) = \{j \in V \mid (j, n) \in E\}$$

Where $A(n)$ represents the set of nodes that can be reached from node n while $B(n)$ is the set of nodes that can reach node n with their transmission. The objective in solving this routing problem is to maximize the fraction of the total commodities that is routed through the network from source to destination. This can be expressed introducing a parameter α that represents for all commodities the fraction of F^k admitted in the network. Therefore, the objective is to find the optimum α, denoted as α^*, such that for each one of the given K commodities it is possible to route in the network exactly $\alpha^* \cdot F^k$ units of flow from the source node s^k to destination t^k. The optimum α^* is that particular value of α that satisfies the following objective function:

$$Max \left\{ \sum_{k=1}^{K} \alpha \cdot F^k \right\} \tag{2.24}$$

Further, conservation equations and non-negativity must be satisfied by every flow:

$$\sum_{j \in A(n)} f^k_{n,j} - \sum_{j \in B(n)} f^k_{j,n} = \begin{cases} \alpha \cdot F^k & \text{if } s^k = n \\ -\alpha \cdot F^k & \text{if } t^k = n \\ 0 & \text{otherwise} \end{cases} \tag{2.25}$$

$$f^k_{i,j} \geq 0 \text{ for all } (i,j) \in E \tag{2.26}$$

$$\alpha \in [0,1] \tag{2.27}$$

Equation (2.25) states for each commodity k that the difference between the incoming flow and the outgoing flow on a given node n is positive if node n is the source of the commodity; it is negative if node n is the destination, and is equal to zero if node n is an intermediate node of the k's path. Equation (2.26) states that negative commodities cannot exist in the network.

To model the capacity and interference constraints, new sets of nodes are introduced: the set made up by all links that are adjacent to node i is given by:

$$S^1_i = \{(i,j), j \in V\} \tag{2.28}$$

and for each node $j \in A(i)$ (i.e., nodes u,v,z in Figure 2.18) the set of all links that have one of their end in $A(i)$ and do not belong to the first set [see (2.28)] is given by (2.29):

$$S^2_{i,j} = \{(j,k) \mid k \neq i\} \tag{2.29}$$

For the given node i, exactly $|A(i)|$ set of this type exist.

From these sets, it is possible to build up the set G_i formed by all relays within two hops from relay i, this set can be expressed as:

$$G_i = \left\{ S^1_i \cup \underset{j \in A(i)}{Y} S^2_{i,j} \right\} \tag{2.30}$$

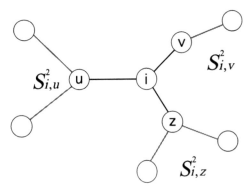

Figure 2.18 Two-hop constraints of node i [21].

It is possible to characterize each set G_i by a theoretical capacity C_i that represents the maximum aggregate flow that can be routed over the whole set of links belonging to that group. All flows routed on links and belonging to set S_i^1 contribute in consuming the common resources associated to the whole group G_i and to bound possibilities for new transmissions/receptions of relay i. Related to the various set $S_{i,j}^2$ built on each neighbor of node i, only the heaviest loaded one should be considered for writing the new QoS constraint, because it is the one that gives the more restrictive condition about the resources consumption of G_i and about the limitation for further transmissions/receptions of node i. It is, therefore, possible to write the constraint to be inserted in the mathematical model as follows:

$$\sum_{k=1}^{K} \sum_{(i,j)\in S_i^1} f_{i,j}^k + \max_{j\in A(n_i)} \left\{ \sum_{k=1}^{K} \sum_{(j,l)\in S_{i,j}^2} f_{j,l}^k \right\} \leq C_i \forall G_i \qquad (2.31)$$

The non linear constraint of (2.31) can be transposed to a set of linear constraints without any approximation giving a linear model for the QoS problem. This is done eliminating the **max** operator that gives the nonlinearity and extending the constraint to the all sets $S_{i,j}^2$ related to relay i comprising in this way also the heaviest loaded one that gives the more restrictive condition. This set can be compacted in the following expression:

$$\sum_{k=1}^{K} \sum_{(i,j)\in S_i^1} f_{i,j}^k + \sum_{k=1}^{K} \sum_{(j,l)\in S_{i,j}^2} f_{j,l}^k \leq C_i \forall S_{i,j}^2 \in G_i, \forall G_i \qquad (2.32)$$

With this it becomes possible to write the mathematical model used to describe the problem of QoS routing in a network composed by fixed/movable relays, equipped with omni-directional antennas, and a unique radio interface:

$$\text{Max} \left\{ \sum_{k=1}^{K} \alpha \cdot F^k \right\} \qquad (2.33)$$

so that:

$$\sum_{j\in A(n)} f_{n,j}^k - \sum_{j\in B(n)} f_{j,n}^k = \begin{cases} \alpha \cdot F^k & \text{if } s^k = n \\ \alpha \cdot F^k & \text{if } t^k = n \\ 0 & \text{otherwise} \end{cases} \qquad (2.34)$$

$$f_{i,j}^k \geq 0 \text{ for all } (i,j) \in E$$
$$\alpha \in [0,1]$$

and

$$\sum_{k=1}^{K} \sum_{(i,j)\in S_i^1} f_{i,j}^k + \sum_{k=1}^{K} \sum_{(j,l)\in S_{i,j}^2} f_{j,l}^k \leq C_i \forall S_{i,j}^2 \in G_i, \forall G_i \qquad (2.35)$$

The QoS algorithm described above is *wireless fixed relay* (WiFR) routing and has the following features [21]:

- A centralized algorithm, where the route computation is performed in a central entity;
- The route choice is based on a precise knowledge of the global network state and an accurate global information;
- Flow-by-flow bases (i.e., packets belonging to different flows are treated separately even if the flows have the same source and/or the same destination);
- End-to-end bases (i.e., considering the source-destination parameters of all flows in the network and not only taking a greedy decision hop-by-hop);
- Automatic load balancing through the network;
- Full functionalities with any lower layer technologies.

To ensure that the QoS algorithm avoids saturation of one or more links a penalty function of the network can be introduced:

$$PF(\underline{S}) = \sum_{(j,k)\in E} \left(\frac{1}{1 - S_{j,k}} \right) \tag{2.36}$$

where $\underline{S}$ is a new matrix, which is introduced whose generic element $S_{j,k}$ is a gauge of saturation level of the radio link from j to k. Because $S_{j,k} \in [0,1]$ even one fully saturated single link would cause the penalty function to diverge. The penalty function is also representative of the overall network condition, in fact the higher its value, the heavier the network load. The WiFR algorithm sets up routes trying to keep as low as possible the penalty function (i.e., it tries to distribute as uniformly as possible the traffic in the network, while avoiding congestion) [21].

When a route request from relay s to relay t with a request bandwidth $\tilde{b}$ (normalized to the provided bandwidth B) is delivered to the central entity, the central entity applies the route searching routine that selects, just among the feasible paths between the source node and the destination one, the path that has the minor impact on the network global saturation level. As basis for the route searching routine the mechanism of Dijkstra's algorithm can be used [21], where the metric can be modified in order to search the route just among the feasible paths and in a greedy manner, so that the optimal route is selected avoiding the exhaustive exploration of all existing feasible paths. This can be done thanks to the following exploring routine that was developed in [21] to determinate if a potential next hop relay, say k, should be explored by the Dijkstra 's algorithm or not. The following are the fall-through controls executed by the routine; j is supposed to be the relay actually selected by the routine:

1. (Preliminary controls, only if j is the source). For a given relay j the parameters stored in its record are:

$$userband(j) = ub \subset [0,1]$$
$$freerx(j) = fr \subset [0,1]$$
$$freetx(j) = ft \subset [0,1]$$

These parameters represent, respectively, the fraction of provided bandwidth B, normalized to 1, that relay j "sees" as yet consumed either for its own transmissions or for receptions of other signals both addressed to it and both not addressed to it; the fraction of provided bandwidth B, normalized to 1, that relay j has still free to receive without having collision with its own transmissions or with other received signals, this parameter depends only on what happens in relay j itself and in the set, the last parameter is the fraction of provided bandwidth B, normalized to 1, that relay j has still free to transmit without causing collision with other relays transmissions and respecting all the constraints introduced in the mathematical model.

If j is the source node then it is immediately discarded and connection is refused if it $freetx(j) < \tilde{b}$.

a. Otherwise, if j is the source node and k is not a destination node, this means that node k will transmit either to the destination or to a next hop towards the destination. It must be controlled so that j can receive the signal retransmitted by node k without violating any constraints. This means that j should not have usedband (j); this latter condition is the necessity that j should receive the signal retransmitted by node k, that consumes a fraction b of its still not in use capacity, after that j itself has transmitted the signal thus having yet consumed a fraction b. If the last equation is true, then connection is refused.

2. If connection is not refused preliminarily, then the routine controls if node k is the destination node, if so node k could be selected since node j can surely transmit directly to the destination having passed all controls before being selected.

3. The algorithm routine arrives at this step if and only if j is not the source node and k is not the destination node. If j is not the source node, then it can receive a signal from its previous hop, transmit to its next hop and receive the possible next hop's retransmission.

a. A preliminary condition that must stand not to discard immediately node k as potential next Dijkstra's selection, is that node k can forward packets of the new connection. Therefore if node k has $freetx(j) < \tilde{b}$, then it is immediately discarded because it is useless to make all other controls. If node k is not discarded here, it means that potentially it could forward packets of the new connection but other controls must be passed before selecting it.

b. Then the routine controls if one or more common neighbors exist for node j and node k, such common neighbor will receive transmission both of j and of k. If such common neighbors cannot receive both transmissions without violating the constraints, then node k is discarded.

c. If node k has not been discarded before, then the routine controls if the destination node t is a neighbor of k or not. If t is a neighbor, then it is controlled that k could receive a signal from its previous hop j and that can retransmit to destination. So if $usedband\,(k) + 2\tilde{b} \leq 1$ then node k is selected, otherwise discarded. If destination node t is not a neighbor of node k, this means that node k must be able to receive from its previous hop j, transmit to a next hop different from the destination and finally receive retransmission of its next hop. So if $usedband\,(k) + 3\tilde{b} \leq 1$ then node k is selected, otherwise discarded.

If the routine states that node k can support the new flow then it is explored by Dijkstra's algorithm as in the usual version (i.e., relay k is labeled with a temporary mark that indicates the overall distance from the source node). If the routine instead states that relay k is not able to support the new flow no further actions are taken on relay k. In both cases, after having executed proper actions based on the routine's response, a new neighbor of the actual selected node j is chosen, if existing, and the routine is called again. Once all neighbors of j have been controlled by the exploring routine and, if necessary, by Dijkstra's algorithm, then the algorithm selects the new node to be marked as definitive exactly as in the usual version of Dijkstra's algorithm and on this node the algorithm reiterates itself. All these actions are repeated until the destination is reached or until no node can be marked as definitive and so Dijkstra's algorithm cannot iterate and stops. If a feasible path is found, the central entity reads the sequence of relays forming the path and performs operations in order to update the information about the global network state [21].

The heuristic QoS routing algorithm described here finds a *suboptimal solution,* which implies that when no feasible path is found for a new flow, it is not necessarily true that the network capacity is insufficient to support the new flow, but a *different* suboptimal solution can exist that leaves room also for the new flow. Based on this assumption, [21] developed an optimization routine.

The central entity selects, according to a specified criterion one of the yet to be routed flows and tries to reroute it along a different feasible path using the same route searching routine described above and respecting the algorithm constraints in order to maintain the QoS required by all flows in the FRN. If it is not possible to reroute the selected flow, the *optimization* routine marks the selected flow and selects a new flow on which it reiterates itself. If rerouting is possible, then the optimization routine tries to admit the new flow; if this is possible, then rerouting is effectively performed and the new flow admitted; otherwise the optimization routine reiterates itself.

Optimized Scheduling

The WiFR mathematical model embeds new constraints, which allow for the QoS routing algorithm to be completely independent from the particular MAC layer effectively used in the network and for not needing any information about the scheduling if it is performed. This is a great advantage, also in terms of a commercial use of the algorithm it could be used upon any kind of MAC and physical layer without any adjustment or modifications. Further, even if no information about the scheduling is needed, the algorithm allows for the existence of a proper scheduling that supports the routing scheme for the relay transmissions at MAC layer [21].

Offering QoS at routing level and then coupling this with a non-QoS MAC layer (e.g., one from the 802.11 family) is a very suboptimal choice and many advantages conquered through the proposed routing algorithm are lost through the MAC layer incompatibility, which decreases the overall network performance. In [21] a *Time Division Multiple Access* (TDMA) MAC layer optimized for use in FRN with the WiFR as the routing algorithm was developed.

The problem of slot assignment was modelized as a variation of the *Minimum Order Frequency Assignment Problem* (MO-FAP) where a maximum given number

N_s of slots may form the MAC frame and each relay would require a given number of time slots according to the capacity it needs for transmission. The same slot is assigned to all relays, whose transmissions are not conflicting. The assignment is resolved through a heuristic algorithm of graph multicoloring that keeps the overall number of slots used minimal.

Integration of Optimization Routine in the WiFR

Figure 2.19 shows how the different routines interact with one another and how the routing algorithm accesses the external input it needs and the information stored in its internal structures.

The blocks and information regarding the MAC layer are effectively performed only either when the optimized TDMA MAC layer or another slotted MAC layer are used in the FRN.

Figure 2.20 to show the performance results for the WiFR algorithm. Figure 2.20 shows an evaluation, where the WiFR was benchmarked to a QoS routing algorithm based on pure hop-count metric named *Wireless Shortest Path First* (WSPF). The simulations were performed with an NS2 simulator [32]. A two-ray (ground reflection) channel model was used, and the provided bandwidth was 2 Mbit/sec, packets of 1 kbyte, with exponential *On/Off* traffic sources with differentiated bandwidth requests resembling voice traffic [21].

Figure 2.20 shows the throughput obtained in a FRN of 60 relays, randomly deployed and following an uniform distribution over a 1000m × 1000m area for various values of the radio range of the relays. The network is fully connected (i.e., each source reaches directly its destination) for the highest values of the radio range and both algorithm gives the same throughput. For a network with low/medium connectivity degree, the WiFR routing algorithm not only outperforms the WSPF with about 13%, but also increases the network capacity with about 19% with respect to the fully connected situation. As the radio range increases, the throughout initially is reduced until a minimum because every single transmission, impacts on the set G_i by rising the resources consumption. Increasing radio range means that the route length is diminished and, hence, also the retransmissions, in order to avoid this, after reaching the minimum, the throughput rises to a steady value.

The results show that the introduced metric is really efficient and able to distribute the load in the FRN and that mesh topology may increase the overall network performances if the resources are efficiently managed [21].

Figure 2.21 (left to right) shows a basic grid of a Manhattan type of topology with 4x4 relay used as sources and/or destinations of the connections for which the number of additional relays is zero ($n = 0$), and particular topologies with $n = 2$, and $n = 5$, respectively.

Figure 2.22 shows that WiFR outperforms WSPF in all Manhattan topologies by 6% to 10%; further, the additional relays allows for increasing the throughput thanks to the possibility of increasing the spatial reuse of the shared radio resources. The capacity gain is about 16% for only one additional relay, and about 20% for two additional relays.

It can be noticed that with three and four additional relays the throughput decreases because the route becomes longer and the retransmissions causing a performance decrease are not completely balanced by the higher spatial reuse. Therefore,

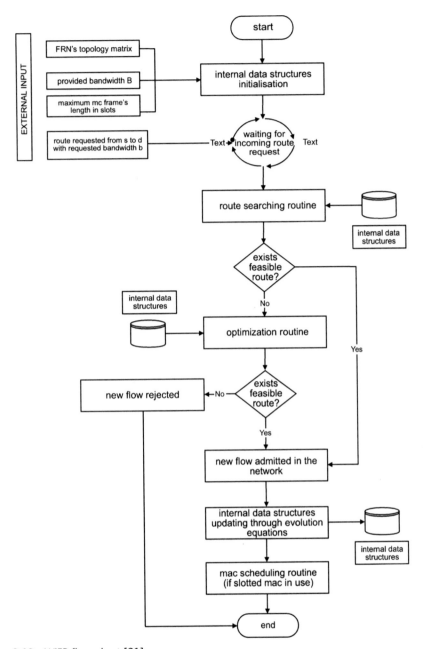

Figure 2.19 WiFR flow chart [21].

one or two additional relays can be the optimal compromise between performance and costs.

Figure 2.23 compares the achievable performance when using different MAC layers, namely, a TDMA MAC layer and an IEEE 802.11 MAC layer with RTS/CTS/Data/ACK mechanism.

For the simulations with IEEE 802.11, the overall load admitted in the FRN was kept below 20%, to ensure an initial better performance than the TDMA MAC layer. The centralized access mechanism results in no loss for the TDMA MAC

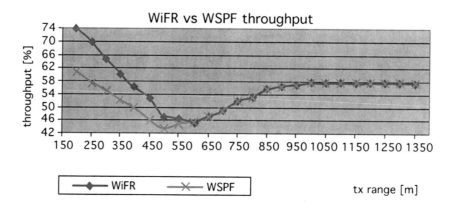

Figure 2.20   Performance of WiFR benchmarked to WSPF [21].

layer, whereas the packet loss for the 802.11 MAC layer, due only to the distributed way of accessing, is in the range from 20% to 25%. A nonslotted MAC layer, such as 802.11 is deeply inadequate to support the QoS guaranteed on the other hand by WiFR, because many of the advantages gained by the routing algorithm are lost during the channel access [21].

WiFR shows that it can keep the network conditions very stable even with an extremely varying offered load. Figure 2.24 shows the end-to-end delay when the connection requests follow a Poisson process with different inter-arrival times.

As the frequency of the connection requests increase, the end-to-end delay remains stable with a slight variation of about 50 msec on an average delay of 0,1-0,2 sec. This shows that WiFR is able to offer QoS and maintain stable the network, independently from the external offered load.

Figure 2.25 shows the performance of WiFR compare to that of ad hoc routing protocols, such as DSR, DSDV, and AODV.

The ad hoc routing algorithms do not offer QoS, therefore these admit all offered load in the FRN. WiFR outperforms all ad hoc routing protocols of at least 10% with a maximum of 55% with respect to DSDV and Manhattan topologies with five additional relays. This means that an FRN should not be considered as a particular case of an ad hoc network, but requires new routing algorithms to be developed taking into consideration the full specifics of such a scenario [21].

For a randomly generated topology, WiFR gives advantages for networks of medium, and large dimensions, in terms of number of nodes, and can enhance the overall network performances admitting a greater amount of traffic with low network connectivity with respect to a fully connected network. Further, a trade-

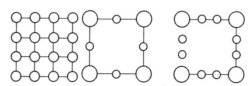

Figure 2.21   Manhattan basic grid (*n*=0) and particular for *n*=2 and *n*=5 topologies [21].

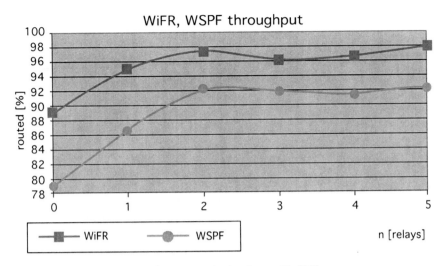

Figure 2.22   WiFR vs WSPF, Manhattan topologies, low traffic [21].

off between the network connectivity in terms of the possibility of finding several feasible routes and the number of relays influenced by each single transmission is possible and is most favorable for low network connectivity. For a Manhattan topology, the WiFR performs best for two additional relays inserted between two neighboring basic relays that act as a source and/or destination of the connections.

WiFR is able to support with QoS both *constant bit rate* (CBR) traffic and exponential traffic sources generating packet bursts even when each connection requires a bandwidth equal to its average rate. Further, WiFR can support dynamic traffic with requests following a Poisson process and with randomly generated source and destination. QoS is always preserved, also, under very heavy traffic and the end-to-end delay can be kept stable under different network conditions.

WiFR can be optimized further by a different kind of path search based on the residual capacity of each link of the network. A local search can be designed to better

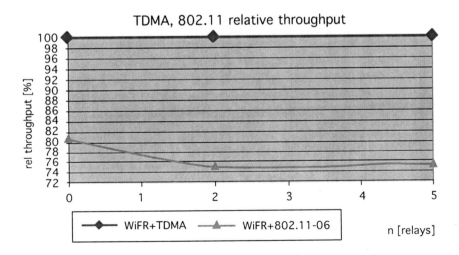

Figure 2.23   TDMA vs 802.11 relative throughputs, Manhattan topologies [21].

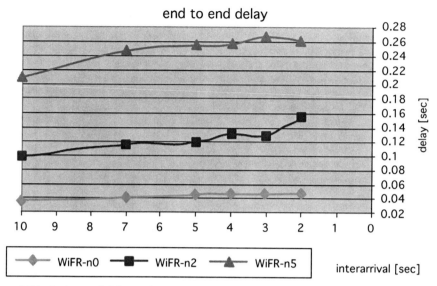

Figure 2.24   End to end delay with WiFR for Poisson arrival of connection requests [21].

distribute the routed flows and to create free space for the new connections [46]. Such improvements lead to an increased network throughput.

The WiFR with optimized path search computes a sort of residual matrix, from which it works out the new topological matrix used by the Dijkstra algorithm, in order to find the best route, using the weight as a metric. The novelty is that the Dijkstra's algorithm does not act on the whole topological matrix but on a modified topological matrix, which takes in consideration only the links, on which the desired amount of traffic can flow with respect to the mathematical constraints.

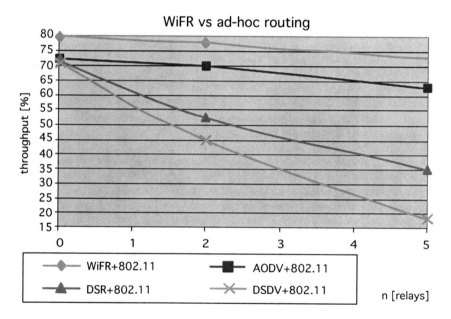

Figure 2.25   Performance of WiFR compared to ad hoc routing protocols [21].

When a route request arrives at the central entity, a routine updates the *flow_unit* structure by computing for each relay of the network the amount of unit of traffic than can be routed through the selected relay taking into account the amount of free resources of the node and the role of the node in the path. In fact, as explained previously, when updating the value of *usedband* for all the relays, it is to notice that a relay forming path p has its *usedband* updated up to three times as it may receive packets from the relay behind in the route, may transmit to its next hop and finally may receive the (useless for it) transmission of the next hop. The resulting matrix is a sort of a residual matrix because it contains for each link the amount of traffic, in relation to a particular couple source-destination, that can flow through the selected link and in that particular direction. It assures that the amount of traffic can flow across the link with respect to the available network resources with the only exception of the presence of common neighbors along the path [46]. In this case, it is not guaranteed that there are sufficient available resources, however, this control is made at the end.

As soon as this matrix has been computed a modified version of topology matrix is worked out without considering the links, which cannot form the path because of the lack of resources. Now the FRN can be treated as a fixed network described by the modified topology matrix and the path is selected by a routine based on the classical Dijkstra algorithm, which stops as soon as the destination is reached and labeled [18]. In this way the computational time is reduced, and when a route is not found because of problems such as scheduling or lack of resources due to common neighbors, the links, which cause these problems can be deleted so that the Dijkstra's algorithm can be recalled.

Another additional step to enhance the WiFR performances is a *local search* node by node, which starts as soon as the attempt to route all the given connection is ended and at least one connection has been rejected because of insufficient network resources. The purpose of this local search is to better distribute the routed flows and to create available resources to route some of the previously rejected flows. Thus, for each rejected flow a routine starts. This routine orders the nodes by their used band. Then starting from the relay with the lowest available resources the algorithm searches which of the already routed flows are responsible for the node resources consumption, either because the node belongs to those paths or because the paths flow near the selected node. As soon as the routed flow to work on is found, the algorithm tries to find another path for this flow, which consumes in general less network resources in the nodes which should be interested in the attempt to route the designated rejected flow. If a new path with those features is not found, the routine examines, if it exists, another flow related to the selected node; otherwise, the next node of the list ordered by the *usedband* is selected and the procedure is repeated. All the nodes of the list can be examined by routine or only those nodes with *usedband* upper than a fixed quantity.

As soon as a flow is moved on another path, which guarantees a total network weight lower than the previous one, an attempt to route the rejected flow is done and, even if a path cannot still be found, all the network structures are updated taking in account the modified path just found and the routine starts again from the beginning creating the new list of nodes ordered by used band and examining them one by one.

When a route for the rejected flow is found the routine stops and the following rejected flow is considered.

When an alternative path is found for an already routed flow, this one is accepted and all the structures updated only if less network resources have been consumed, that is, if the sum of the weights of each node is lower than the actual network weight. For this reason, after the choice of the rejected flow to try to route, a weight is given to each node with regard to its value of used band and to its position with respect to the couple source-destination of the flow under examination. In order to better assign the weight to each node, three different classes of nodes have been adopted. The three classes are divided as follows.

The *first* class consists of the destination node, the source node, and all the nodes at one hop away from the source. The *second* class groups, between the remaining relays, all those nodes, which are situated on a portion of the network described by a virtual circumference (or ellipse) built on the junction between the source and the destination. The *third* class contains the rest of the nodes.

A range of five different weights was adopted in [46] and these were assigned, starting from the higher and decreasing to each node in the following way:

- Nodes of the first class with available resources lower than the required bandwidth of the rejected flow and lower than two times the required bandwidth if the node is the source one and the destination is not one hop away;
- Nodes of the second class with available resources lower than the required bandwidth of the rejected flow;
- Nodes of the first class with available resources upper than the required bandwidth of the rejected flow and upper than two times the required bandwidth if the node is the source one and the destination is not one hop away;
- Nodes of the second class with available resources lower than the required bandwidth of the rejected flow;
- Nodes of the third class.

The division in five classes avoids that nodes, which should be involved in the finding of the path for the rejected flow will not be too much loaded causing the well-known blocking effect. For example, a great weight is given to the neighbors of the source, in fact, if one of these has a free band lower than the band required by the rejected flow that should be routed, the path would never be found because this node cannot receive the signal, which all the neighbors receive, when the source transmits, because of "physical broadcast." For the same problem the source, when it is not one hop away from the destination, should be led to have at least two times the required bandwidth of the available band because if a path is found this node will surely transmit and will surely receive back at least the signal of the next hop, thus with a high weight it is avoided to load too much this node reserving the available network resources to try to find a route. In general terms, the higher the risk of the blocking effect for the new flow, the higher the weight assigned to that node; and the closer the position of the node with respect to the couple source-destination, the higher the weight assigned to that node. In order to establish, which nodes are closer to the couple source-destination, if the distance between this couple is quite low with respect to the range of coverage, a circumference is chosen with a diameter equal to that distance while, if the distance is rather high, an ellipse is used with the major axis equal to the distance.

The performance was evaluated by using two ray (ground reflection) channel, a provided bandwidth of 2 Mbit/sec, packets of 1 Kbyte, a given traffic matrix with a number of sources, which is the 20% of the number of relays for random topologies and is equal to 16 for the Manhattan topologies, and CBR traffic sources with different random data rates. Given the traffic matrix, the routing was defined trying 300 times to route the given connections picking them up in random order and maintaining the best attempts as the final routing, while for the new version only 30 attempts were done selecting, for every 10 simulations, the best attempt, and running over this one the local search optimization [46].

Figure 2.26 shows the throughput obtained in a FRN of 90 relays randomly deployed and following an uniform distribution over a 1000m × 1000m area for various values of the radio range below 600 meters, which is the radio range under which WiFR outperforms the other algorithms (see Figure 2.25 and [47]).

With the only exception of the lowest values of the radio range (where the connectivity is quite low), the optimized version of the WiFR algorithm (*WiFR_new*) outperforms WiFR algorithm without modifications (WiFR). The throughput is increased on the average of about 5% with a peak of about 10% for a particular radio range. *WiFR_1step* represents the network throughput obtained only with the first step of the modifications.

Figure 2.27 shows the results obtained from simulations on a Manhattan topology starting from a basic grid of 4 × 4 relays used as sources and/or destinations of the connections and adding a number *n*, from *n* = 0 (basic grid) to *n* = 5, of additional relays that have as the only task the forwarding of the packets.

The curves differ clearly from one another showing the two steps of the algorithm enhancement. The reference curve was obtained with the basic version of the WiFR algorithm. A first improvement is observed with the modified path search

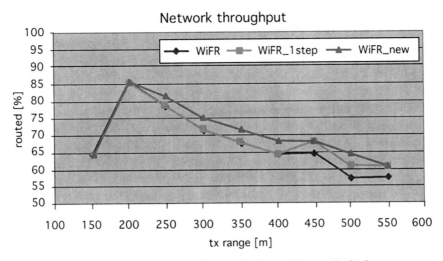

Figure 2.26   Throughput in an FRN for random topologies and low traffic [46].

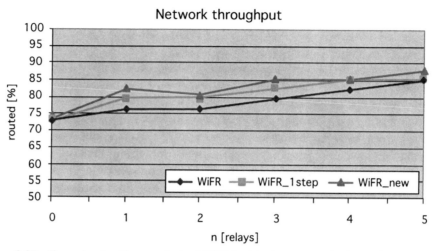

Figure 2.27 Throughput with an optimized WiFR for Manhattan topologies and medium traffic loads [46].

(*WiFR_1step*) and a final enhancement is obtained with the new version of the algorithm, which consists of both steps.

The best results are obtained for a high number of relays because for a couple source-destination the number of paths available is higher. A higher gain is achieved when the network is loaded with low/medium traffic (i.e., the connection data rates are up to 10/20% of the provided bandwidth), because with such data rates the network resources to be freed in order to route the rejected flow are lower.

2.2.3.5   Multiconstrained QoS Routing

QoS routing is based on the resource reservation process that consists of the finding the resources and making reservations [48, 49]. Before making a reservation, the routing algorithm must select the appropriate path, which meets the given requirements. To this end multiple metrics such as bandwidth, delay, loss probability, hop count, link load, or other should be considered. Finding a route subject to multiple constraints, however, is in general an NP-complete problem [46].

The network may be modeled as a weighted, directed graph $G = (V, E)$, where the vertices V represent the network nodes, and the edges E correspond to the communication links. Each of many possible weights of an edge is equivalent to one particular metric of a link. There may not exist a link either in one of or in both directions if there is no physical radio link available between any two radio interfaces. An example network graph is shown in Figure 2.28.

In the example graph of Figure 2.28, there are three weights next to each edge corresponding to the bandwidth, delay, and loss probability metrics, respectively. For the purposes of QoS provisioning, the routing algorithm must take all of them into account when selecting the best paths between any two distinct nodes in the network. Those metrics, however, are of different types and cannot be mixed.

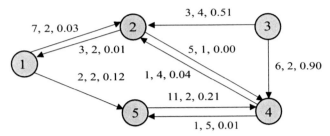

Figure 2.28 An example of a network graph [21].

In general, if $w(i, j)$ is a metric for the link (i, j) then for any path $p = (i, j, k, ..., l, m)$ the metric $w(p)$ is [48]:

- Additive if: $w(p) = w(i, j) + w(j, k) + ... + w(l, m)$,
- Multiplicative if: $w(p) = w(i, j)w(j, k)... w(l, m)$,
- Concave if: $w(p) = min[w(i, j), w(j, k), ...,w(l, m)]$.

It means that metrics such as delay, delay jitter, and cost flow are additive, loss probability is multiplicative, and bandwidth is concave [21]. The NP-completeness theorems for additive and multiplicative metrics are presented in [48] where the authors additionally explain why it is not recommended to define a single metric as a function of multiple parameters.

Let assume that the following metrics: bandwidth B, delay D, and loss probability L are put together into the formula metric of (2.37):

$$f(p) = \frac{B(p)}{D(p)L(p)} \tag{2.37}$$

Equation (2.37) does not allow to assess, whether the QoS requirements for a specific route are met or not. If a path $p = (a, b, c)$ is assumed, then in the case of delay (additive metric), $f(ab + bc) = f(ab) + f(bc)$, and in the case of a bandwidth (concave metric) $f(ab + bc) = min[f(ab), f(bc)]$. In the case of the metric in (2.37), there may not exist a composition rule as it comprises of three different types of metrics (additive, multiplicative, and concave) [21].

Despite the NP-completeness problem, computationally feasible algorithms are known for bandwidth and hop count optimization, some of them are the following [50]:

- *Widest-shortest path algorithm*: the path with the minimum number of hops is chosen. If several paths have the same number of hops, then the one with the maximum available bandwidth is selected.
- *Shortest-widest path algorithm*: the path with the maximum available bandwidth is chosen. If several paths have the same available bandwidth, then the one with the minimum number of hops is selected.
- *Shortest-distance algorithm*: the shortest path is chosen, where the distance of a *k-hop* path p is defined as:

$$\text{dist}(p) = \sum_{i=1}^{k} \frac{1}{r_i} \tag{2.38}$$

where r_i is the bandwidth of the link i.

Those algorithms are not sufficient for the purposes of efficient QoS routing in the context of next generation multihop radio systems [21] where the radio network architecture may be based on fixed and mobile/movable nodes with different capabilities, which would result in special requirements for the routing protocol.

The FP6 IST project WINNER [21] proposed a multiconstrained QoS routing algorithm, which is based on the generalized Dijkstra's algorithm. In this approach k shortest paths are selected between the source and destination nodes. Hop count was proposed to be the main criterion for this selection. Once the selection procedure is completed, every single application may select a path from the resulting set, which best meets any other criteria. Either a shortest path might be chosen, or a longer one with more bandwidth available. It is also important to provide some balancing procedures for the network traffic. To this end there may be a third criterion taken into account, which guarantees that out of the paths meeting the first and second criteria the one is chosen, which additionally will allow to distribute the load between distinct routes more evenly. If there is a need, also other criteria may be applied to the initial set of paths, which has been selected with the use of the main criterion.

The classical Dijkstra's solution [51] is applicable to the single shortest path problem, whereas the generalized version [52] allows for finding k shortest paths leading from the source node s to the target node t. Those paths are the elements of the set P, which contains all the possible paths between the aforementioned nodes. Let the *ith* shortest path in the graph G, which is leading from s to t, be denoted by p_i. The weight $w(p)$ associated with this path is defined as follows:

$$w(p) = \sum_{(i,j) \in p} w(i,j) \tag{2.39}$$

The generalized algorithm searches for the set of paths $P_k = \{p_1, p_2, ..., p_k\} \subseteq P$ under the following conditions:

1. The path p_i is selected prior to p_{i+1} for any $i = 1, ..., k$-1.
2. $w(p_i) \leq w(p_{i+1})$ for any $i = 1, ..., k$-1.
3. $w(p_K) \leq w(p)$ for any $p \in P - P_K$.

In the classical Dijkstra's algorithm [51] there is a single label d_i associated with each node, which is equivalent to the distance between that particular node and the source node. The distance is expressed by the means of the sum of the weights corresponding to the individual edges. In the case of the generalized version for selecting k shortest paths, there may be one or more labels associated with one node, depending on the number of paths passing the specific node. Each such label, connected with the node i, corresponds then to the distance between this node and the source node s on the k-th path.

Associating many labels to one node is usually computationally ineffective [21]. That is why a modified version of the generalized Dijkstra's algorithm was proposed in [52] concerned with applying the extended set of nodes. It means that each node may appear many times, depending on the number of the shortest paths that were built with the use of this particular node. As a result there is always a single label assigned to each node. For the purposes of unambiguous identification of the nodes that are used repeatedly, the algorithm uses a function *node*: $N \rightarrow V$. The aim of this function is to assign the next natural number to each node that was accepted as an element of the path leading from s to t. In consequence, many numbers may be associated with the nodes that are elements of distinct paths. The generalized Dijkstra's algorithm is shown in Figure 2.29.

The following notation is used:

- V – the set of nodes;
- V^* – the extended set of nodes;

1: **for** $\forall i \in V - \{s\}$ **do**
2:     $count_i \leftarrow 0$
3: **end for**
4: $count_s \leftarrow -1$
5: $elm \leftarrow 1$
6: $h(elm) \leftarrow s$
7: $d_{elm} \leftarrow 0$
8: $X \leftarrow \{elm\}$
9: $h^{-1}(s) \leftarrow \{elm\}$
10: $P_k \leftarrow \varnothing$
11: **while** $(count_t < k)$ and $(X \neq \varnothing)$ **do**
12:     $l \leftarrow$ element of $X \mid d_l \leq d_x \forall x \in X$
13:     $X \leftarrow X - \{l\}$
14:     $i \leftarrow h(l)$
15:     $count_i \leftarrow count_i + 1$
16:     **if** $(i = t)$ **then**
17:         $p \leftarrow$ path from 1 to l
18:         $P_k \leftarrow P_k \cup \{h(p)\}$
19:     **end if**
20:     **if** $(count_i \leq k)$ **then**
21:         **for** $\forall arc(i, j) \in E$ **do**
22:             $elm \leftarrow elm + 1$
23:             $\pi_{elm} \leftarrow l$
24:             $h(elm) \leftarrow j$
25:             $d_{elm} \leftarrow d_l + w(i, j)$
26:             $X \leftarrow X \cup \{elm\}$
27:             $h^{-1}(j) \leftarrow h^{-1}(j) \cup \{elm\}$
28:         **end for**
29:     **end if**
30: **end while**

Figure 2.29 The generalized Dijkstra's algorithm [21].

- X – the subset of V^*, used for storing information about the nodes that will be used for building the set of the k shortest paths;
- $count_i$ – the number of paths leading from the source node s to the node I;
- elm – the number of elements in the set V^*, whereas the $\{elm\}$ is the element of index elm;
- d_i – the distance of the i-th element (representing the $node(i)$) from the source node s;
- π_i – the predecessor (father) of the node i on the path from the source node s;
- P_k – the number of the shortest paths between nodes s and t.

Another approach towards a multiconstrained QoS routing problem is a modification of the OLSR protocol (see Section 2.2.1.3) [53]. The classical OLSR routing protocol stems from the *Open Shortest Path First protocol* (OSPF), which is one of the best solutions for wired networks. As a link-state class protocol it is very stable and due to its proactive nature it offers the instant availability of the routing information [21].

One of the main advantages of the OLSR routing protocol is that it uses the selected nodes only for transmission of the control messages. Those nodes are called *multipoint relays* (MPRs) and are chosen by a given node out of its all 1-hop neighbors, to which bidirectional links exist. In consequence, all other neighbors in the range of the node N, which do not belong to the *MPR* selected by this node, also receive and process the broadcast messages sent by the node N, but do not retransmit them. This is shown in Figure 2.30.

Such an approach is aimed at minimizing the number of redundant retransmissions and therefore optimizing the global control traffic level. There are two types of the control messages: *HELLO* and *topology control* (TC) messages. The first type is used for neighbor discovery and the second one for topology dissemination [54]. *HELLO* messages are broadcasted by nodes on their interfaces in a periodic manner for the purposes of link sensing, neighbor detection, and MPR selection signaling [53]. These messages are generated on the basis of the information stored in the *local link set, neighbor set,* and *MPR set.* Link sensing is necessary

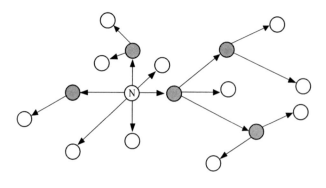

● MPR node

Figure 2.30   MPR selection in OLSR [21].

for detecting whether the radio link exists in both directions, just one, or maybe none of them. There is a direct association between the existence of a link and a neighbor. *HELLO* messages are received by all one-hop neighbors but must not be forwarded. Nevertheless, they allow each node for discovering its one-hop as well as two-hop neighbors. By the means of *HELLO* messages, each node also selects a set of MPRs out of its one-hop neighbors. Each selected node maintains a set containing those nodes, by which it was selected as the MPR. Those nodes are called the MPR selectors and each selected node may broadcast messages on behalf of one of them only.

MPR nodes broadcast TC messages only for building the topology information base. Those messages are flooded to all nodes and take advantage of the MPRs [53]. As a result the information about the MPR selectors for all MPR nodes is advertised throughout the entire network in a way that minimizes the number of retransmissions. Routes are constructed on the basis of the advertised links and links to the neighbors. To this end a node must disseminate the information about the links between itself and all the nodes in its *MPR* selector set, so that there was sufficient information provided to other nodes. The TC message follows the pattern of the general message format, where the *Message Type* field is set to *TC_MESSAGE* [53].

The information concerning the neighbors and the topology is updated periodically, in a proactive way. Usually, due to the constraints of the battery powered mobile nodes, reactive approaches are rather recommended for ad hoc-like networks. Nevertheless, if assumed that only no-power constrained relays are chosen as the *MPRs*, the *OLSR* protocol can be used without implications of the above constraint [21].

Another advantage of the OLSR protocol is that every single node may specify its willingness to forward traffic coming from its neighbors. By default all the nodes *WILL_DEFAULT* forward packets. Later on, some of them may decide that they *WILL_NEVER* do that, and so on. In consequence, a mobile terminal node may be willing to take part in routing as far as it does not detect too low battery power level. It was proposed in [21] in the context of the WINNER system characteristics that all the relay stations without power constraints were of the type *WILL_ ALWAYS*.

The *OLSR* protocol has also some drawbacks [21]. First of all, it provides no QoS functionality and uses the classical Dijkstra's algorithm for the routing purposes. The *QOLSR* protocol proposition could be based on the *shortest-widest path algorithm* [51]. However, only two types of metrics (bandwidth and delay) are considered in this solution, which might not fulfil the requirements of the communication system/scenario.

The modified *OLSR* protocol was proposed as the optimum solution [21]. The modification is connected with replacing the classical Dijkstra's algorithm with its generalized version, which is suitable for selecting the k shortest paths. There is the need for collecting more information about different parameters of the network links, but in consequence, the *QoS* routing will be feasible. The inherent features of the *OLSR* protocol can be very useful for the purposes of mobility management. It can be assumed that only nodes without power constraints (fixed or mounted on a vehicle) are used for the forwarding of the traffic, whereas other nodes are not excluded from taking part in this process.

2.2.3.6   Routing for Heterogeneous Relays

The *heterogeneous relay station* (HERS) was defined in [21] as a network node with relaying capabilities that is wirelessly connected to an AP, other relay station and/or user terminal, using different radio interface technologies for its connections. Here an access point is understood like a network node terminating the physical layer, (parts of) the link layer, and possibly also parts of the network layer of the radio interface technology from the network side. Moreover, the AP is the node closest to the core (backbone) network that a user terminal (UT) may be (directly) connected to.

In the case of heterogeneous relays it is important that the QoS parameter will be maintained on the transition. Namely, the links on both side of the heterogeneous relay node can have different requirements for RRM because those will reflect different scenarios and often would require interworking between different RAT modes [21]. For example, for a HERS, the difference of the two links can be described as follows [21]:

- **Hop 1 (from AP to HERS):**
    - No change in link quality (static and well known link conditions);
    - Point to point connection from HERS's point of view.
- **Hop 2 (from HERS to UTs)**
    - Dynamic link conditions up to loss of connection;
    - Resource has to be shared by one or more connections.

This means that on one side a HERS has to distribute the resources between the UTs based on their demands, and on the other hand, it has to provide a mechanism to release resources on the HOP 1 to achieve an efficient resource use on both sides of the relay. Another possibility would be if both hops of the relay share one radio resource. In this case a common MAC can be introduced, for example, to be in charge of the shared medium access between both systems [21]. The possible protocol stack for the HERS is shown in Figure 2.31. The common MAC is given as *generic link layer* (GLL).

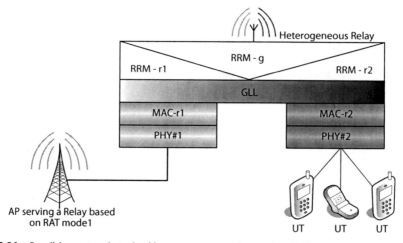

Figure 2.31   Possible protocol stack of heterogeneous relay station [21].

Figure 2.32 shows a scenario where one HERS fed by one AP is serving two UTs. The link of UT1 to the HERS is assumed to be interfered. It is further assumed that UT1 will adapt its physical layer (PHY) mode to the changed situation.

In the downlink (DL), it is assumed that UT_1 changes to a PHY mode with less data rate, which means that the HERS has to inform the AP to reduce the amount of data for UT_1 in order to avoid a buffer overflow. The consequence is that unused resources occur on the first hop. These resources could be used by the AP, (e.g., to feed some other relays or UTs in the coverage area of the AP. On the other side the UT_1-HERS connection could be only temporarily interfered, therefore, the HERS would release the resources on the first hop also only temporarily. In case, that the link between UT_1 and the HERS is broken, the HERS has two options one is to assign the free resources to UT_2 requesting more data for UT_2 from the AP and the other one is to release the resources on the first hop.

For the uplink (UP), the situation is quite similar with the only difference that the HERS is not endangered to run in a buffer overflow. This means no end to end signaling is required. But, also, here the HERS has to release the radio resources on the first hop temporarily.

This means that an end-to-end signaling is required in order to allow for efficient use of the radio resources for both the UL and the DL.

[21] proposed to perform forwarding at the GLL level instead of my means of more conventional routing procedures (e.g., IP-level routing) in addition to the development of new novel schemes to be able to exploit the added diversity gain due to the presence of multiple modes. Most developed multihop packet radio network (MPRN) routing protocols are IP-based and may be used over any underlying radio access technologies, however, when one device contain more than one radio access technology it normally implies that for every interface a unique address is used for routing purposes.

Consider a case with three nodes: one combined node (with access router-AR, border gateway-BG, and an AP, which provides access to the Internet); one RN and one UT. Depending on how many modes the different nodes support, different scenarios may be envisioned for the use of multimode capable nodes. In Figure 2.33, the UT only incorporates one mode (m2), whereas the AP and the RN incorporate two modes (m1 and m2).

The DL traffic from the AP to the UT may follow three different wireless paths: one direct path using mode m2 and two two-hop paths (wherein the first path uses mode m2 for both hops and the second one uses mode m1 for the first hop and mode m2 for the second hop). Usually, a path is made up of the nodes traversed from source to destination, but in the scenario here, a path is made up of the nodes and modes traversed/used from source to destination.

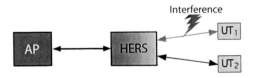

Figure 2.32   Interference situation in HERS [21].

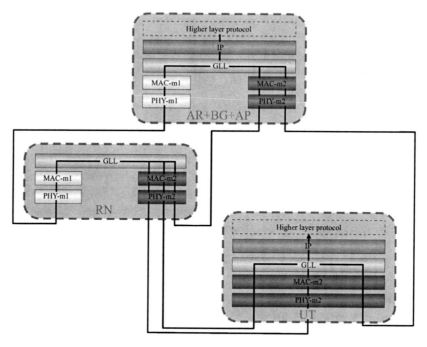

Figure 2.33 Relaying scenario involving two modes [21].

Figure 2.34 and Figure 2.35 show two scenarios were two different paths and four different paths may be used from source to destination respectively. From this it was concluded [21] that in a scenario encompassing multimode capable nodes several different paths may be used for any flow between the AP and UT.

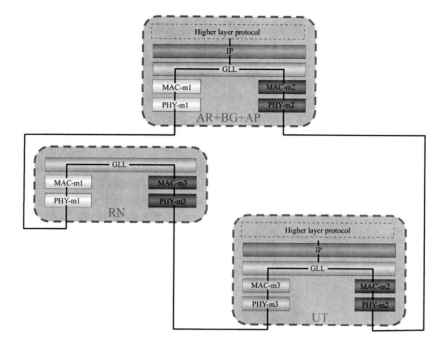

Figure 2.34 Example of a relaying scenario involving three modes [21].

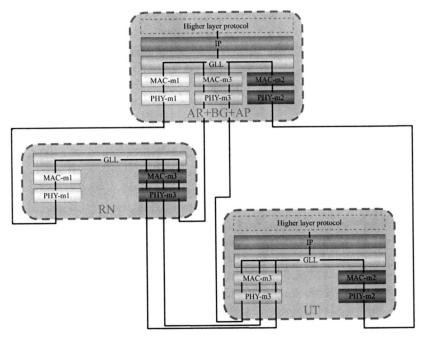

Figure 2.35   Example of another possible relaying scenario involving three modes [21].

By letting the transmitter choose one of the available modes for data transmission, the overall throughput may be enhanced and may be seen as an extension of transmission diversity techniques [21]. This form of diversity is denoted as *multimode transmission diversity* (MMTD) [21, 55, 56]. In a relaying scenario, another form of diversity may also be accounted for, namely the multiple paths outlined above. This form of combined diversity may be defined as the dynamic selection of one (or more) mode(s) to be used on the one (or more) path(s) from source to destination and was denoted in [21] as *multimode path diversity* (MMPD). As an example, a case where several modes and routes are used in parallel can be seen as MMPD. This case may be further refined to reflect whether the multiple parallel transmissions are used for increasing the robustness or data rate depending on if data is duplicated or not.

Applying MMPD in the AP for DL traffic would require (1) the knowledge of which RNs and which modes the AP may use to reach the UT; and (2) some form of (mode-dependent) routing metric to be able to make the route and mode selection (e.g., capacity estimations on the links between AP and UT). The rate, at which the nodes within the network may switch paths and modes will affect the possible gains of performing MMPD. The greatest gains (at least in theory) will be encountered when nodes are able to switch path(s) and mode(s) on a packet-by-packet basis. In this case, one may exploit the variations in radio channel quality over short time intervals to optimize the selection decisions. However, being able to select path(s) and mode(s) on a packet-by-packet basis is also accompanied by the greatest overhead since the nodes in the network will have to rely on up to date information on several paths and modes. Another possibility is to restrict the switch-rate and only allow the GLL to switch the mode and path once every x-th second or even not allow for mode and path switching for the whole duration of a communication session.

Figure 2.36 shows a scenario where diversity is provided by heterogeneous relays.

In the two-hop scenario of Figure 2.36 with multiple relays, the source sends signals via interface A to the relays; these forward the information on interface B to the destination.

The first phase solely uses interface A to convey information to the relays. The relays then independently perform conversion to interface B. The second phase is similar to the soft handover transmission known in 3GPP systems [12]: two stations, which (i.e., the two RSs), simultaneously transmit to the destination.

There the following challenges associated with such a scenario, namely:

- Coordination of transmissions (i.e., assigning relays), and conveying this information to the receiving mobile station;
- Orthogonal transmissions in phase two are required to achieve separation of the signals of the RS. This can be done by using different spreading codes as in 3GPP, or by using orthogonal space-time codes such as the Alamouti's scheme [57]. In the latter case, symbol-level synchronization must be employed [21].

The Alamouti scheme was studied in [58, 59] by assuming that the powers of two relays are constrained individually. [21] proposed a cooperative relaying scheme based on the Alamouti diversity where the power allocation optimization is done under aggregate *relay power* constraints. The rationality for a relay power constraint is that the aggregate relay transmit power should be kept at the lowest level, for a given performance, as interference can be minimized. The optimization criteria of maximizing the *signal-to-noise-ratio* (SNR) and/or the average capacity were used.

Cooperative relaying brings together the worlds of multiantenna systems and relaying. By allowing cooperation of mobile terminals, distributed antenna arrays can be built that overcome the drawbacks of correlation and spatial limitations. At

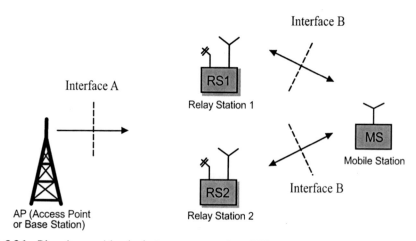

Figure 2.36   Diversity provision by heterogeneous relays [21].

the same time, placing a relay between communicating nodes is beneficial from the viewpoint of propagation losses.

In general, cooperative networks employ relay stations that receive signals from a source and resend a processed version of these signals to the intended destination. In a *cellular* or any other *infrastructure-based* network, this third node could be a BS, while in an ad hoc context, the third node would just be the common destination node or the next hop for the two source nodes.

An example of cooperative relaying scenario is shown in Figure 2.37.

Such a scenario has the following two advantages [21]:

- Cooperative systems are able to explicitly make use of the *broadcast nature* of the wireless medium as the signal transmitted by the source can in principle be received by the relays as well as by the destination;
- The relay channel offers *spatial diversity*, as the signals from the source and the relays propagate to the destination through essentially uncorrelated channels.

In a conventional, direct system, both source nodes would transmit independently and directly to the destination. Assuming a TDMA protocol with K symbol periods per slot, node 1 would transmit in times $[0, ..., K]$ to the destination node, which is followed by node 2s transmission in the second slot $[K+1, ..., 2K]$ such as shown in Figure 2.38.

In a cooperative relaying scenario, the two nodes may decide to relay parts or all of their information via the other node in order to take advantage of the spatial diversity. This cooperative transmission is restricted by the so-called *orthogonality constraint*: a terminal cannot transmit and receive simultaneously at the same frequency. This requires that the stations receive and transmit at orthogonal subchannels, which could in practice be achieved by splitting the available resources in the frequency or time domain. If the total time for sending data is equal for both the direct systems and the cooperative scheme of the example scenario in Figure 2.38, a *simple static cooperative relaying protocol* that yields diversity for both sources splits the two frames of total duration $2K$ symbols into *four* subslots of duration $K/2$ [21]. The fundamental tradeoff of cooperation and relay protocols is that the diversity gains are challenged by the constraints imposed by the orthogonality restrictions and the repetition-coded nature of the protocols.

Relaying schemes and distributed multi-antenna systems can be categorized according to a variety of parameters. Figure 2.39 gives the most important parameters for such a categorization.

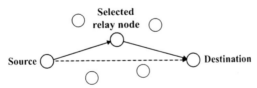

Figure 2.37 A two-hop relaying scenario: conventional relaying occurs along the solid arrows; for cooperative relaying, the destination additionally takes the signals sent by the source into account (dashed-line arrow) [21].

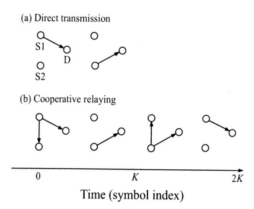

Figure 2.38 A simple cooperative relaying scheme. Two nodes may either send consecutively and directly to the destination (a), or the slots are subdivided to allow for cooperation in time-division manner (b).

Forwarding Schemes

Amplify-and-Forward: In *amplify-and-forward* schemes, the relays amplify the signal in the analog domain, without decoding the received and re-encoding the newly transmitted message. Such retransmission is characterized by the *noise amplification*. While *amplify-and-forward* protocols are simple in nature, they may be hard to implement: the use of time-division schemes for relaying requires large amounts of samples to be stored. Similarly, frequency-division approaches require the relay to retransmit the received signal on a different frequency, thus calling for the hardware that performs such frequency conversion. Amplify-and-forward schemes are also known as *analog relaying* or *nonregenerative relaying*.

Decode-and-Forward: In *decode-and-forward* schemes, also referred to as *digital relaying* or *regenerative relaying*, the relays are required to demodulate and decode the signal prior to the re-encoding and retransmission. While these schemes do not suffer from noise amplification and complex implementation constraints, they pose the danger of *error propagation* that may occur if the relay incorrectly decodes a message and retransmits this wrong information "newly" encoded. One can further subdivide *decode-and-forward* schemes according to the coding scheme as follows [21]:

- *Repetition coding* schemes have the relay repeat the source message.
- Additional coding gains can be realized by advanced *spatially distributed coding* schemes, such as the distributed Alamouti coding or distributed turbo coding. These gains, however, come at the cost of increased complexity.
- Schemes providing *incremental redundancy* have been proposed in connection with feedback. Protocols similar to ARQ can be designed, where the redundancy is delivered by relays instead of the original source node.

Protocols

Almost independently of the above classification, the relaying schemes can be of a *static* or *adaptive* nature with respect to their transmission rate and protocol.

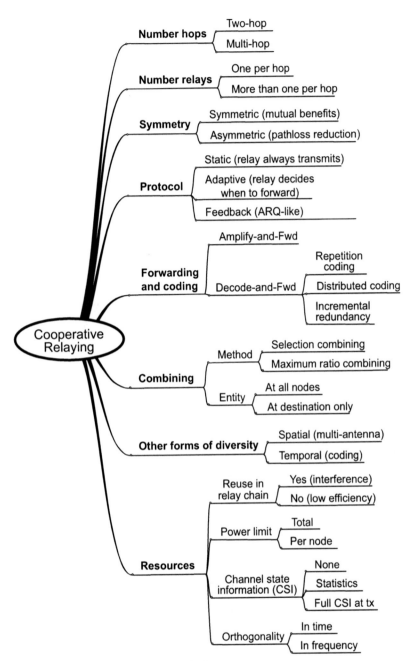

Figure 2.39   Classification of cooperative relaying.

Static Protocols: In static protocols, transmissions from the source and the relays follow a fixed pattern. Usually, a two-phase protocol would be employed
that makes the source broadcast in the first phase, and relays and destination receive this broadcast signal. In the second phase, the relays would retransmit their
signals, using either *amplify-and-forward* or *decode-and-forward* schemes. Such
static schemes may suffer from the drawbacks of their repetition-coded nature: for
some channel conditions, the destination may have successfully decoded the source

message in the first phase, thus making the second phase a waste of resources. Similarly, when the relays were not able to receive the broadcasted message with a quality that allows for retransmission, then the second phase is useless as well. On the other hand, the static protocols are attractively simple.

Adaptive Protocols: In adaptive protocols, the relays can prevent error propagation by deciding whether or not to retransmit signals. A variety of protocols can be designed.

Protocols Based on Feedback: Feedback protocols can overcome the drawbacks of the fix two-phase nature by exploiting the variable channel conditions. Retransmissions occur only if the destination indeed requires additional information. Such adaptive protocols generally yield a variable rate and may require some form of feedback, thereby complicating protocol design.

Symmetry
A symmetrical network scenario is shown in Figure 2.40 (a).

In this scenario, the two source terminals are assigned in turns the roles of a source and relay. This can be seen as a truly *cooperative relaying* scenario, and can be useful when both transmitting terminals experience similar pathloss to the common destination. In symmetric networks, the average channel fading (or pathloss, respectively) is equal for all the links from the cooperating source terminals to the destination. In asymmetric networks [see Figure 2.40 (b)], the pathloss reduction comes at the cost of loss of symmetry (i.e., the roles of relay and source can no longer be exchanged for the mutual benefit of both stations).

The performance studies of the above strategies for single and multiple antennas scenarios are described in Chapter 3 of this book.

2.2.3.7   Centralized and Distributed Routing

The objective of the routing function in the context of a next generation multihop communication system based on relays is to optimize the system performance (e.g., throughput) [60]. The routing functionality also provides input to, or has interoperations with other RRM functions such as the *access-point selection, flow admission, load control,* and *handover.*

From an architectural perspective, two basic routing strategies have been proposed for relay-based systems: *centralized* and *distributed.* As a *centralized* strategy, the route computation is performed in a *central controller,* which normally possesses powerful processing capabilities and has knowledge on the global network status. Following a *distributed* strategy, all network nodes between the source and

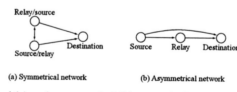

(a) Symmetrical network   (b) Asymmetrical network

Figure 2.40   Symmetrical (a) and asymmetrical (b) networks [21].

the destination (inclusive) *jointly* perform route determination. This strategy can function when no central controller is reachable, but its performance is normally limited by the processing capabilities of the network nodes and its knowledge of the network status.

The two strategies were analyzed in [60] for a fixed-relay based air interface technology with a maximum of two hops. The routing is done per user; however, *per-flow* routing can be also considered as an option to facilitate more advanced load balancing and radio resource management schemes. The communication between the BS and the UTs follows a *star* topology (i.e., there are no links between the FRNs), and each FRN communicates only with one BS (no meshing). It is assumed that the UL and DL communications follow the same route. The routing decision can be made based on either UL or DL. Packet forwarding is performed in the MAC layer.

Centralized Routing
Figure 2.41 shows an example of a centralized routing architecture [60] initially proposed for the WINNER radio system [55].

The route calculation (routing) resides at the *Access Control Server* (ACS), which is the logical node that controls the access to the air interface radio resources.

For the purpose of packet forwarding, the centralized route information needs to be distributed to the *Forwarding Information Bases* (FIBs) of the APs and the RNs. The UTs are always sources in the upstream and destinations in the downstream, (i.e., the UTs do not need to forward packets), therefore, no FIB is needed in at the UT.

Generally, within the centralised routing functional module, there are three basic functional blocks, namely, the *route discovery*, the *proactive routing* and the *in-use routes updating*, as well as a database, called *candidate routes database*. These blocks are shown in Figure 2.42.

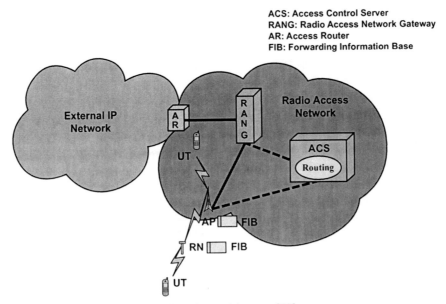

ACS: Access Control Server
RANG: Radio Access Network Gateway
AR: Access Router
FIB: Forwarding Information Base

Figure 2.41   Example of a centralized routing architecture [60].

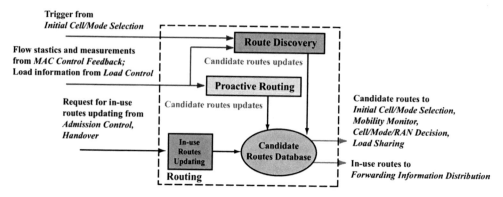

Figure 2.42 An internal structure of the routing functional module [60].

The *route discovery* is responsible for calculating the initial candidate routes under the request of *initial cell/mode selection* during the flow establishment. This is basically an initialization to a proactive routing function. The *proactive routing* can be executed periodically or on demand in order to maintain the overall quality of the candidate routes in the database. It is able to perform cross-optimization (if necessary) for the routes in the system. The main task of the *in-use routes updating* is to update the "in-use" flag of corresponding routes in the database under the requests of other RRM functions (e.g., *admission control*) and *handover*. Each active flow has one in-use route but the *candidate routes database* holds up-to-date candidate routes for all the active flows. Each flow can have multiple candidate routes in the database, and the in-use flag associated with each candidate route indicates whether the route is in use or not.

The measurements and flow statistics from the *MAC control feedback*, and load information from the *load control*, provides the required information to the *route discovery* and the *proactive routing*. The execution time scale of the *proactive routing* generally depends on the system dynamics (e.g., user mobility, flow traffic volume fluctuations), and normally is fairly slow, (e.g., once per several seconds). Moreover, if necessary, an adaptive time scale could be considered for the *proactive routing* (i.e., different time scales are adopted for users with different mobility or traffic characteristics).

The *route discovery* and the *proactive routing* are only responsible for providing the candidate routes of flows. Other RRM functions such as *admission control* and *handover* will eventually decide, which of these candidates should be used, and then send requests back to the *in-use route updating* to update the in-use flag of corresponding routes in the database.

The candidate routes in the database can be the output to other system functions, such as *initial cell/mode selection, mobility monitor,* and *load sharing*. The in-use routes in the database should be signalled to APs and RNs by means of *forwarding information distribution*, to update their FIBs for packet forwarding purposes.

Generally, a FIB is a database purely for the purpose of packet forwarding. Each entry of it consists of the minimum amount of information necessary to make a forwarding decision on a particular packet. The typical components within an FIB entry are the address of the destination (or flow ID) and the address of the next hop.

FIBs are maintained by means of *forwarding information distribution* based on the corresponding in-use routes in the centralized *candidate routes database.*

For a system with maximum of two-hops for a connection, in the downstream, the BS and RNs are supposed to do the packet forwarding, whereas in the upstream, only the RNs act as packet forwarders. In the upstream case, the RNs always forward packets to the BS and as such, the upstream FIB of each RN essentially has only one fixed entry, and therefore can be omitted. One major advantage of the centralized routing strategy is that cross-optimization can be performed in order to improve the system performance. As a result, the centralized route calculation algorithm usually involves some degree of algorithm complexity. Nevertheless, owing to the powerful process capability of the central node, sophisticated routing algorithms are normally feasible to implement. Moreover, heuristic algorithms can also be devised to notably bring down the algorithm complexity with small performance penalty.

The route calculation algorithm needs some input measurements, (e.g., channel qualities between users and BSs/RSs, and the loading status of the BSs and RSs). These are normally available already in the central node to be used by other RRM functions such as handover, admission control, and load sharing. As a result, the *centralized* routing strategy is not likely to induce much extra complexity due to the input information gathering.

User candidate routes calculated are stored in the *candidate routes* database, and the corresponding forwarding information is distributed to the relevant network nodes once the flow is established. Hence, some system complexity will be brought to maintain the *candidate routes* database. Based on the consideration that the route updates are happening in a fairly slow time scale, this is normally not a major problem.

Distributed Routing
In the distributed routing architecture, the UT makes the decision on the RAP (i.e., BS or FRN), it is going to be connected to, which is implicitly a *routing* decision. The routing decision is made in the UT based on the information, which is broadcasted by the RN and BS in the network coverage area [60]. This would mean that only a respective routing algorithm has to be placed in the UT, which is relaying on the information conveyed by each RAP during the broadcast phase. The resulting logical nodes architecture is shown in Figure 2.43.

The FRNs and BSs include *link cost information* in their *broadcast channel* (BCH). Link cost information for the BS-RN link indicates the radio channel status (e.g., path-loss, interference) as well as the BS or FRN use (e.g., their traffic load). For the RAP-UT link, the link cost only indicates the utilizations cost of the RAP as the UT can extract the radio channel status from its measurements.

The UTs listen to the BCH and evaluate the optimal access point. Implicitly, the selection of the optimal server also results in a route to be followed toward the access point into the core network. Here, the selection of the optimal route involves the radio channel status and traffic load.

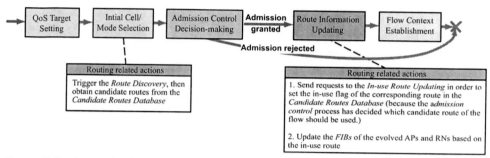

Figure 2.43 Interaction between routing and admission control at the flow establishment in a distributed architecture [60].

Based on the downlink pilot signal measurements, the UT evaluates the cost of the air interface link UT-RAP for the received pilots. From the BCH, the UT also extracts the costs related to the RAPs utilization as well as the cost of the RAP-BS link. By selecting the RAP with the lowest total cost, the UT also selects an appropriate or optimal route. Hence, it can be seen that the UT selects not only the *best server*, but actually the *best access route*. The algorithm can be enhanced to consider additional parameters such as the user subscription attributes, and so fourth.

One solution could be that the radio resources in each relay enhanced cell, including the BS and the first tier of FRNs, are administered from the BS. In this way, for example, if the traffic load increases within the coverage area of a RN, the BS can assign it additional resources at the expense of other RNs in the cluster. The BS would not have visibility into resource assignment/ scheduling within a RN (i.e., between flows / UTs). Similarly, the assignment of resources for competing BSs will be arbitrated by the central ACS entity. The ACS will not have visibility into the cluster level resource assignments, which are controlled by the BSs.

In the distributed routing architecture, the signaling overhead related to the routing is limited to the cost information broadcasted by BCHs. The UTs then make the routing decision based on a simple decision making procedure to find the route with the lowest cost. Therefore, the routing complexity here is mainly a function of the number of the received BCHs. For cases with more than two hops, the signaling overhead would be an important factor that should be further investigated. In such cases, the decision making procedure would be also more complex.

Routing Interactions with Other RRM Functions
Routing and Admission Control: In order to make an appropriate *admission control* decision at the flow establishment, apart from the flow QoS parameters, the network load status, and so fourth, the possible routes for the new coming flow are also necessary. Therefore, prior to the *Admission Control* decision making, the *route discovery* needs to be performed to find out candidate routes for the flow (see Figure 2.43). If the flow is admitted, the in-use route for the flow is therefore chosen, and hence, *route information updating* will be executed to send requests to the *in-use route updating* in order to set the in-use flag of the corresponding route in the *candidate routes database*, and update the *FBIs* of evolved APs and RNs based on the in-use route.

Routing and Handover: When the areas covered by the BS and RN essentially comprise separate cells (i.e., the RNs will have their own broadcast channels), the handover can, therefore, take place between the APs and RNs. The *proactive routing* is executed periodically or on demand to maintain the candidate routes in the *candidate routes database*. The *mobility monitor* keeps monitoring the quality of the candidate routes for the individual flows. Once the *mobility monitor* finds the in-use route of a flow needs to be updated, it triggers a handover process. If the handover is successful, requests will be sent to the *in-use route updating* in order to modify the in-use flag of corresponding routes in the *candidate route database*, and the *FBIs* of evolved APs and RNs will be updated based on the in-use route. The interaction between *routing* and *handover* is shown in Figure 2.44.

Routing and Load Sharing: Load sharing is an important congestion avoidance control mechanism, and is responsible for the prevention of the system overloading by means of appropriately sharing loads between the RNs and the APs [60]. The *load sharing* periodically checks the candidate routes of the flows. If it finds it beneficial for the system (e.g., the overloading of some RNs or APs can be avoided) to change the route of a flow, it will trigger a handover process. The routing related actions after the handover are the same. The interaction between *routing* and *load sharing* is shown in Figure 2.45.

In summary, under the centralized strategy which is *network-oriented*, the route computation is performed in a central controller, which normally possesses powerful processing capabilities and has knowledge on the global network status. In the distributed strategy, which is *user-oriented*, all network nodes between the source and the destination (inclusive) involve in performing route determination. This strategy can function when no central controller is reachable, but its performance is normally limited by the processing capabilities of the network nodes and its knowledge of the network status.

It can be feasible to implement both approaches in a single radio system [60]. The technical comparison must consider the advantages and disadvantages of the centralized and distributed decision making procedures, which involves complexity,

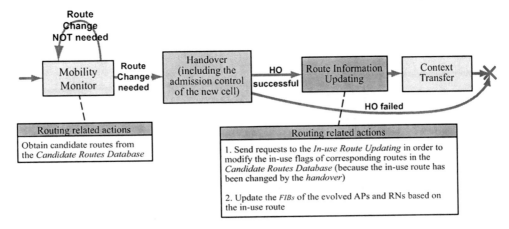

Figure 2.44 Interactions between routing and handover [60].

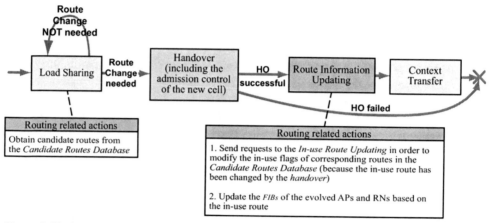

Figure 2.45 Interaction between routing and load sharing [60].

signaling overhead, response time to variations in different time-scales, and so forth. In the case of having at most two hops with no meshing and very slow network dynamics, it is simple to show that given a perfect link cost information, the two presented methods result in a unique routing solution [60]. If the objective is to optimize the system performance (e.g., throughput), instead of to maintain the network connectivity, the centralized strategy is potentially more capable. However, the computational complexity and signaling overhead of centralized approach should be taken into consideration. The signaling overhead and delay of the centralized strategy (caused by the information gathering) would be acceptable for a two-hop system. In addition, some other RRM functions require lots of measurements gathering, (e.g., link qualities), and some of these measurements can be shared with the centralized routing function, and thus further mitigate its problem of information gathering.

The distributed routing approach reduces the signaling overhead. Because the routing decision in this approach is made locally, potentially this approach can react with a lower latency in responding to the changing conditions of the air interface. The distributed routing however may require a higher level of computational complexity in the UTs. Such approach, because of its distributed nature, may also make the implementation of the advanced traffic management functionalities harder than in a centralized architecture. The signaling overhead for this approach would be increased for cases with more than two hops and/or meshing [60].

2.3 Protocols and Algorithms for Self-Organizing and Opportunistic Networks

Opportunistic networking constitutes a medium-term application of a general-purpose MANET for providing connectivity opportunities to pervasive devices when no direct access to the Internet is available. Pervasive devices, equipped with different wireless networking technologies, are frequently out of range from a network but are in the range of other networked devices, and sometime cross areas where some type of connectivity is available (e.g., Wi-Fi hotspots). Thus, they can opportunistically exploit their mobility and contacts for data delivery. Opportunistic networks aim at building networks out of mobile devices carried by people, possibly without relying on any pre-existing infrastructure. Moreover, opportunistic networks look at mobility, disconnections, partitions, and so fourth as "features" of the networks rather than exceptions. Actually, mobility is exploited as a way to bridge disconnected "clouds" of nodes and enable communication, rather than a drawback to be dealt with. More specifically, in opportunistic networking no assumption is made on the existence of a complete path between two nodes wishing to communicate. Source and destination nodes might never be connected to the same network, at the same time. Nevertheless, opportunistic networking techniques allow such nodes to exchange messages. By exploiting the "store-carry-and-forward" paradigm, intermediate nodes (between source and destination) store messages when no forwarding opportunity towards the final destination exists, and exploit any future contact opportunity with other mobile devices to bring the messages closer and closer to the destination. This approach to build self-organizing infrastructure-less wireless networks has been recognized as much more practical than the conventional MANET paradigm.

2.3.1 Protocols and Algorithms for Sensor Networks

Sensor networks developments are one of the driving forces towards pervasive and self-organizing systems. The field of sensor networking research has been gathering a large interest, because it has demonstrated that concepts for multihop, self-organizing networks with energy-efficiency as a primary constraint and high levels of the miniaturization of network devices, even up to smart dust, are feasible [61].

2.3.1.1 MAC Protocols and Link Layer Techniques

The design of an appropriate MAC protocol is very relevant for the behavior of sensor networks. Such design can solve a number of the problems that can occur in WSn due to the shared wireless medium [61]. An efficient operation for these MAC protocols can be achieved by carefully considering the cross-layer dependencies (e.g., with routing, or the antennas at the PHY layer). In addition, a well-designed MAC protocol for a WSN should be adapted to the overall objective imposed to the network, (e.g., a long-lasting energy-efficient operation, the specific task that the network should perform, and so forth). In the following, the design of some generic MAC protocols is described.

Energy Efficiency of P-Persistent CSMA in Nonsaturated Traffic Conditions
Contention-based MAC schemes offer a number of advantages for use in WSNs, including minimal synchronization requirements between the nodes and making the need of a central scheduler redundant. As a result, the deployed WSNs can be more robust. However, contention-based schemes suffer from many sources of energy wastage, namely, collisions, transmission overhearing, and idle listening. The various sources of energy wastage for *P*-persistent CSMA MAC were thoroughly analyzed in the FP6 IST project CRUISE [7] and a closed formula for calculating the expected energy efficiency as a function of the various design parameters, for a fully

connected, one hop network topology, was derived [61]. Previous analysis concerning the energy efficiency of CSMA-based protocols had been performed considering the IEEE 802.11 protocol family and focusing on laptop applications, rather than WSNs [62]. As a result, the analysis had been performed assuming saturated conditions, in which, every node has a packet to transmit at any time. While a saturated conditions model can be used for typical ad hoc networks, consisting of laptop computers, it proves inadequate for the design of CSMA-based MAC protocols for sensor networks where the nodes are expected to require a more infrequent access to the channel [61]. At the same time, optimization of the energy efficiency becomes very important due to the scarce energy resources of the nodes.

Following the approach for slotted p-persistent CSMA/CA, for a finite number of nodes [63, 64], forming a fully connected, one-hop wireless network, a model was derived in [61].

Each node can be either a transmitter or a receiver with equal probability. All transmitted packets have the same length, and are transmitted over an assumed noiseless channel. Each terminal can sense any transmission that occurs within the *carrier sensing* (CS) area and delays its own transmission. All the transmissions inside the *transmission range* (TR) are successful, (i.e., the SIR is greater than a threshold, which allows error-free reception).

For simplicity, it is assumed that a packet transmission duration (T) equals unity (this includes the interruption time), therefore all the values are normalized to the packet duration. The slot duration denoted by a is chosen to be equal to the propagation delay. Furthermore, $1/a$ is assumed to be an integer.

All the users are synchronized to start their transmissions at the beginning of a slot. Each terminal has periods, which are independent and geometrically distributed (in order to take advantage of the memoryless property), in which there are no packets.

A terminal is called "empty" if it has no packets in its buffer awaiting transmission, and ready otherwise. A terminal switches from the ready state to the active state at the moment that it starts a transmission.

In each slot, an (empty) terminal generates a new packet with probability g $(0 < g < 1)$, comprised of both new and rescheduled packets. For the p-persistent protocol each ready terminal starts transmitting in the next slot with probability p $(0 < p \leq 1)$.

Each user is assumed to have at most one packet requiring transmission at any one time (including any previously blocked packet). This means that all users make room for new packet arrivals from the beginning of a packet transmission by putting aside the already buffered packets.

The system state consists of a sequence of regeneration cycles composed of consecutive busy and idle periods. An idle period (I) is defined as the time, at which the channel is idle and all the terminals are empty. In contrast, a busy period (B) is defined as the period, at which there is a transmission, successful or not, or where at least one of the M terminals is ready. A busy period ends if no packets have accumulated at the end of a transmission. The time sequence of the events is shown in Figure 2.46.

The system is regenerative with regards to the successful transmissions from the tagged node. A *virtual transmission period*, which starts at the end of a successful

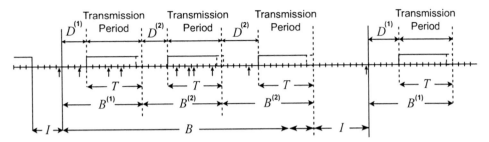

Figure 2.46   Time sequence of events in a one-hop wireless network [61].

transmission by the tagged node and finishes at the end of its next successful transmission, can be defined.

The virtual transmission periods are identical and include a number of subperiods as well as idle period(s). In every subperiod, the tagged node will successfully transmit a packet with a probability $\overline{P_s}$ and fail to transmit with probability $1 - \overline{P_s}$, where $\overline{P_s}$ is the average success probability in any subperiod, given by:

$$\overline{P_s} = \frac{1}{\overline{J}} P_s^{(1)} + \frac{\overline{J} - 1}{\overline{J}} P_s^{(2)} \tag{2.40}$$

J is the average number of subperiods within a busy period, and $\overline{J}$ is the expectation of J, given by:

$$\overline{J} = 1/(1-g)^{(1+1/a)M} \tag{2.41}$$

The expected number of subperiods until the first successful transmission by the tagged node (including the transmission) is given by:

$$N_s = \sum_{i=1}^{\infty} i(1 - \overline{P_s})^{i-1} \overline{P_s} = \frac{1}{\overline{P_s}} \tag{2.42}$$

On the average, every $\overline{J}$ subperiods, there is an idle period $\overline{I}$, then the average idle time in every virtual transmission period is given by:

$$\overline{I}_v = \frac{\overline{I}}{\overline{P_s}\overline{J}} \tag{2.43}$$

The virtual transmission period is shown in Figure 2.47.

From an energy consumption point of view, the node can be in one out of four different states at any time, namely the following:

- *Transmit*: the node is transmitting, successfully or not, consuming energy of W_T per timeslot.
- *Receive*: the node is receiving a packet, successfully decoding the enclosed information bits and consuming energy of W_R per timeslot.
- *Idle*: the node is sensing the medium to detect a carrier, but is not attempting to decode any information bits. In this state, the node is consuming energy

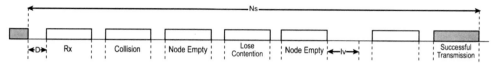

Figure 2.47 Virtual transmission period [61].

of W_I per timeslot, which is slightly smaller than W_R since decoding is not performed.

- *Sleep*: In this state, the transceiver is switched off and the sole energy (W_S) that is consumed is from the DSP clock.

Assume that whenever a transmission begins the nontransmitting nodes try to decode the header of the packet to identify if they are the intended receiver and read the *network allocation vector* (NAV), indicating the length of the packet. The length of the header is T_H and that T_H/α is an integer. If there is no collision all the listening nodes will successfully decode the above information and the intended receiver will continue decoding the remaining of the packet, while the rest of the nodes will go to sleeping state until the NAV expires. If there is a collision, none of the listening nodes can decode the NAV, so they keep sensing the channel until the transmissions are over, when they can contend for the channel. The average number of the successful transmissions in one virtual transmission period is $N_s - N_{cs} = M$.

From definition, the tagged node has exactly one successful transmission in one virtual transmission period. Because all the nodes become receivers with the same probability, the average subperiods that the tagged node will be a receiver in a virtual transmission period is given by:

$$N_R = \frac{N_s - N_{cs} - 1}{M - 1} = 1 \tag{2.44}$$

The energy consumption of the tagged node during transmission, reception, idle listening, contention, overhearing, and collision is calculated as follows:

$$E_{Tx} = \frac{1}{\alpha} W_T$$

$$E_{Rx} = \frac{1}{\alpha} W_R$$

$$E_{idle} = \left\{ N_s + \frac{\overline{I_v} + N_e \overline{D}}{\alpha} \right\} W_I$$

$$E_{contend} = \frac{(N_I + N_c + 1)\overline{D}}{\alpha} W_I \tag{2.45}$$

$$E_{overhearing} = (M - 2) \left\{ \frac{T_H}{\alpha} W_R + \frac{1 - T_H}{\alpha} W_S \right\}$$

$$E_{coll} = N_c E_T + (N_{cs} - N_c) \left\{ \frac{T_H}{\alpha} W_R + \frac{1 - T_H}{\alpha} W_I \right\}$$

The total energy spent by the tagged node in a virtual transmission period (E_{Total}) is given by the sum of the above amounts of energy, and corresponds to the total energy spent (on average) for the successful transmission of one information packet. The energy efficiency (η) is defined as the ratio of the energy spent on the payload transmission over the total energy:

$$E_{\text{Total}} = E_{Tx} + E_{Rx} + E_{\text{idle}} + E_{\text{contend}} + E_{\text{overhearing}} + E_{\text{coll}}$$

$$\eta = \frac{\dfrac{1 - T_H}{a} W_T}{E_{\text{Total}}} \tag{2.46}$$

Figure 2.48 shows the percentage of the total energy that is spent on different states of the tagged terminal (E_x/E_{Total}) as a function of the contention probability p. For lower values of p most of the energy is spent on contending for the channel, due to the very long transmission delays $\bar{D}$. As p increases, the collisions become the dominant cause for the energy expenditure. Besides the colliding nodes, the overhearing ones are also affected by the collisions because they are not able to decode the NAV and thus cannot go to a sleeping state.

Figure 2.49 shows the energy efficiency, η, as a function of and the total offered load $G = gM/a$.

There is, generally, an optimum p for given traffic conditions. If the total offered traffic is below a certain value most energy is spent in idle listening to the channel, reducing η to zero. A straightforward way to increase the efficiency would be to introduce sleeping periods, during which the sensors collect data to forward to the

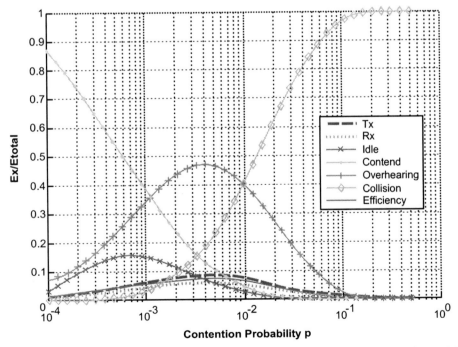

Figure 2.48 Percentage of the total energy spent on different states (M = 50, G = 50, a = 0.01, TH = 0.1) [61].

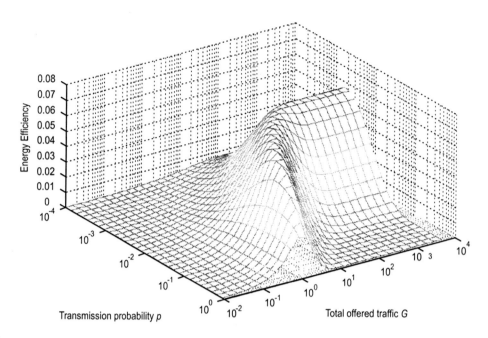

Figure 2.49   Energy efficiency (η) as a function of G and p (M=50, a=0.01, TH=0.1) [61].

MAC when it is awake. Effectively, the sleeping patterns have the same effect when increasing the total offered load.

The provided framework gives the expectation of the energy expenditure from a node in any possible communication state in for the transmission of one information packet [61]. A variety of parameters can be considered, including the number of terminals sharing the channel, the total offered traffic, the contention probability, and the packet header and payload lengths, making this framework a very valuable tool in the design of CSMA-based MAC protocols.

Energy Efficiency of WiseMAC in Low-Traffic Conditions

WiseMAC is based on the preamble sampling technique [65, 66]. This technique consists in regularly sampling the medium to check for activity. By sampling the medium, listening to the radio channel for a short duration is realized, (e.g., the duration of a modulation symbol). All sensor nodes in a network sample the medium with the same constant period T_W. Their relative sampling schedule offsets are independent. If the medium is found busy, a sensor node continues to listen until a data frame is received or until the medium becomes idle again. At the transmitter, a wake-up preamble of size equal to the sampling period is added in front of every data frame to ensure that the receiver will be awake when the data portion of the packet arrives. This technique provides a very low power consumption, when the channel is idle. The disadvantages of this protocol are that the (long) wake-up preambles cause a throughput limitation and a large power consumption overhead in transmission and reception. The overhead in reception is not only bared by the intended destination, but also by all other nodes overhearing the transmission. The WiseMAC protocol was designed to reduce the length of this costly wake-up preamble [61].

The novel idea introduced by WiseMAC consists in learning the sampling schedule of one's direct neighbors to use a wake-up preamble of minimized size. This provides a significant improvement compared to the basic preamble sampling protocol, as well as to S-MAC and T-MAC. WiseMAC basically works as follows [66].

Because the wireless medium is error prone, a link level acknowledgement scheme is required to recover from packet losses. The WiseMAC ACK packets are not only used to carry the acknowledgement for a received data packet, but also to inform the other party of the remaining time until the next sampling time. In this way, a node can keep a table of sampling time offsets of all its usual destinations up-to-date. Using this information, a node transmits a packet just at the right time, with a wake-up preamble of minimized size, as shown in Figure 2.50.

The duration of the wake-up preamble must cover the potential clock drift between the clock at the source and at the destination. This drift is proportional to the time since the last resynchronization (i.e., the last time an acknowledgement was received). The synchronization mechanism of WiseMAC can introduce a risk of systematic collision. Indeed, in a sensor network, a tree network topology with a number of sensors sending data through a multihop network to a sink often occurs. In this situation, many nodes are operating as relays along the path towards the sink. If a number of sensor nodes try to send a data packet to the same relay, at the same scheduled sampling time and with wake-up preambles of approximately identical size, there are high probabilities to obtain a collision. To mitigate such collisions, it is necessary to add a medium reservation preamble of randomized length in front of the wake-up preamble. The sensor node that has picked the longest medium reservation preamble will start its transmission sooner, and thereby reserve the medium.

A very important detail of the WiseMAC protocol, which is also found in the IEEE 802.11 power save protocol, is the presence of a more bit in the header of data packets. When this bit is set to 1, it indicates that more data packets destined to the same sensor node are waiting in the buffer of the transmitting node. When a data packet is received with the more bit set, the receiving sensor node continues to listen after having sent the acknowledgement. The sender will transmit the following

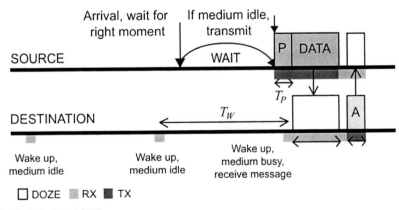

Figure 2.50 WiseMAC [61].

packet right after having received the acknowledgement. This scheme permits to use a sampling period that is larger than the average interval between the arrivals for a given node. It also permits to reduce the end-to-end delay, especially in the event of traffic bursts.

The performance of the WiseMAC protocol was studied in [61, 65] using both a regular lattice topology with traffic flowing in parallel and a typical sensor network topology with randomly positioned sensors forwarding data towards a sink. The interest of a lattice topology with traffic flowing in parallel is that it allows for exploring the behavior of a MAC protocol without inserting aspects linked to routing, load balancing, and traffic aggregation. The regularity of the topology allows deriving mathematical expressions to approximate the power consumption. Once the parameters defined, the analysis starts considering the lattice topology and then it addresses the random network topology [61].

WiseMAC presents many appealing characteristics. It is scalable as only local synchronization information is used. It is adaptive to the traffic load, providing an ultra low power consumption in low traffic conditions and a high energy efficiency in high traffic conditions. Thank to the "more" bit, WiseMAC can transport bursty traffic, in addition to sporadic and periodic traffic. This protocol is simple, in the sense that no complex signaling protocol is required. This simplicity can become crucial when implementing WiseMAC on devices with very limited computational resources [61].

MAC Design for WSN Using Smart Antennas
WSNs differ from wireless ad hoc networks because of more energy constraints: nodes employed in WSNs are characterized by limited resources such as storage, processing, and communication capabilities. To cope with these impairments, there has been a lot of interest in the design of new protocols to facilitate the use of smart antennas in ad hoc networks [67, 69]. Smart antennas allow the energy to be transmitted or received in a particular direction as opposed to disseminating energy in all directions. This helps in achieving significant spatial reuse and thereby increasing the capacity of the network. However, the MAC and the *network* layers must be modified and made aware of the presence of the enhanced antennas in order to exploit their use. This might be accomplished by means of the cross-layer principle, as widely adopted in recent wireless networks design [61, 70].

The use of smart antennas in WSNs is highly desirable for several reasons: higher antenna gain might compensate the reduced coverage range due to higher frequencies (for realizing small size nodes) or to preserve connectivity in networks and efficiently use the node energy thus increasing its lifecycle. Besides, the adoption of smart antennas allows the gain maximization toward the desired directions by concentrating the energy in a smaller area, with a transmitted power decreasing, a received power increasing, a power consumption reduction, a coverage range increasing, a network lifetime increasing and an error probability reduction. Moreover, as the joint application of smart antennas and channel access management protocol permits to reduce the undesired power radiation, it is possible to reduce the interference caused by other transmissions and collision probability. There are several protocol design solutions in the presence of directional antennas proposed in the literature.

[71] proposes several extensions of the IEEE 802.11 protocol. The basic protocol, named *directional MAC* (DMAC) [72], operates in two phases. The first phase
(based on directional RTS/CTS exchanges) is used for tuning the receiver antenna on
the sender direction. During the first phase, the receiver listens to the channel omnidirectionally. After this phase, DATA and ACK messages are exchanged directionally.
Overhearing nodes update their directional network allocation vectors as in IEEE
802.11. This protocol suffers from several problems due to the directional nature
of the communications and the increased directional range: hidden terminals due to
unheard RTS/CTS messages, hidden terminals to asymmetry in gain and deafness.

Multihop MAC (MMAC) [72] extends the basic DMAC protocol by using multihop RTSs to establish a directional link between the sender and the receiver. Then
CTS, DATA, and ACK are transmitted over a single hop by exploiting the directional antennas gain. The neighbors of a node are divided into two types:

- The direction-omni (DO) neighbors, which can receive transmissions from
  the node even if they are in the omnidirectional reception mode;
- The direction-direction (DD) neighbors, which can hear from the node only
  if beam formed in the direction of the node.

The idea behind MMAC is to form links between the DD neighbors; the advantage of doing so, is to reduce the hop-counts on routes and in bridging possible
network partitions. The basic problems with hidden terminals and deafness still exist with the MMAC protocol, however, the benefits due to the use of the increased
range somewhat compensates for the other negative effects.

The previous shortcomings for DMAC and MMAC are reduced in random
topologies of networks. Those protocols do not study neighbor discovery and the
tracking of neighbors in mobile scenarios.

The *receiver-oriented multiple access* (ROMA) [73] protocol was proposed
for a scheduled access with the presence of *multi-beam antenna arrays* (MBBA).
It exploits the multilink feature of the MBBA so a node can commence several
simultaneous communication sessions by forming up to K links, where K indicates
the number of antenna beams. Besides, the system is capable of performing omnidirectional transmissions and receptions. This protocol relies on a time-slotted system, where time is divided into contiguous frames. Thus, ROMA creates a certain
number of links in every time-slot through a two-hop topology of transmission.
Resorting to the capacity of MBBA, in transmission and in reception, good results
in terms of throughput and delay can be achieved.

In [74], a protocol MAC with multiple selectable, fixed oriented antennas was
proposed. Each node has the option to use all antennas simultaneously, resulting in
an omnidirectional transmission/reception; alternatively, a node can, while sending
or receiving, decide to turn off some of the antennas. The signal is then only trans-
mitted/received over the active antennas in their respective sectors. After a sensor
node is back to an active state, it listens omnidirectionally to the channel with all of
its antennas. Then, when a signal on one of its antennas is strong enough to mark
the beginning of a packet reception, the node determines the antenna, which provides the highest signal level and shuts down the other antennas. At the end of the
reception the packet is marked with the direction it came from and the node returns

to omnidirectional listening mode. Before a sensor node is going to send any data, it has to know which antennas are involved; as this information is neighborhood information and in our case stored in the MAC layer, the MAC layer has to mark every outgoing packet with the antennas this packet has to be radiated on. Based on this behavior for reception/transmission, this protocol works as follows: before the transmission of a DATA packet, RTS/CTS messages are transmitted using all free antennas; conversely, DATA and ACK packets are only sent on the antenna, which with the strongest signal when receiving the RTS or CTS, respectively. The nodes which hear the RTS/CTS exchange, instead of abstaining from transmission during the packet only turn off the sector from which the control had been received by entering the information in the related DNAV (every antenna has its NAV). These nodes, thus, can pursue the communication with other nodes using any remaining free antenna. This protocol is energy saving thanks to the reduced number of re-transmissions in the presence of dense networks.

In [75], a new protocol for WSNs was proposed; it tries to maximize the efficiency and minimize the energy consumption by favoring certain paths of local data transmission towards the sink using switched beam antennas at the nodes. It is suited for those cases, where unexpected changes to the environment must be propagated quickly back to the BS without the use of complicated protocols that may deplete the network from its resources. During the initialization phase of the network, the BS transmits a beacon frame with adequate power to reach all the network nodes. Each node switches among its diverse beams and finds the one that delivers the best signal. Then, the nodes will use this beam only for transmitting data and they will use the beam lying on the opposite side of the plane only for receiving data. The protocol is highly dependent on the antenna beam-width; by carefully selecting the appropriate beam-width one obtains a tradeoff between robustness and load incurred in the network.

[76] introduced an opportunistic MAC scheme for controlling the uplink message transfer from sensor nodes to a central controller in WSN. The WSN controller is equipped with multiple antennas to communicate with single-antenna sensors. Using multiple antennas, the controller forms a beam that sweeps in an iterative, pseudorandom manner across the entire deployment area and it broadcasts a beacon signal. Each node will discover the beam of the controller because it will receive its beacon with a larger SNR than other nodes. If a node has a packet to transmit, it can send it. This protocol benefits from diversity of reception at the controller and leads to a considerable reduction of battery energy consumption in sensor nodes.

The use of directional antennas implies higher capacity thanks to an increased spatial reuse, lower latency, better connectivity, longer network lifetime, and greater security. But each protocol grants a privilege to certain positive aspects of these antennas because the efficiency of a MAC protocol depends on several aspects: density and traffic of the network, type of application, and topology of antennas used [61].

Comparison of MAC Protocols
The following parameters can be used to make a comparison between different MAC protocols proposed for sensor networks, including the ones based on IEEE 802.15.4 standard [i.e., personal area networks (PANs)] (Table 2.2):

Table 2.2 Comparison Between Different Existing MAC Protocols for Sensor Networks [61]

Services	Traffic Load Adaptation	QoS Guarantee	Scalability	Network Sync.	Low Power / Sleep Operation	Mobility Support	Interface for Higher Layers
Protocols							
CSMAC	NO	NO	YES (weak)	N/A	YES	N/A	NO
S-MAC	NO	NO	NO	YES	YES	NO	NO
DMAC	YES	YES (slightly)	YES (weak)	YES (weak)	YES	N/A	NO
TMAC	YES (weak)	NO	NO	YES	YES	NO	NO
EMAC	NO	NO	YES	YES	YES	YES	NO
TRAMA		YES	NO	YES	YES	NO	NO
MMAC	YES	YES	NO	YES	YES	YES	NO
BMAC	NO	NO	NO	NO	YES	NO	YES
Z-MAC	YES	YES (weak)	YES (weak)	YES	YES	NO	YES
LR-WAPN	YES	YES	YES	YES	YES	YES	YES
WiseMAC	YES	NO	YES (weak)	YES (locally)	YES	YES	YES

- *QoS Guarantees*: this is the ability of a protocol to give medium access priority to a node or a group nodes when required.
- *Traffic load adaptation*: This is the ability of the protocol to dynamically adapt to varying traffic characteristics of the network.
- *Scalability*: this is the capability of a protocol to divide the network into virtual or real clusters, such that medium access activities are restricted by the cluster boundaries irrespective of the number of nodes in the network.
- *Low power mode operation*: this is the ability of the protocol to support low power mode operation (sleep) during periods of inactivity.
- *Network synchronization*: node sleep and active schedule is central to the operation of low power operation. In order to support it, the nodes in the network must be synchronized with one another (at least locally).
- *Mobility support*: the protocol must be able to handle channel access issues related to the node movement, redeployment, or in the event of death, of a node. In effect, it should efficiently handle cluster associations and disassociations.
- *Interface for higher layers*: this is the property of giving higher layers the ability to set certain MAC layer parameters, depending on the upper layer needs.

The choice of a protocol will depend on the type of WSn and the application scenario, which would define the type of expected performance. In general, the WSNs introduce the following requirements to the MAC layer services:

- Network synchronization in the absence of a global synchronization mechanism;
- Extreme scalability due to high node density in some cases;

- Energy efficient communication protocols, (which should deal with over-hearing, data retransmission, idle listening, etc.);
- Adaptability to changing radio conditions;
- Low-footprint implementation due to limited capabilities of sensor nodes;
- Power-efficient channel utilization at high and low data rate;
- Tight coupling with higher and lower layers.

2.3.1.2   Routing for Sensor Networks

Routing techniques face a number of challenges, which are specific to sensor networks, such as a unique identifying scheme, the substantial asymmetry in traffic patterns, the redundancy in the information transmitted, as well as the processing time. Another distinctive aspect of communications in the WSNs domains is that WSNs are usually data-centric, with the objective of delivering the collected data in a timely fashion. The characteristics of the data-centric protocols are their ability to query a set of sensor nodes (rather than an individual device), attribute-based naming, and data aggregation during relaying; they route each piece of data according to its own characteristics, thus typically requiring some sort of data naming scheme. On the other hand, the location of the nodes is employed by, for example, routing protocols that use spatial addresses, and by signal processing algorithms that are used for tasks such as target tracking [61].

Sensors can be considered in most of the cases [77] as quasi-static devices, which are mostly limited by resource energy, and whose consumption needs to be optimized. Routing has to be adapted to the network topology/architecture, and has to consider the way, in which the sensor devices generate the traffic (in a significant number of cases, traffic is either broadcast or multicast). The basic idea is to add information to the data, so that the lifetime of data can be used in combination with the local time and the arrival time for evaluating the relevance of the data at the moment when it is used. Another possibility is to evaluate the temporal and spatial consistency of the data transmitted from different locations. This concern has been considered for a long time in the fieldbus and industrial/real-time communication domain [78, 80].

On the other hand, typical performance metrics, (e.g., throughput) are relatively less important in the case of WSNs. Different routing protocols proposals, both reactive and proactive, have been proposed and analyzed for the identified WSNs application portfolio [77].

In the following the most important types of protocols for the optimal performance of WSNs [61] are described.

Hierarchical Protocols
[81] proposed a hierarchical architecture for sensor networks, which can affect the underlying routing protocols. In addition, it is also important to address the impact of clustering techniques on the data aggregation and fusion. Some of the most relevant examples of hierarchical routing protocols are the Low Energy Adaptive Clustering Hierarchy (LEACH) protocol [83], the Power Efficient Gathering in Sensor Information Systems (PEGASIS) [84], and the Threshold Sensitive Energy Efficient Sensor Network (TEEN) [85].

Data Centric Networking

One of the most outstanding features of WSN is that the focus should be normally put on the data that the devices collect, rather than on the node identity (as opposed to other types of networks, where the identity–address–of the node is the distinguishing aspect); this normally leads to different routing strategies [61].

The data-centricity of WSNs gives rise to new challenges in their information processing and data management. In many applications, users may frequently query information in the network. The trade-off between updates and queries needs to be addressed. In-network data processing techniques, from simple reporting to more complicated collective communications, such as data aggregation, broadcast, multicast, and gossip should be developed. In data centric protocols sources send data to the sink, but routing nodes look at the content of the data and perform some form of aggregation/consolidation function on the data originating at multiple sources.

The characteristics of *data-centric* protocols are the ability to query a set of sensor nodes, attribute-based naming, and data aggregation during relaying. Some examples are briefly described below.

The *sensor protocol for information via negotiation* (SPIN) [86] sends data to the sensor nodes only if these are interested; it uses three types of messages: ADV (metadata), REQ (metadata), and DATA. In this case, the topological changes are localized. Each node needs to know only its one-hop neighbors. SPIN halves the redundant data in comparison to pure flooding. However, it cannot guarantee the data delivery and thus, it is not appropriate for applications that need reliable data delivery.

The *directed diffusion* [87] sets up gradients for the data to flow from the source to the sink during the interest dissemination phase; it has several elements: interest messages, data messages, gradients, and reinforcements.

The *rumour routing* [88] is intended for contexts, in which the geographic routing criteria are not applicable. Routers send queries to the nodes that observed a particular event rather than flooding the entire network to retrieve the information. This protocol consists of a trade-off between *query and event flooding*. A long-lived packet is generated when events happen, being the agents in charge of propagating the event to distant nodes.

Other protocols are the *gradient-based routing* (GBR) [89], the *constrained anisotropic diffusion routing* (CADR) [90], COUGAR [91], and the *active query forwarding in sensor networks* (ACQUIRE) [92].

Location Aware Protocols

There are two different aspects [82] to be considered under this class of routing protocols. First, there exists sometimes the need to address nodes according to their physical location, while on the other hand being aware of the positions of the nodes (both the destination and the intermediate ones) could also leverage some optimization in routing procedures.

Node location is employed by routing protocols that use spatial addresses, and by signal processing algorithms (e.g., beam-forming) that are used for tasks such as target tracking. The underlying algorithm problem is that of localization whereby the nodes in the network discover their spatial coordinates upon the network boot-up. When the sensor nodes are deployed in an unplanned topology, there is no a priori knowledge of location. The use of the *Global Positioning System* (GPS) in sensor nodes is ruled out in many scenarios because of the power consumption, antenna size, and overhead obstructions such as dense foliage [61]. The ad hoc nature of the deployment rules out infrastructure for many scenarios of localization. It is critical that the sensor network nodes are able to estimate their relative positions without assistance, using means that can be built-in.

The localization problem in itself is a good example of a signal processing task that the sensor network needs to solve. The basic approach would be for sensor nodes to gather sufficient number of pairwise distance estimates via some suitable mechanism, and then use multilateration algorithms to estimate the positions of the nodes [61]. To begin with, a few nodes might know their position via other means (beacon nodes), but at the end of the localization process every node would hopefully know its position.

A key problem, however, is that in conventional formulations of multilateration [93, 94] one needs to estimate the location of an entity given estimates of its distance to three or more beacons with known positions. In sensor networks, a very high density of beacons nodes would be needed. To keep the required beacon density and energies low, a preferred method would be to jointly estimate the positions of all the non-beacon nodes via a collaborative multilateration formulation based on a criterion, such as the *least-square error minimization*. Besides being computationally hard for a large number of nodes, doing this would require a centralized node, where all the distance estimates would be collected at significant communication and associated energy costs. A more scalable solution is the *locally distributed iterative multilateration* [95], whereby a node calculates its position and is promoted to a beacon as soon as enough of its 1-hop neighbors are beacons. Starting with a critical density of beacons, a percolation-like phenomenon would result in all the nodes gradually discovering their position. With a sufficient beacon density, a small number of successive multilateration steps leads to the rapid convergence of location estimates. The communication overhead is much lower than in a centralized approach as all the message exchange is strictly local and is easily piggybacked on the routing messages.

Another challenge in localization is the estimation of the distance between a pair of nodes. Using time-of-flight of radio signals (as in GPS) is ruled out when the distances are too tiny and the radio frequencies are not very high. A readily available method would be to use the *received signal strength indication* (RSSI) provided by the radio. The RSSI data can be cheaply piggybacked on regular routing and data. The accuracy of this approach can be improved by using a parameterized channel, a path loss model whose parameters are also estimated together with the position [95]. However, in practice, the RSSI-based approach works only in the absence of significant multipath effects. In most environments other than open spaces, multipath is an issue. A promising alternative technology is to estimate the distance by the time of flight of acoustic or ultrasound signals, and using the much faster radio signal to establish the time reference [96–98].

Location awareness may be the only difference between traditional MANETs and WSNs that can actually be used for optimization reasons [61]. There is a trade-off between scalability, mobility, and location-awareness in the sense that as the network size and mobility increases, the location information bulk becomes larger and the necessary updates more frequent. In this context, an adaptive lightweight location service supporting routing over a hierarchy that enforces balanced energy conservation across the whole network, to ensure prolonged lifetime would be beneficial [61].

QoS-Aware Protocols

QoS-aware protocols consider the end-to-end requirements of the network link and normally aim to maximize a given quality metric across the entire link. Energy-aware protocols are a special case of the latter, in which the energy cost is the main metric. Another stringent QoS requirement is the reliability. Another metric that is commonly used to determine the QoS in sensor networks, is the quality or reliability of collected data. A wireless network of sensor nodes is inherently exposed to various sources of unreliability, such as unreliable communication channels, node failures, malicious tampering of nodes and eavesdropping. Approaches adopted to improve the reliability of sensor data are also included with the proposed QoS-aware protocols [61].

The *maximum lifetime energy routing* [99] aims to maximize the network lifetime by defining the link cost of each link as a function of the energy remaining at the node and the energy required to perform data forwarding. The Bellman-Ford shortest path can be used to determine the path across the network that maximizes the residual energy. This differs from the *minimum transmitted energy* (MTE) algorithm by the consideration of the remaining energy at a node in the route path determination—as a node exhausts its energy supply it is less likely to take part in the data forwarding.

The Bellman-Ford algorithm is likely to consume a lot of energy in the setup state for a large sensor network.

The *minimum cost forwarding* (MCF) [100] is used to provide a simple and scalable solution to the problem of finding the minimum cost path in a large sensor network. The cost function at each link is a function of the delay, throughput, and energy consumption at that node. A fixed sink node initiates the protocol by broadcasting. This is diffused through the network, with each node adding its cost information and retransmitting using a backoff-based algorithm. The length of the backoff interval is proportional to the cost being transmitted—in this way, a node that is about to transmit a large cost value waits a long time to see if it receives a lower cost, while a low-cost path is propagated quickly.

Once this has been established, each node knows, which neighbor to transmit to in order to send data to the sink with the lowest cost. This protocol is simple and easy to implement on a sensor node.

Sequential assignment routing [101] is a sensor network routing protocol that includes QoS directly in its routing decisions. It creates multiple spanning trees rooted at the sink node, considering QoS, energy and packet priority into account. Using these trees, multiple paths may exist from a particular node to the sink node; the path that is the most appropriate to the desired QoS metric is chosen. The *path failures* automatically trigger *path updates*. The protocol suffers from having to maintain a lot of state information in terms of a sensor node.

Energy-aware QoS routing [102] combines different QoS requirements in a network. The real-time traffic is routed along least-delay paths, while the nonreal time traffic is routed to minimize the energy consumption. A clustered approach is used for scalability, and a class-based queuing scheme at each node. Dijkstra's algorithm [18] is used at each node to choose the best path from those available.

According to [103], the main sources of unreliability in sensor data can be classified into two categories: (1) faults that change behavior permanently and (2) failures that lead to transient deviations from normal behavior. The permanent faults include failures due to unavailability of energy resources, calibration errors after prolonged use, and loss of wireless coverage. The transient errors are caused by the noise from various sources, such as the thermal noise at the receiver, the channel interference, and the multipath fading effects. A number of strategies have been proposed in the literature to combat the effects of data unreliability that lead to improved QoS.

An application level error correction strategy was proposed in [103], in which a data prediction algorithm is used to filter out the errors caused by the transient faults. While the data predictions filter out the majority of the errors in the observed values, it is possible that the predictions may not always track the data process variations correctly. In such cases, to increase the effectiveness of the algorithm, a delay in reporting the data is introduced within the application's delay constraints. The delayed reporting allows the use of the observed values in the next few samples to guide the choice of correct value between the predicted and observed value. The advantage of this approach is that is does not require any overhead in the transmission by the sensor nodes. Instead, most of the error correction functionalities are pushed to the receiver, which is less resource constrained than the sensor nodes.

A learning algorithm for determining the quality of the sensor data in the data fusion process in presented in [104]. In this approach each sensor actively learns the quality of information from the different sensors and updates their reliabilities using the *weighted majority* technique. In the strategy, a sensor node forwards data to one of its neighbors once it detects a new target. Any sensor that fuses all the relevant data for a target can stop the data propagation if the confidence of the fused data meets a threshold.

A data collection *protocol with expected reliability guarantees* (PERG) was proposed in [105]. PERG provides expected reliability guarantees, while minimizing the resource consumption by adaptively adjusting the number of retransmissions based on the current network fault conditions. The objective is to minimize the communication overhead involved in maintaining the desired reliability. The basic idea of PERG is to use retransmissions to achieve the user required reliability. At the end of each collection period, the sensor forwards to the server information about the number of the data items and messages sent in the collection period. Based on this information, the server estimates the current network situation, as well as the actual reliability achieved in the current collection period. The sensor derives retransmission times for data items generated in the next period based on the feedback from the server.

Cross-Layer Routing
Routing techniques may benefit from information provided by the different layers within the protocol stack. Cross-layer algorithms aim to derive efficiency by dispensing with the traditional layered network stack and considering the information from

different layers in the decision process. Each layer is not developed in isolation and is dependent on the other layers. Routing is one independent layer in a traditional network stack but in cross-layer systems this independence no longer applies [61].

The two layers lending themselves most to such approaches are the MAC and the network layer. The information from one layer can prove very useful to the other layers, particularly in sensor networks, where the two layers may not be distinct at all.

In the following, a *link adaptive transport protocol (LATP)* is described that was proposed in the frames of the FP6 IST project E-SENSE [70, 106] to combat the negative effect of the traditional transport protocols (i.e., UDP, TCP) in a WSN environment. The traditional transport layer protocols not only worsen the MAC contention in the network, but also trigger routing dynamics in the network due to the increased MAC contention and eventually worsen the transport layer performance. In general, the routing protocols do not distinguish between MAC contention loss, buffer overflow, channel errors, and mobility induced errors [106]. Therefore, if the MAC contention loss is frequent and persistent during the entire course of the transport connection, the routing protocols will be triggered unnecessarily for routing maintenance. This would further affect the performance of the end-to-end connection. Thus, provisioning of reliable or real time (audio/video) data transfer over multi hop wireless sensor networks (MWSN) is extremely difficult.

The congestion (MAC contention) in *mobile WSN (MWSN)* has direct impact on the energy efficiency and the QoS metrics, such as latency, jitter, packet loss ratio, and bandwidth. For example, the packet losses due to the MAC contention result in increased latency, degrade the link utilization, and also waste energy, due to the MAC level back-off and retransmission in the random access based (or contention-based) networks. The project E-SENSE focused on the problems arising from the MAC contention and the routing dynamics induced by mobility in the MWSN and proposed a *transport layer* solution based on cross-layer information, to support the real-time applications in MWSN [106].

The proposed protocol, LATP is suitable for A/V streaming applications in outdoor environmental sensor networks (ESN), where the ambience sensors (A/V sensors) are connected to the sink via multihop wireless links.

LATP is based on a cross-layer approach as shown in Figure 2.51.

In LATP, the MAC layer estimates the degree of the MAC contention (i.e., permissible throughput) experienced by each intermediate node (sensor) in the path for the end-to-end connection and provides this information to the source sensor to adjust its sending rate [107]. Meanwhile, the routing protocol (network layer) notifies the sender when it detects route failures. Because the route failures may be triggered by the MAC contention in the path, the sender verifies the route failure notification with the MAC contention information and determines whether the route has actually failed due to mobility.

When the sender detects the actual route failure due to mobility (i.e., reception of route failure notification when the degree of MAC contention is low), it freezes the transmission and probes for the path less frequently in order to avoid unwanted transmissions and save energy.

Once the route is re-established, the sender will restart the transmission for the connection. On the other hand, if the sender receives a route failure notification

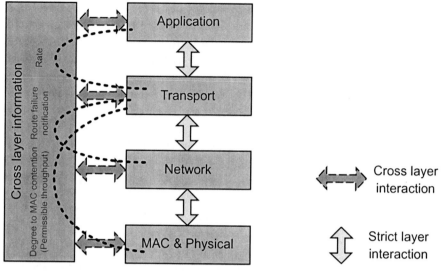

Figure 2.51   Cross layer approach for LATP [106].

when the degree of MAC contention is high, it will take an appropriate action to reduce the MAC contention in the network as it assumes the route failure has been triggered by the MAC contention. The architecture of the LATP protocol is shown in Figure 2.52.

[106] evaluated LATP for the problems induced by the MAC contention in a MWSN with a simple chain topology, using *ns*2 simulations. The underlying MAC protocol is assumed to be CSMA/CA with *request-to-send and clear-to-send* (RTS/

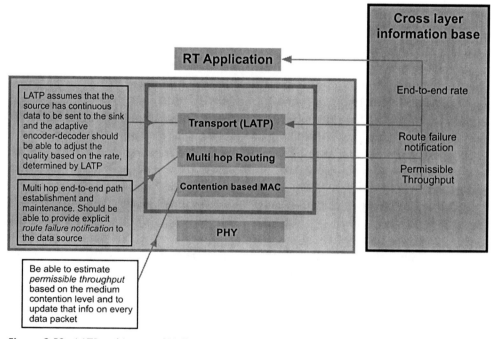

Figure 2.52   LATP architecture [106].

CTS) frames. The simulations considered fixed nodes (no mobility) and, therefore, the effects of mobility (i.e., route failures) were not evaluated. Because LATP was proposed for real-time applications over MWSN, the primary evaluation metrics were in terms of end-to-end delay and jitter. LATP was evaluated based on a simple linear chain deployment and with and without competition (i.e., a single flow or multi flows in the network).

The following simulations assumptions were made for the assessment:

- Number of nodes: n=1 to 10 sensor nodes;
- Deployment: linear chain, equally spaced nodes; 75m distance between nodes;
- Sink position: right end of the chain.
- Channel rate: 2 Mbps is assumed for preliminary assessment
- Data rate: Adaptive rate; however maximum rate is limited to 100 pkts/s for single flow and 25 pkts/s for multi flows in the network;
- Packet size: 500 bytes;
- Range: 120m;
- Energy level: infinite;
- Energy model: not analyzed;
- Simulation duration: 400s;
- Traffic model: CBR application with adaptive encoder-decoder at left end of the chain (it is assumed that the application has always data to send);
- Duty cycle: none;
- Buffer size: 10, 20, and 50 pkts for single flow and 20 pkts for multi flows;
- Observed metrics (as defined in IR Sim): End-to-end delay (latency) and jitter.

The CBR application with an appropriate encoder-decoder was used with the *TPC friendly rate control* (TFRC) protocol and LATP. An FTP application was used with the TCP-Few [108] for the comparison analysis, from the source to the sink. Initially, a single flow was established from the source to the sink at 10s and the measurements were taken over a steady period of 50s to 400s for different buffer sizes (Q) of 10, 20, and 50 packets at each node. The end-to-end delay and the jitter obtained with the increasing number of hops connections are presented in Figure 2.53 and Figure 2.54, respectively. All the results presented were taken over 10 simulation runs, unless otherwise specified.

Further simulations were performed over a 4-hop linear chain with the same simulation assumptions as before, in order to evaluate the LATP performance with competing flows. In this scenario, it was assumed that the source can support multiple applications at the same time [106]. The buffer size of the nodes was fixed at 20 packets. Simulations were performed for 400s for 1, 2, 4, 6, 8, and 10 competing flows with LATP, TFRC, and TCPFeW protocols, separately. The first flow was started at 10s and the subsequent flows were started in 10s interval. All the flows were stopped at 400s. The measurements were taken over a steady period of 150s to 400s. Figure 2.55 and Figure 2.56 show the results for the delay and jitter performance, respectively.

With increasing the number of flows, LATP provides a lower average end-to-end delay per flow compared to TFRC as shown in Figure 2.55. Although TCP-FeW

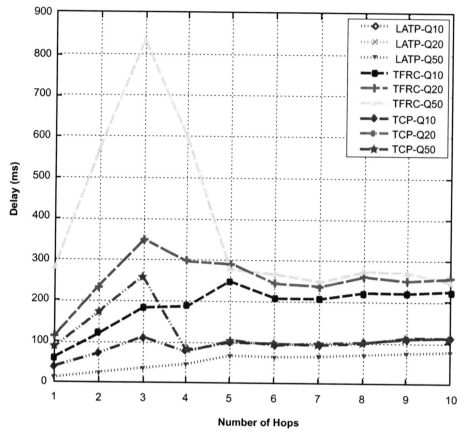

Figure 2.53 LATP, TFRC and TCP-FeW delay performance with increasing number of hops connection [106].

exhibits better delay performance, it shows very poor performance in terms of jitter and throughput. The jitter performance of LATP almost saturates at 40 ms when the number of competing flows increases as shown in Figure 2.56.

The presented version of LATP only provides a solution for problems arising from medium contention for real-time applications in MWSN. In order to combat the full scope of problems as outlined in the beginning of Section 2.3.1.4.5, mobility, energy consumption, and packet delivery ratio must also be implemented [106].

In the following, LATP is evaluated over a variety of scenarios using *ns*2 simulations. LATP is compared with TFRC for end-to-end delay, jitter, packet loss rate, and fairness with competing flows in chain, star, and random deployments [109]. The simulations are performed in multi hop ESNs with static nodes (sensors and sink). In addition, the wireless channel was assumed to be perfect with no bit errors.

The underlying MAC protocol was assumed to be CSMA/CA in basic mode (i.e., without RTS/CTS). The other simulation parameters were configured as follows. The channel capacity 2Mbps, the transmission range 125m, the data packet size 512 bytes and the buffer size 25 packets at each node. The traffic model was set again to CBR application with an adaptive encoder and decoder mechanism, so that the transport protocols LATP and TFRC have always data to send. The simulation

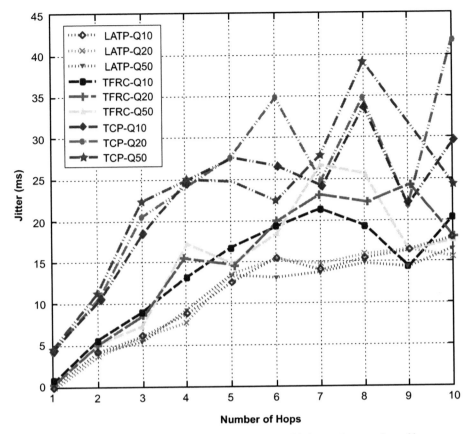

Figure 2.54 LATP, TFRC and TCP-FeW jitter performance with increasing number of hops connection [106].

time was set to 400s and the results presented were taken over 10 simulation runs with 95% confidence interval, unless otherwise specified.

In the chain topology simulations, the sensor nodes are deployed in a linear chain, as shown in Figure 2.57 and only neighboring nodes can directly communicate with each other. LATP and TFRC flows were performed separately from node 0 to node n. The measurements were taken over a steady period and the average results obtained with an increasing number of hops connection are shown in Figure 2.58.

Figure 2.58 (a) shows the delay performance of the LATP and TFRC flows. It can be observed that TFRC experiences higher delay than the LATP flows. In particular, the TFRC flows experience at least 100% more delay than the LATP flows. The reason behind this is, that TFRC loads the network since it produces a sending rate that is above the rate supported by the underlying MAC layer in multi hop networks. Then, for each packet the MAC layer attempts multiple retransmissions and back-offs before transmitting or dropping the packet. This increases the end-to-end delay of the TFRC packet as TFRC waits for the sender to be notified of the packet losses in order to control the sending rate. Although TFRC eventually resolves some packet losses caused by the medium contention, it receives them too late due to the MAC layer retransmission and back-off. Thus, TFRC will experience a maximum

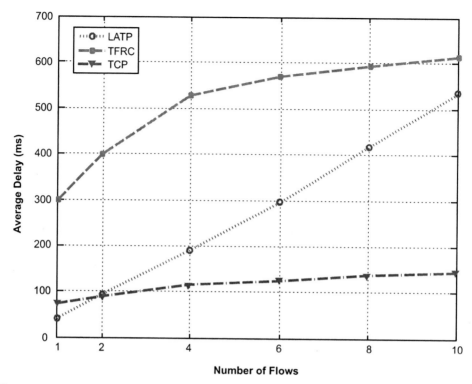

Figure 2.55 Delay performance comparison with different number of competing flows over a 4-hop MWSN [106].

delay if the nodes participating in the connection have a sufficient interface queue (buffer) size. On the other hand, LATP provides much better delay performance than TFRC, for all number of hops connections. Since LATP controls the sending rate based on the degree of the medium contention level in the network, it operates at a rate supported by the MAC layer and does not over load the network.

The jitter performance is shown in Figure 2.58(b). TFRC and LATP both provide considerably good jitter performance for real-time applications over the chain topology, still, LATP outperforms TFRC.

Figure 2.58(c) shows that TFRC has higher throughput in small number of hops connections than LATP and that both, TFRC and LATP, achieve almost the same amount of throughput as the number of hops connection increases.

The *packet loss rate*, PLR (i.e., the fraction of packets sent and not received by the destination) is shown in Figure 2.58(d). LATP exhibits much better performance in terms of PLR than TFRC. LATP prevents the transmission of packets during the excessive medium contention in the network and avoids packet losses. Thus, energy wastage due to the unsuccessful transmission attempts is greatly reduced with LATP.

In another set of simulations, the sensor nodes were placed in a star topology as shown in Figure 2.59.

The node 0 is set as the sink (server), the nodes 5, 6, 7, and 8 are set as the sources, and the other nodes act as routers. First, a single flow was established from node 5 to node 0 and the measurements were taken. Next, two flows from nodes

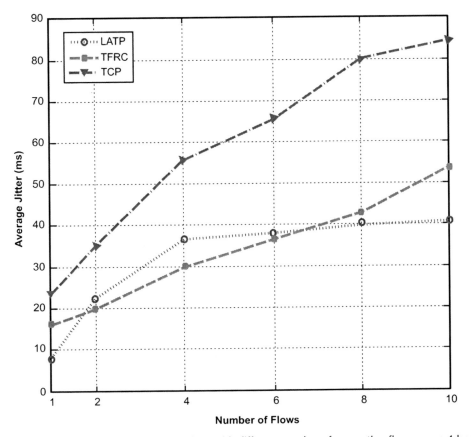

Figure 2.56   Jitter performance comparison with different number of competing flows over a 4-hop MWSN [106].

5 and 6 each were performed simultaneously in order to evaluate the performance with competing flows.

Similarly, in the next phase, 3 flows from nodes 5, 6, and 7 were performed. In the final phase, all four source nodes (4 flows) transmit packets simultaneously. The measurements were taken over a steady period in order to avoid the transient states at the beginning of the simulations.

Figure 2.60 shows the per-flow average end-to-end delay with a number of competing flows in the network.

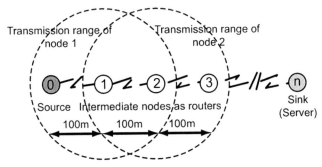

Figure 2.57   Chain topology for LATP evaluation [109].

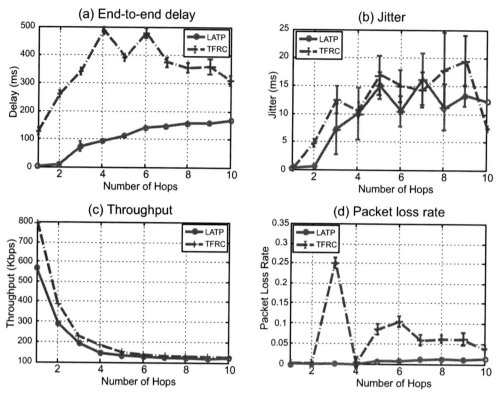

Figure 2.58 LATP performance in chain topology [109].

In all cases, end-to-end delay is significantly reduced with LATP flows, compared to the delay experienced by TFRC flows. The improvement achieved with LATP in terms of jitter and PLR is also highly significant as shown in Figure 2.60(b, c). On the other hand, the aggregate throughput achieved with LATP flows is around 10% less than that achieved with the TFRC flows, as shown in Figure 2.60(d).

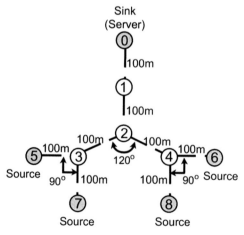

Figure 2.59 Star topology [109].

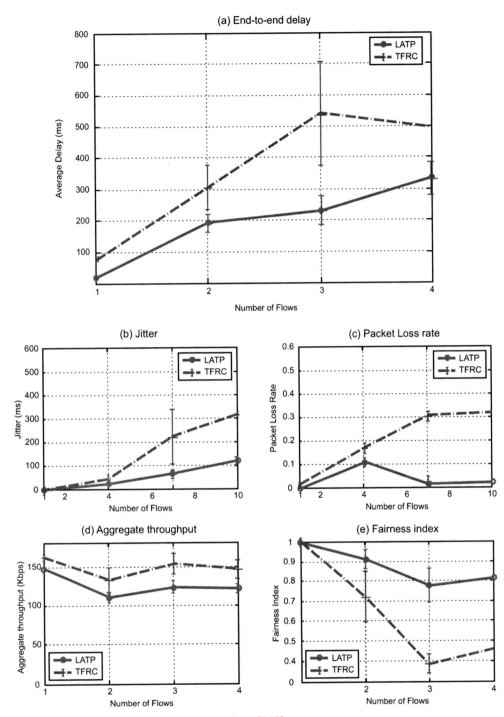

Figure 2.60   LATP performance in star topology [109].

LATP shows an excellent fairness among the competing flows compared to TFRC, as shown in Figure 2.60(e). This means that with the TFRC protocol, only some of the flows achieve high throughput and the others loose their fair share almost completely. LATP, however, shares the available bandwidth fairly with the other competing flows in the network. Although LATP looses some aggregate through-put in the star topology, the improvement achieved with LATP in terms of end-to-end delay, jitter, PLR, and fairness is considerably significant to provide QoS support for the A/V streaming applications in multihop WSNs with star-like topologies.

The *random* topology simulation scenario considered 100 sensor nodes, placed uniformly and randomly in an 800-m × 800-m area. Ten streaming flows were sent simultaneously between ten randomly chosen source and destination pairs, with a minimum hop distance (between a source and its destination) of at least two hops. A random topology creates a complex simulation scenario, which results in only slightly better performance of LATP compared to TFRC [109].

The simulation results over chain, star, and random topologies confirm that LATP can support the QoS performance in terms of end-to-end delay, jitter, and packet loss rate, while providing excellent fairness to the competing flows. The LATP protocol element requires few processing cycles and overhead for its func-tions and does not maintain any per-flow state table at intermediate nodes. This makes it less complex and more cost effective for multihop wireless sensor network applications.

2.3.2 Self-Organization and Topology Control

Self-organization is realized by organizing a sensor network into clusters where a *Connected Dominant Set* (CDS) is formed by the *cluster-heads* [109]. A CDS is defined as a subset of the network nodes that are connected in such a way that the nodes that are outside of the subset connect to at least one node, which belongs to that subset. By *clustering*, a network is logically changed from a *flat* topology to a *hierarchical* topology.

Self-organization can be implemented in different layers. The simplest CDS selection algorithm was introduced in a *low-energy adaptive clustering hierarchy* (LEACH) [83]. In LEACH, the nodes are assumed to have the ability to transmit to the remote sink node directly. Hence, every node can listen to other nodes. A node makes the decision to be in the CDS based on a pre-defined percentage ([83] observed that 5% is optimal by simulations) of cluster-heads for the network and the number of times the node has been a cluster-head so far.

The IEEE 802.15.4 standard [110] enables the formation of a *clustered tree*. A cluster-tree topology combines point-to-point communications from device to router with point-to-point multipoint communications between the routes and gateways. If the direct, single-hop transmission path between a router and gateway is interrupted, messages can multihop through other routers to reach their destina-tion. This is shown in Figure 2.61.

The fact that both network layer and MAC layer protocols implement self-organization may result in functions overlapping. For example, if both network and MAC protocols collect the neighbor information for the cluster forming, more overheads are produced.

[106] proposed to realize the self-organization and topology control of a WSN with the following three mechanisms:

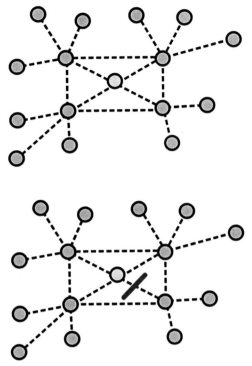

Figure 2.61   A cluster-tree topology.

- Decision-making mechanism;
- Topology construction mechanism;
- Self-recovering mechanism.

Figure 2.62 shows a proposed architectural realization of self-organization and topology control.

The *neighborhood information base* is a neighbor list, which is shared by several layers. The MAC and the network layers can update the neighbor list. The MAC layer provides each neighbor with the receiving power and link quality indicators. Both the MAC and the network layers may add and remove the neighbor entries from the neighbor list. However, the operation of adding or removing should be agreed by the other protocol layer, (e.g., if the MAC removes a neighbor entry from the list, the neighbor list manager should first confirm the removing operation with the network layer protocols and then remove the entry. An approach for the implementation of the neighbor list and its manager can be found in [111]. Although the self-organization and topology control manager take advantage of the neighbor list, the list is not a part of the manager.

2.3.2.1   Decision Making

The decision-making mechanism decides if a network should be changed to a hierarchical topology. The decision can be made based on the information collected from the application layer and the neighbor list. The network conditions such as the

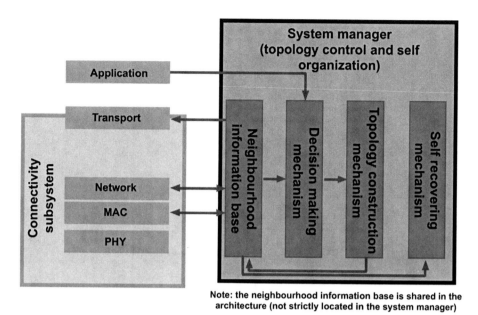

Note: the neighbourhood information base is shared in the architecture (not strictly located in the system manager)

Figure 2.62 Self-organization and topology control architectural elements [107].

number of nodes and network dimension should be considered. In general, the decision relies on whether the change to hierarchical topology saves energy compared to the flat topology. To find the energy gain of clustering, the energy consumption of the flat topology and the clustered topology in different network conditions must be analyzed [106].

Assume that in both topologies, the nodes maintain an awake/asleep cycle of t_c seconds. In the flat topology, a CSMA/CA MAC protocol with RTS/CTS is applied, while the clustered topology uses two MAC schemes. Specifically, in a cluster, a *cluster member* (CM) has to transmit its sensed data to its *cluster head* (CH) at the time assigned in the beacon message in a TDMA-like manner. This assumption may not be true for all the MAC mechanisms in a clustered topology; however, here it can ease the analysis. Among the clusters, the CHs use the CSMA/CA MAC. The CH aggregates its CMs' packets and transmits them to a sink node in a burst via a route made up only of CHs. It can be assumed that the neighboring clusters active periods do not overlap because scheduling is out of the scope of this analysis. The MAC protocols of both scenarios are assumed to be ideal, therefore, no collision happens. Some of the above assumptions may be too strong, such as no-collision MAC and non-overlapping active period, but this simplifies the analysis.

Assume that in both scenarios, n nodes are uniformly distributed in a network and that the maximum transmission range of a node is R [106]. For a clustered topology, k nodes are selected to be the CHs. Therefore, each cluster has one CH and $(n-k)/k$ CMs. The network occupies an $mR*mR$ [m²] area, where m is a constant larger than 1.

In the case that all the nodes transmit with full power, the average hop-count, $E[h]$ can be approximated by $2m/3$ [112], provided that n is large and the network is connected. Assuming the first order energy model proposed in the LEACH protocol [83], to transmit I bits at a distance of R, the energy consumption is given by:

$$E_t = E_{elec}I + \varepsilon_{amp}IR^2 \tag{2.47}$$

and to receive I bits, the energy consumption is:

$$E_r = E_{elec}I \tag{2.48}$$

In both scenarios, every node sends a data packet of L bits with intensity λ [packet/s]. The length of an RTS/CTS (denoted as L_{rts} and L_{cts}, respectively) packet is ηL bits, and a beacon message length (denoted as L_{beacon}) is γL bits, where η, γ < 1 are constant (because RTS/CTS and the beacon packets are normally smaller than the data packets). The nodes in a flat topology have to wake up every cycle to listen to the channel. To ease the derivation, the energy consumption in this period was equaled to that of receiving βL bits. The data compression ratio in a clustered topology is α. For a node, the energy consumption includes the energy spent on transmitting, receiving, idle listening, and sleep, which is denoted as:

$$E = E_{tx} + E_{rx} + E_{listen} \tag{2.49}$$

where the energy consumption of sleeping is neglected in this phase.

Compared to sending and receiving, the power consumption of sleeping can be neglected [111].

In a flat topology, the average total traffic per node is the sum of the node's own traffic and the whole relay traffic. The own traffic that each node generates is $\lambda (L + L_{rts})$ bits. In order to relay a packet for another node, a node has to send out a CTS and an RTS packet to the previous hop and the next hop, respectively. Therefore, in total the node sends $L + L_{rts} + L_{cts}$ bits to relay a packet of L bits and a node relays $\lambda (E[h] - 1)(L + L_{rts} + L_{cts})$ bits/s.

A *minimum transmission power* means that the nodes transmit with the smallest power sufficient to maintain the connectivity of the network. A key to analyze the gain numerically is to know the size of the CDS [106]. Different algorithms may results in CDS of different size. Finding a *minimum CDS* (MCDS) in a unit disk graph is an NP-hard problem [113].

Moreover, the CDS size does not depend on the number of nodes but on the geographical size of the network. In [114], the authors iteratively try each node combination of a proposed MCDS size in a network until they find a set of nodes of that size, which is connected and dominant. The size of MCDS is always $k = 8$, while the node number varies from $n = 20$ to $n = 110$.

In the following the two topologies are compared in a network where $R = 50$m, which is a typical transmission range of most commercial wireless sensors [115]. The size is 175x175 m^2.

The number of nodes, n, , the length of the awake/asleep cycle, t_c and the aggregation ratio, α, are varied during the simulation. Each data packet is chosen to be 80 bytes. The size of the RTS, CTS, and the beacon packets is 20 bytes. Each node sends a data packet every 30 seconds. The value of β is selected based on the maximum clock drift per second, which in IEEE 802.15.4 is found to be 0.08

ms [110]. If the neighboring nodes synchronize with each other every 10s, then a node has to sample a channel for at least 0.8 ms to be able to cope with the drift. The goal of the simulations is to observe the energy gain of a *clustered* versus a *flat* topology.

Figure 2.63 is plotted with the length of an awake/asleep cycle of one second. The aggregation ratio and the node number are varied from zero to one and from 100 to 1000, respectively. Both the clustering gain from E_{flat} and E'_{flat} are plotted. Approximately, 10% more gain is obtained with the longest transmission range, which means that $E_{flat} > E'_{flat}$. The aggregation introduces a larger gain when the node number is small. When the network is very dense, the effect is not so significant. It can be observed that clustering saves at least 50% energy. The gain increases over 90% when the node number is large and the aggregation ratio is small [106].

Figure 2.64 shows the results for the gain for $\alpha = 0.5$.

It can be observed that the increase of t_c does not result in an increase of the gain. The reason is that when there are many CMs in a cluster, the gain from other factors, such as less RTS/CTS and data aggregation, is more dominant, and the energy saving from the less frequent beacon message is not significant.

The above calculations allow to compute the actual gain of forming a clustered WSN topology. The analysis shows that it is beneficial to organize a network into a clustered topology whenever possible. The analysis does not include the cost of topology construction and maintenance, which can be quite high in some conditions, thus decreasing the gain of clustering. For accurate results this should be part of the analysis.

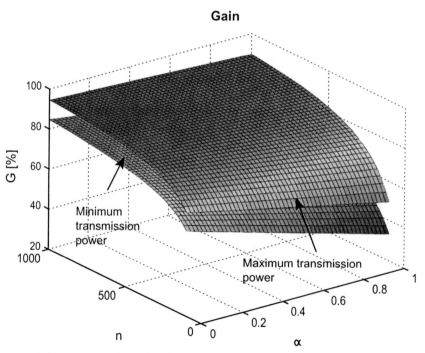

Figure 2.63 Clustering gain G as a function of the aggregation ratio α and the number of nodes n ($t_c = 1$s) [106].

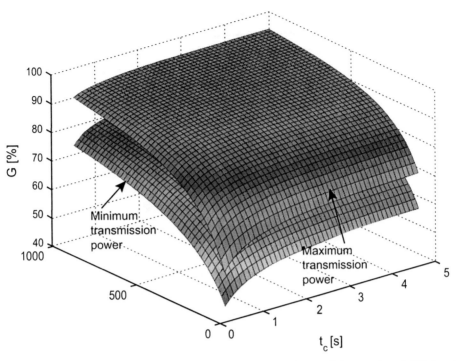

Figure 2.64   Clustering gain as a function of the cycle length t_c and the number of WSN nodes n for $\alpha = 0.5$ [106].

2.3.2.2   Topology Construction

After the clustering decision is made, the decision-making mechanism informs the topology construction mechanism to change the topology accordingly.

The clustered topology can be generated by a clustering algorithm, which partitions the network into several self-organized clusters, while maintaining the connectivity between the clusters. In each cluster, there is one CH and several CMs. The algorithm should consider the lifetime of the CHs, because it has a direct effect on the maintenance cost of the clustered topology [106]. This problem can be solved when selecting the CHs. For the purpose, a selection criterion C can be defined, such as:

$$C = \frac{E}{E_{\text{max}}} F + \frac{D}{D_{\text{max}}} (1 - F) \qquad (2.50)$$

where C is the criterion, E is the remaining energy of the node, E_{max} is the maximum energy when a node is fully charged, D is the node's degree, D_{max} is the maximum degree of the node's neighbors, and $F \in [0, 1]$ is a predefined weighting parameter [106]. When F is large, the energy level of the nodes is considered more important, thus the selected CHs should have more energy. When F is small, the node degree is more dominant and, thus, the nodes having more neighbors are selected. Therefore, more nodes can be covered by a CH and the resulting CH set can be small.

Each node on power up selects a random time in a predefined interval to broadcast the first round of the HELLO message to its neighbors. It carries the selection criterion and its direct neighbor's ID. The messages have to be sent at least three times [106]. Thus, a node knows the IDs of its neighbors in a two-hop range and what is the connectivity between its direct neighbors. The degree and remaining energy represented by C are also collected.

When every node has decided on its status in the network, the CHs will broadcast their CH status to all their direct neighbors in the beacon messages. On receiving the status indication message from all neighboring CHs, a non-CH node sends an association request message to associate with the closest CH. The CH will reply with an association reply to acknowledge. It is possible that some CHs do not have associated CM. In this case, the CH just connects the neighboring CHs but does not serve as a coordinator of a cluster. Therefore, it does not have to send the beacon message periodically.

To evaluate the impact of the CHs selection cost function, the size of the CDS and the average energy level are observed.

Figure 2.65 shows the number of selected CHs. A small F, which implies that the nodes degree is considered more important than the energy level, results in a small number of CHs. With the increase of the number of nodes in the network, the curve raises more rapidly with a large value F.

Figure 2.66 shows the average energy level of the selected CHs for different values of F.

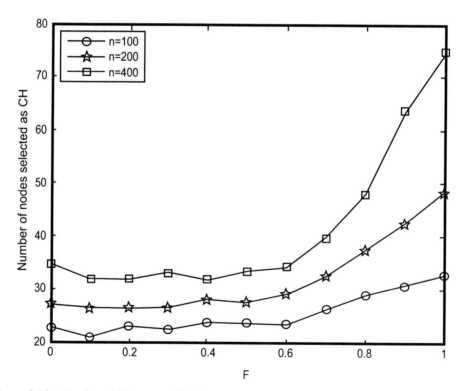

Figure 2.65 Number of CHs versus F [106].

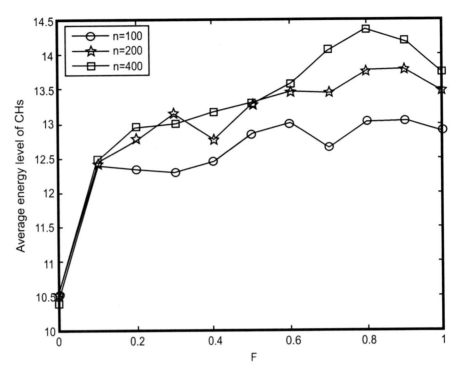

Figure 2.66 Energy level of CHs versus F [106].

The average energy level reaches its peak when *F* is 0.8. When *F* equals to 1, more CHs are selected (see Figure 2.65). In this case, although the energy level is dominant in the selection, the large number of CHs makes a larger divider, which decreases the average energy level.

It can be concluded that there is a trade-off in the number of CHs and their energy level when there is a change in the value of *F*. When F is increased until 0.7, the number of CHs doest not increase significantly and the average energy level of CHs almost reaches the peak. [106] suggests that to obtain a small set of CHs, which have relatively high energy level, F should be chosen around 0.7.

2.3.2.3 Self-Recovering

The self-recovering mechanism should detect any changes in the topology from the neighbor list and take the appropriate action to cope with these changes. When a node is added into a network, it will first listen to the medium for a specified period. In the case that it could receive beacons from the neighboring CHs, the nodes simply would send *association request* messages to the closest CH and associate with it. However, there is a possibility that the node is in radio range of the network but all its direct neighbors are CMs. Because the CMs do not transmit, unless requested by their CH, the newly added node cannot find them. Such a node is isolated by clustering.

The solution is to let the newly added node to send a preamble to interfere with the beacon messages sent to the CMs. To guarantee that at least one beacon message is interfered, a preamble has to be longer than one superframe. After transmitting a preamble, the newly added node keeps listening to the medium. The CMs interfered by the preamble will identify that a new node is added and report it to their CHs. Then the CHs will trigger the CH reselection process.

The failure of a CM does not have any impact on the clustered structure of the network. The only thing that it has to do is disassociate from its CH before dying. However, when a CH dies, other CHs must be reselected to maintain the connectivity. The CH may switch to a CM status and keep functioning in the network. The exact value of the threshold depends on several parameters, such as the traffic level, the number of CMs that a CH has, the power of the transceiver, and even the MAC protocol. When a CH decides to give up the CH status, it informs all its CMs and neighboring CHs to perform a CH reselection. It is a *quasi-cluster* formation procedure, in which the CH does not participate. On receiving the message, CHs connected with this resigning CH inform their respective CMs for holding on for five superframes of its absence, which is enough for a reselection action. Its CMs then delete their association with the CH. All the CMs and CHs linked with the resigning CH redo the whole cluster construction procedure. The difference here is that the original CHs are still CHs, irrespective to the result of the CH reselection. It is easy to prove that the resultant small CDS is connected to the whole network, since the newly selected CHs are certainly connected to the original CHs. The CH reselection procedure when a new node is added is the same than when a CH resigns, except that the CHs trigger the reselection and also participate in the reselection procedure.

After the new clusters are constructed, the self-recovering mechanism should inform the neighbor list to update the neighbor information accordingly.

2.3.2.4 Summary

The described architecture and algorithms clarify the basic functions of the self-organization and topology control in a mobile WSN network and can benefit the design of cross-layer optimization strategies.

2.4 Conclusions

Future communication systems will support multiuser applications, in which various users, using various devices, can dynamically change devices and continue their tasks in applications that dynamically enable and disable the available functions. Dynamic networks, networks that are mobile and can be created in an ad hoc manner, are core to the pervasive computing paradigm.

This chapter described basic protocols and algorithms including their interaction for the support of high-level performance in different types of ad hoc networks and network topologies. Although, there are a number of basic protocols that can be applied to any ad hoc network, one must identify the limitations that come from the architectural specifics of each type of network.

Next generation radio systems need to comply with a variety of requirements, not the least of which is ubiquitous coverage. Multihop communications is one way to increase the coverage but it brings about the challenge of designing appropriate routing, QoS, and mobility protocols.

As a way to ensure seamless mobility, integration and coexistence of various types of communication networks and architectures, has become quite important. Interoperability between solutions proposed for the support of inherently different scenarios create another challenge for the optimal design of protocols and increasing the complexity of algorithms.

The lack of ready made WSN communication protocols/mechanism forces developers to provide a solution for each particular need. Research towards self-organization requires that common and interoperable solutions are designed and deployed, making the communication framework more generic and allowing the autonomous action of the system. Topology control and self-organization are very important for a well-performing WSN.

Communications in WSNs are usually data-centric, with the objective of delivering collected data in a timely fashion. Also, such networks are resource-constrained, in terms of sensor nodes processing power, communication bandwidth storage space, and energy. This gives rise to new challenges in information processing and data management in WSNs. In many applications, users may frequently query information in the network. The trade-off between updates and queries needs to be addressed.

The classical layered protocol architecture is slowly giving way to design and solutions that are based on information form different protocol layers.

The future vision for the global information society is characterized by a massive proliferation of intelligent devices, where billions or even trillions of small electronic devices will be embedded in everyday objects to support us in our daily activities. This in turn will entail a massive increase in the complexity and heterogeneity of the network infrastructure characterized by convergence (fixed and mobile), and support for sensors, mobility and a variety of new highly dynamic services. This requires a different approach to the design of protocols and algorithms based on the concepts of uncertainty and autonomicity.

References

[1] FP6 IST Project DAIDALOS II Deliverable D211, "Concepts for Networks with Relation to Key Concepts, Especially Virtual Identities," September 2006, at http://cordis.europa.eu/ist/ct/proclu/p/mob-wireless.htm.

[2] FP6 IST Projects in Mobile and Wireless Beyond 3G, at http://cordis.europa.eu/ist/ct/proclu/p/mob-wireless.htm.

[3] FP6 IST Projects in Broadband for All, at http://cordis.europa.eu/ist/ct/proclu/p/broadband.htm.

[4] FP6 IST Projects MAGNET and MAGNET Beyond, at http://cordis.europa.eu/ist/ct/proclu/p/mob-wireless.htm.

[5] FP6 IST project DAIDALOS and DAIDALOS II, http://cordis.europa.eu/ist/ct/proclu/p/mob-wireless.htm.

[6] FP6 IST Project ADHOCSYS, at http://cordis.europa.eu/ist/ct/proclu/p/broadband.htm.

[7] FP6 IST Project CRUISE, at http://cordis.europa.eu/ist/ct/proclu/p/mob-wireless.htm.

[8] FP7 ICT Project E3, Deliverable 3.1, "Requirements for Collaborative Cognitive RRM," August 2008, at https://ict-e3.eu/.

[9]   FP6 IST project CRUISE, Deliverable 113, "Future Needs, Research Strategy and Visionary Applications for Sensor Networks," August 2008, at http://cordis.europa.eu/ist/ct/proclu/p/mob-wireless.htm.

[10] FP6 IST project DAIDALOS II, Deliverable D241," Architecture and Design: Routing for ad-hoc and moving networks, " October 2007, at http://cordis.europa.eu/ist/ct/proclu/p/mob-wireless.htm.

[11] Media Independent Handover (MIH) 802.21, at http://ieee802.org/21/index.html.

[12] 3GPP 1999, "Opportunity Driven Multiple Access," Technical Specification TR 25.924 V1.0.0, 1999.

[13] Rouse, T., I. Band, and S. McLaughlin, "Capacity and Power Investigations of Opportunity Driven Mulitple Access (ODMA) Networks in TDD-CDMA Based Systems," in *Proc. of the IEEE International Conference on Communications*, Tel Aviv, Israel, April 2002.

[14] CELTIC EU-funded Project WINNER+, Deliverable 1.3, "Innovative concepts in Peer-to-Peer and Network Coding," January 2009, at www.ist-winner.org.

[15] FP6 IST Project FIREWORKS, Deliverable 1D1 "Service Requirements and Operational Scenarios," May 2006, at http://cordis.europa.eu/ist/ct/proclu/p/mob-wireless.htm.

[16] FP6 IST Project FIREWORKS, Deliverable 1D2, "System Requirements," May 2006, http://cordis.europa.eu/ist/ct/proclu/p/mob-wireless.htm.

[17] NEXT WAVE Project FORWARD, Deliverable D3, "Correctness of Routing in Wireless Communications Environments," October 2003, at www.forward-project.org.uk.

[18] Dijkstra, E. W., "A Note on Two Problems in Connexion with Graphs," *Numerical Mathematics*, Vol. 1, 1959, pp. 269–271.

[19] Ford, L. R., Jr., and D. R. Fulkerson, *Flows in Networks*, Princeton University Press, 1962.

[20] Perlman, R., *Interconnections: Bridges and Routers*, Addison-Wesley, 1992.

[21] FP6 IST Project WINNER, Deliverable 3.1, "Description of Identified New Relay Based Radio Network Deployment Concepts and First Assessment," November 2004, at http://cordis.europa.eu/ist/ct/proclu/p/mob-wireless.htm.

[22] Johnson, D. B., and D. A. Maltz, "Dynamic Source Routing in Ad Hoc Networks," in *Mobile Computing*, T. Imielinski, and H. Korth, (eds.), Kluwer, pp. 152–181, 1996.

[23] Johnson, D. B., and D. A. Maltz, "The Dynamic Source Routing Protocol for Mobile Ad Hoc Networks," October 1999, IETF Draft, http://www.ietf.org/internet-drafts/draft-ietf-manet-dsr-06.txt.

[24] Perkins, C. E., and E. M. Royer, "Ad Hoc On-Demand Distance Vector (AODV) Routing," IETF Internet Draft, draft-ietf-manet-aodv-13.txt.

[25] Spirent Communications, White Paper, "Multicast Routing, PIM Sparse Mode and Other Protocols," November 2003, at http://www.spirentcom.com/.

[26] Prasad, R. (ed.), *Towards the Wireless Information Society*, Volumes 1 and 2, Artech House, 2005.

[27] Zhang, H.-K., B.-Y. Zhang, and B. Shen, "An Efficient Dynamic Multicast Agent Approach for Mobile IPv6 Multicast," in *Academic Open Internet Journal*, 2006, ISSN 1311-4360, Vol. 17.

[28] Couto, D. S. J. D., et al., "A High-Throughput Path Metric for Multihop Wireless Networks," in *Proceedings of ACM MOBICOM*, September 2003, San Diego, California.

[29] Haas, Z. J., and M. R. Pearlman, "The Zone Routing Protocol (ZRP) for Ad Hoc Networks," Internet draft, 1997.

[30] Chakrabarti, S., and A. Mishra, "QoS Issues in Ad Hoc Wireless Networks," in *IEEE Communications Magazine*, Vol. 39, No. 2, February 2001, pp. 142–148.

[31] Crawley, E., et al., "A Framework for QoS-Based Routing in the Internet," RFC 2386, August 1998.

[32] Sobrinho, J. L., and A. S. Krishnakumar, "Quality-of-Service in Ad Hoc Carrier Sense Multiple Access Wireless Networks," in *IEEE Journal on Selected Areas of Communications*, Vol. 17, No. 8, August 1999, pp. 1353–1414.

[33] Iwata, A., et al., "Scalable Routing Strategies for Ad Hoc Wireless Networks," *IEEE Journal on Selected Areas of Communications*, Vol. 17, No. 8, August 1999, pp. 1369–1379.

[34] Lin, C. R., "On-demand QoS Routing in Multi-Hop Mobile Networks," in *Proceedings of IEEE INFOCOM*, Anchorage, Alaska, Vol. 3, pp. 1735–1744, April 2001.

[35] Chen, X., Z., and M., S., Corson, "QoS Routing for Mobile Ad Hoc Networks," *Proceedings of IEEE INFOCOM*, June 2002, New York, USA, Vol. 2, pp. 958–967.

[36] Hongxia, S., and H. D. Hughes, "Adaptive QoS Routing Based on Prediction of Local Performance in Ad Hoc Networks," in *Proceedings of IEEE WCNC 2003*, Vol. 2, pp. 1191–1195, March 2003.

[37] Garey, M., and D. Johnson, *Computers and Intractability: a Guide to the Theory of NPCompleteness*, W. H. Freeman, 1979.

[38] Mauve, M., J. Widmer, and H. Hartenstein, "A Survey on Position-Based Routing in Mobile Ad-Hoc Networks," *IEEE Network*, Vol. 15, No. 6, 2001, pp. 30–39.

[39] RECOMMENDATION ITU-R M.1645, "Framework and Overall Objectives of the Future Development of IMT 2000 and Systems Beyond IMT 2000," at www.itu.int.

[40] Lott, M., et al., "Hierarchical Cellular Multihop Networks," in *Proceedings of Fifth European Personal Mobile Communications Conference (EPMCC 2003)*, April 2003, Glasgow, Scotland.

[41] Kosch, T., C. Schwingenschlögl, and L. Ai, "Information Dissemination in Multihop Inter-Vehicle Networks—Adapting the Ad-hoc On-demand Distance Vector Routing Protocol (AODV)," in *Proceedings of the IEEE International Conference on Intelligent Transportation Systems*, Singapore, September 2002.

[42] Ko, Y.-B., and N. H. Vaidya, "Location-Aided Routing (LAR) in Mobile Ad Hoc Networks," in *Proceedings of ACM/SIGMOBILE 4th Annual International Conference on Mobile Computing and Networking (MobiCom'98)*, Dallas, October 1998.

[43] Li, H., et al., "Multihop Communications in Future Mobile Radio Networks," in *IEEE PIMRC 2002*, September 2002.

[44] Li, H., et al., "*Hierarchical Cellular Multihop Networks*," in Proceedings of EPMCC 2003, March 2003.

[45] Assad, A. A., "Multicommodity Network Flow—A Survey," in *Networks*, Vol. 8, pp. 37–91, John Wiley & Sons, Inc.

[46] FP6 IST Project WINNER, Deliverable 3.4, " Definition and Assessment of Relay-Based Cellular Deployment Concepts for Future Radio Scenarios Considering the 1st Protocol Characteristics," November 2005, at http://cordis.europa.eu/ist/ct/proclu/p/mob-wireless.htm.

[47] FP6 IST Project WINNER, Deliverable 3.2, "Description of Identified New Relay Based Radio Network Deployment Concepts and First Assessment by Comparison against Benchmarks of Well Known Deployment Concepts using Enhanced Radio Interface Technologies", February 2005, http://cordis.europa.eu/ist/ct/proclu/p/mob-wireless.htm.

[48] Wang, Z., and J. Crowcroft, "Quality-of-Service Routing for Supporting Multimedia Applications," in *IEEE Journal on Selected Areas in Communications*, Vol. 14, No. 7, September 1996, pp. 1228–1234.

[49] Chen, S., and K. Nahrstedt, "An Overview of Quality-of-Service Routing for the Next Generation High-Speed Networks: Problems and Solutions," in *IEEE Network*, November/December 99, pp. 64–79.

[50] Chao, H. J., and X. Guo, *Quality of Service Control in High-Speed Networks*, New York: John Wiley & Sons, 2002.

[51] Cormen, T., H. Leiserson, C. E., and R. L. Rivest, *Introduction to Algorithms*, 2nd ed., The MIT Press, September 2001.

[52]   Martins, E., M. Pascoal, and J. Santos, "Labeling Algorithms for Ranking Shortest Paths," 1999, http://citeseer.ist.psu.edu.

[53]   Clausen, T., and P. Jacquet, "Optimized Link State Routing Protocol (OLSR)", RFC 3626, IETF, October 2003, http://www.ietf.org.

[54]   Benzaid, M., P. Minet, and K. Al Agha, "A Framework for Integrating Mobile-IP and OLSR Ad-Hoc Networking for Future Wireless Mobile Systems," *Wireless Networks*, Vol. 10, No. 4, July 2004, pp. 377–388.

[55]   FP6 IST Project WINNER, Deliverable 3.4.1, "The WINNER Air Interface: Refined Spatial-Temporal Processing Solutions," December 2006, at http://cordis.europa.eu/ist/ct/proclu/p/mob-wireless.htm.

[56]   Prasad, R., and A. Mihovska, *New Horizons in Mobile Communications: Radio Interfaces*, Artech House, 2009.

[57]   Alamouti, S. M., "A Simple Transmit Diversity Technique for Wireless Communications," *IEEE Journal on Selected Areas in Communications*, Vol. 16, No. 8, October 1998, pp. 1451–1458.

[58]   Anghel, P. A., G. Leus, and M. Kaveh, "Multi-User Space-Time Coding in Cooperative Networks," in *Proceedings of the IEEE ICASSP*, April 2003, Hong-Kong, China.

[59]   Anghel, P. A., G. Leus, and M. Kaveh, "Distributed Space-Time Coding in Cooperative Networks," in *Proceedings of the Nordic Signal Processing Symposium*, October 2002, Norway.

[60]   FP6 IST Project WINNER, Deliverable 3.5, "Proposal of the Best Suited Deployment Concepts and Related RAN Protocols," December 2005, at http://cordis.europa.eu/ist/ct/proclu/p/mob-wireless.htm.

[61]   FP6 IST Project CRUISE, Deliverable 220.1, "Protocol Comparison and New Features," August 2007, at http://cordis.europa.eu/ist/ct/proclu/p/mob-wireless.htm.

[62]   Zanella, A., and F. Pellegrini, "Statistical Characterization of the Service Time in Saturated IEEE 802.11 Networks," *IEEE Communications Letters*, Vol. 9, No. 3, pp. 225–227.

[63]   Gkelias, A., et al., "Average Packet Delay of CSMA/CA with Finite User Population," *IEEE Communications Letters*, Vol. 9, No. 3, March 2005, pp. 273–275.

[64]   MacKenzie, R., and F. T. Oapos, "Throughput Analysis of a p-Persistent CSMA Protocol with QoS Differentiation for Multiple Traffic Types," in *Proceedings of IEEE ICC*, May 2008, pp. 3220–3224.

[65]   El-Hoiydi, A., and J.-D. Decotignie "WiseMAC: An Ultra Low Power MAC Protocol for the Downlink of Infrastructure Wireless Sensor Networks," in the *Proceedings of the Ninth IEEE Symposium on Computers and Communication*, June 2004, Alexandria, Egypt, pp. 244–251.

[66]   El-Hoiydi, A., "Spatial TDMA and CSMA with Preamble Sampling for Low Power Ad Hoc Wireless Sensor Networks," in *Proceedings of the Sixth IEEE Symposium on Computers and Communication*, July 2002, Taormina, Italy, pp. 685–692.

[67]   Melly, T., et al., "WiseNET: Design of a Low-Power RF CMOS Receiver Chip for Wireless Applications," CSEM Scientific and Technical Report, 2002, at www.csem.ch.

[68]   Benedetti, M., et al, "On the Integration of Smart Antennas in Wireless Sensor Networks," *IEEE Antennas and Propagation Society International Symposium*, July 2008, pp. 1–4.

[69]   Boudour, G., et al., "On Designing Sensor Networks with Smart Antennas," in *Proceedings of the 7th IFAC International Conference on Fieldbuses and Networks in Industrial and Embedded Systems*, Toulouse, France, 2007.

[70]   FP6 IST project E-SENSE, at http://cordis.europa.eu/ist/ct/proclu/p/mob-wireless.htm.

[71]   Takai, M., et al., "Directional Virtual Carrier Sensing for Directional Antennas in Mobile Ad Hoc Networks," in *Proceedings of ACM MobiHoc*, June 2002.

[72] Breslau, L., et al., "Advances in Network Simulation," *IEEE Computer*, Vol. 33, No. 5, May 2000, pp. 59–67.

[73] Bao, L., and J., J., Garcia-Luna-Aceves, "Transmission Scheduling in Ad Hoc Networks with Directional Antennas," in *Proceedings of MOBICOM*, September 2002, Atlanta, Georgia, USA.

[74] Kubisch, M., H. Karl, and A. Wolisz, "A MAC Protocol for Wireless Sensor Networks with Multiple Selectable, Fixed-Orientation Antennas," TechnischeUniversitat Berlin, February 2004.

[75] Dimitriou, T., and A. Kalis, *Efficient Delivery of Information in Sensor Networks Using Smart Antennas*, Markopoulo Ave., Athens, Greece: Athens Information Technology.

[76] Coronel, P., S. Furrer, and W. Schott, "An Opportunistic Energy-Efficient Medium-Access Scheme for Wireless Sensor Networks," IBM Research GmbH, Zurich Research Laboratory, Switzerland: IEEE, 2005.

[77] FP6 IST Project CRUISE, Deliverable D112.1, "WSN Applications, Their Requirements, Application-Specific WSN Issues and Evaluation Metrics," May 2007, http://cordis.europa. eu/ist/ct/proclu/p/mob-wireless.htm.

[78] Galara, D., and J.-P. Thomesse, "Groupe de réflexion FIP: proposition d'un système de transmission série multiplexée pour les échanges entre des capteurs, des actionneurs et des automates réflexes," French Ministry of Industry, Paris, May 1984.

[79] Decotignie, J.,-D., and P., Prasad, "Spatio-Temporal Constraints in Fieldbus: Requirements and Current Solutions," in *Proceedings of the 19th IFAC/IFIP Workshop on Real-Time Programming*, June 1994, Isle of Reichnau, France, pp. 9–14.

[80] European Fieldbus Standard WorldFIP EN50170, EN 50170 -volume 3-Part 3-3: DATA LINK LAYER Service Definition and EN 50170 -volume 3-Part 5-3: Application Layer Service Definition (Sub-Part 5-3-1: MPS Definition and Sub-Part 5-3-2: SubMMS Definition) and EN 50170 -volume 3-.

[81] FP6 IST Project CRUISE, Deliverable D210.1, "Sensor Network Architecture Concept", November 2006, at http://cordis.europa.eu/ist/ct/proclu/p/mob-wireless.htm.

[82] Holger, K., and A. Willig, *Protocols and Architectures for Wireless Sensor Networks*, Wiley, April 2005.

[83] Heinzelman, W. B., A. P. Chandrakasan, and H., Balakrishnan, "An Application-Specific Protocol Architecture for Wireless Microsensor Networks," *IEEE Transactions on Wireless Communications*, Vol. 1, No. 4, October 2002, pp. 660–670.

[84] Lindsey, S., and C., S., Raghavendra, "PEGASIS: Power Efficient Gathering in Sensor Information Systems," in *Proceedings of the IEEE Aerospace Conference*, March 2002, Big Sky, Montana.

[85] Manjeshwar, A., and D. P. Agrawal, "TEEN: A Protocol For Enhanced Efficiency in Wireless Sensor Networks," in *Proceedings of the International Workshop on Parallel and Distributed Computing Issues in Wireless Networks and Mobile Computing*, April 2001, San Francisco, California.

[86] Kulik, J., W. R. Heinzelman, and H. Balakrishnan, "Negotiation Based Protocols for Disseminating Information in Wireless Sensor Networks," *ACM Wireless Networks*, Vol. 8, March-May 2002, pp. 169–185.

[87] Intanagonwiwat, D., et al., "Directed Diffusion for Wireless Sensor Networking," *IEEE/ACM Transactions on Networking*, Vol. 11, February 2003.

[88] Braginsky, D., and D. Estrin, "Rumor Routing Algorithm for Sensor Networks," in *Proceedings of the First Workshop on Sensor Networks and Applications (WSNA)*, Atlanta, Georgia, October 2002.

[89] Schurgers, C., and M. B. Srivastava. "Energy Efficient Routing in Wireless Sensor Networks," in *MILCOM Proceedings on Communications for Network-Centric Operations*,

2001. MILCOM Proceedings on Communications for Network-Centric Operations: Creating the Information Force.

[90]   Chu, M., H. Haussecker, and F. Zhao, "Scalable Information-Driven Sensor Querying and Routing for Ad hoc Heterogeneous Sensor Networks," *International Journal of High Performance Computing Applications*, Vol., 16, No., 3, August 2002.

[91]   Yao, Y., and J. Gehrke, "The Cougar Approach to In-network Query Processing in Sensor Networks", in *SIGMOD Record*, September 2002.

[92]   Sadagopan, N., B. Krishnamachari, and A. Helmy, "The ACQUIRE mechanism for efficient querying in sensor networks," In *Proceedings of the IEEE International Workshop on Sensor Network Protocols and Applications (SNPA)*, Anchorage, Alaska, May 2003.

[93]   Madden, S., et al, "TinyDB: an Acquisitional Query Processing System for Sensor Networks," *ACM Transactions on Database Systems*, Vol. 30, No., 1, 2005, pp. 122–173.

[94]   Turin, G., W. Jewell, and T. Johnston, "Simulation of Urban Vehicle-Monitoring Systems," in *IEEE Transactions on Vehicular Technology*, Vol. 21, No. 1, February 1972.

[95]   Savvides, A., et al., "Location Discovery in Ad Hoc Wireless Networks," Memorandum, Networked and Embedded Systems Laboratory, UCLA, June 2000.

[96]   Ward, A., A. Jones, and A. Hopper, "A New Location Technique for the Active Office," *IEEE Personal Communications*, Vol. 4, No. 5, October 1997.

[97]   Priyantha, N., A. Chakraborty, and H. Balakrishnan, "The Cricket Location Support System," in *ACM Mobicom*, August 2000.

[98]   Girod, L., and D. Estrin, "Robust Range Estimation for Localization in Ad Hoc Sensor Networks," in Technical Report CS-TR-2000XX, UCLA, November 2000.

[99]   Kalpakis, K., K. Dasgupta, and Parag Namjoshi, "Maximum Lifetime Data Gathering and Aggregation in Wireless Sensor Networks," in *Proceedings of the 2002 IEEE International Conference on Networking (ICN'02)*, Atlanta, Georgia, August 26–29, 2002, pp. 685–696.

[100] Ye, F., et al., "A Scalable Solution to Minimum Cost Forwarding in Large Sensor Networks," in *Proceedings of Tenth International Conference on Computer Communications and Networks*, 2001, Scottsdale, Arizona, pp. 304–309.

[101] Sohrabi, K., et al, "Protocols for Self-organization of a Wireless Sensor Network," in *IEEE Personal Communications*, October 2000, Vol. 7, pp. 16–27.

[102] Akkaya, K., and M., Younis. "An Energy-aware QoS Routing Protocol for Wireless Sensor Networks," in *Proceedings of the 23rd International Conference on Distributed Computing Systems Workshops*, May 2003, Providence, Rhode Island.

[103] Mukhopadhyay, S., D. Panigrahi, and S. Dey, "Data Aware, Low cost Error correction for Wireless Sensor Networks," in *Proceedings of the IEEE Conference in Wireless Communications and Networking*, March 2004, Atlanta, Georgia, pp. 2492–2497.

[104] Yu, B., and K. Sycara, "Learning the Quality of Sensor Data in Distributed Decision Fusion," in *Proceedings of the International Conference on Information Fusion*, 2006, Florence, Italy, pp. 1–8.

[105] Han, Q., et al, "Sensor Data Collection with Expected Guarantees," in *Proceedings of the IEEE International Conference on Pervasive Computing and Communications*, March 2005, Hawaii, Hawaii, pp. 374–378.

[106] FP6 IST Project E-SENSE, Deliverable 3.3.1, "Efficient and Light Weight Wireless Sensor Communication Systems," May 2007, at http://cordis.europa.eu/ist/ct/proclu/p/mob-wireless.htm.

[107] FP6 IST Project E-SENSE, Deliverable D3.2.1, "Efficient Protocol Elements," November 2006, at http://cordis.europa.eu/ist/ct/proclu/p/mob-wireless.htm.

[108] Fu, Z., X. Meng, and S. Lu, "A Transport Protocol for Supporting Multimedia Streaming in Mobile Ad Hoc Networks," in *IEEE Journal on Selected Areas in Communications*, December 2003, Vol. 21, No. 10, pp. 1615–1626.

[109]   FP6 IST Project E-SENSE, Deliverable 3.3.2, "Novel Cross-Optimization," January 2008, at http://cordis.europa.eu/ist/ct/proclu/p/mob-wireless.htm.

[110]   IEEE 802.15.4 standard, "Part 15.4: Wireless Medium Access Control (MAC) and Physical Layer (PHY) Specifications for Low-Rate Wireless Personal Area Networks (LR-WPANs)," http://standards.ieee.org/getieee802/download/802.15.4-2003.pdf.

[111]   Polastre, J., et al., "A Unifying Link Abstraction for Wireless Sensor Networks," in *Proceedings of SenSys 2005*, November 2005, San Diego, California.

[112]   Hekmat, R., *Ad-hoc Networks: Fundamental Properties and Network Topologies*, Springer, 2006.

[113]   Clark, B. N., C. J. Colbourn, and D. S. Johnson, "Unit Disk Graphs," *Discrete Mathematics*, Vol. 86, 1990, pp. 165–177.

[114]   Williams, B., and T. Camp, "Comparison of Broadcasting Techniques for Mobile Ad Hoc Networks," in *Proceedings of the ACM International Symposium on Mobile Ad Hoc Networking and Computing* (MOBIHOC), June 2002, Lausanne, Switzerland.

[115]   CC2420 Datasheet 2006, Chipcon Corporation Available: http://www.chipcon.com/files/CC2420 Data Sheet 1 4.pdf.

[116]   Bahl, P., and V. N. Padmanabhan, "Radar: An In-building RF-based User Location and Tracking System," *Proc. of IEEE INFOCOM 2000*, Tel-Aviv, Israel, March 2000.

[117]   Bulusu, N., J. Heidemann, and D. Estrin. "GPSless Low Cost Outdoor Localization for Very Small Devices," *IEEE Personal Communications Magazine*, Vol. 7, No. 5, October 2000, pp. 28–34.

[118]   Capkun, S., M. Hamdi, and J. P. Hubaux, "GPS-free Positioning in Mobile Ad-Hoc Networks," *Proc. of HICSS 2001*, Maui, Hawaii, January 2001.

[119]   Yu, Y., D. Estrin, and R. Govindan, "Geographical and Energy-Aware Routing: A Recursive Data Dissemination Protocol for Wireless Sensor Networks," UCLA Computer Science Department Technical Report, UCLA-CSD TR-01-0023, May 2001.

[120]   Zorzi, M., and R. R. Rao, "Geographic Random Forwarding (GeRaF) for Ad Hoc and Sensor Networks: Multihop Performance," *IEEE Transactions on Mobile Computing*, Vol. 2, No. 4, October-December 2003.

CHAPTER 3

Technologies for Capacity Enhancement

The main challenges related to the design of optimally performing ad hoc networks comes from the requirements for self-organization, self-configurability easy adaptation to different traffic requirements and network changes. The capacity of ad hoc wireless networks is further constrained by the mutual interference of concurrent transmissions between nodes. Wireless ad hoc networks consist of nodes that communicate with each other over a shared wireless channel. The radio signal attenuation and interference on the shared wireless medium impose new challenges in building large-scale wireless ad hoc networks. Finally, especially in the context of wireless sensor networks (WSN), the provision of energy-saving solutions is yet another challenge.

Chapter 2 identified the fundamental performance limits for ad hoc networks and described the evaluation methodologies for protocols and algorithms (e.g., routing, link adaptation) from a network capacity point of view.

This chapter describes the strategies and their assessment proposed to enhance the performance of the physical network for different types of ad hoc networks. The contributions are based on the achievements of the projects that were active in the area in the scope of the Framework Program Six (FP6) of the Information Society and Technology (IST) research and development initiative [2, 3]. The projects had different objectives and scenarios, therefore, the proposed solutions span areas such as joint scheduling, resource allocation, and power control and cooperative transmissions for multihop and wireless mesh communication systems [4–6], adaptive coding and modulation for opportunistic networks [7]; joint PHY-MAC layer solutions for opportunistic, wireless personal and sensor networks [7–10]; energy-efficiency schemes for WSNs [9–10], capacity enhanced wireless personal area network (WPAN) radio interface technologies [8, 11].

This chapter is organized as follows. Section 3.1 introduces into the main challenges related to enhancing the performance of wireless ad hoc networks and outlines some of the basic strategies to achieve that. Section 3.2 describes strategies for enhancing the performance of mobile relay-based systems. Section 3.3 describes distributed scheduling approaches towards optimized wireless mesh networks. A cross-layer framework for further increase in the performance incorporating the presented scheduling strategies is also described. Section 3.4 describes an approach to optimized inter-WPAN communications. Section 3.5 concludes the chapter.

3.1 Introduction

Wireless ad hoc networks, regardless of the type, must meet a number of stringent technical requirements. Examples of wireless ad hoc networks are the networks used in coordinating an emergency rescue operation, networking mobile users of portable and powerful computing devices (e.g., laptops, PDAs), sensor networks, automated transportation systems, Bluetooth, and home RF. Such networks consist of a group of nodes that communicate with each other over a wireless channel without any centralized control. As every node may not be in direct communication range of every other node, nodes in ad hoc networks cooperate in routing each other's data packets. Thus, the lack of centralized control and possible node mobility give rise to a number of challenging design and performance evaluation issues [12].

First of all, they must meet the high capacity needs of the access nodes which have to forward the accumulated traffic. One important performance analysis issue is to determine the traffic-carrying capacity of multihop wireless networks. In addition, these networks have to cope with the delay and other strict quality-of-service (QoS) requirements of the end user applications. They must be able to self-adapt to varying environment conditions. Future wireless ad hoc networks would integrate a large amount of distributed various types of nodes (including wireless sensors) with various propagation characteristics. Optimizing such wireless networks requires significant time and resource effort to obtain the effective capacity, coverage, and quality of the service gains. A self-optimizing network would adapt and automatically configure its logical topology and its parameters in function of the dynamic changes observed in the network resource state and traffic.

In order to satisfy such stringent technical requirements, a range of novel enabling techniques has been proposed and researched. Opportunistic behavior is highly linked with adaptive coding and modulation to perform a capacity achieving transmission in the case of given opportunities in the dimensions time, frequency, and space [13]. In OFDM-based systems, due to the nature of the waveform, two techniques can be defined for link adaptation: frequency-adaptive transmission and non-frequency adaptive transmission [14]. Adaptive modulation and coding is further used to adapt the physical modulation and coding schemes to a varying environment. Further, resource allocation schemes are also important enabling technologies for the successful transmission of data in the above scenarios. Algorithms for cooperative transmission can allow for adapting the cooperation between the source and relays to the topology of the environment in a cooperative relay-based multihop network [14]. The channel models should be able to include scenarios of many available relays in the communication environment. One-way to capture the multihop behavior is to use the *slotted amplify-forward model*, in which each relay repeats the source transmission in a time slot assigned to itself [15]. A novel concept for high throughput transmissions is the use of *smart* antennas. These techniques include fixed beam antennas, adaptive antennas, and *multiple-input multiple-output* (MIMO) coding. In the context of wireless mesh networks (WMNs), if the independent streams are routed to different relays and at the same time sent directly to the destination, the system throughput can be maximized [16]. This approach is particularly appealing for correlated channels, where the spatial multiplexing gain cannot be fully obtained by sending the data via only one relay node. Further, the design of a cognitive *medium access control* (MAC) protocol can improve utilization of spatial opportunities and spatial efficiency in opportunistic networks [13].

Cross-layer design strategies can effectively exploit the dependencies between layers of the protocol stack to obtain performance gains. In the context of WSNs, appropriate cross-layer protocol design can improve the energy efficiency [17]. Highly optimized, energy efficient PHY layer technologies combined with cross-layer optimized and designed protocols from the PHY to the transport layer, enables a new generation of capable wireless sensory [18].

3.1.1   Channel Network Capacity

Depending of the nature of the network and the assumptions made, channel network capacity is known only in special cases [12, 19]. In general, computing the capacity upper bound is not easily achievable, however, for a relay channel, a simple interpretation of the bound is possible [12]. An upper bound to the capacity of a Gaussian relay network was found in from the cut-set theorem in [19]. It was shown that for a larger class of Gaussian relay networks the typical scaling behavior of capacity is $\log(M)$, where M is the number of relay nodes.

The most well known model is the "water networks" model. A network of water pipes could be seen as a set of nodes or a set of source/destination pairs, with the main node: the source of water, and the final destination: the sink, where all the water has to go [21]. It was shown in [21] that the maximum flow from the source to the sink (the "max-flow") is equal to the minimum flow, minimized over all the cuts (the "min-cut"): *max-flow = min-cut*; often referred to as the *max-flow min-cut* theorem.

Differing from the water network, the *wireless* network has the particularity that every node receives the signal from every other node [12]. The first attempt to address the issue of the maximum amount of information that can be carried over an ad hoc network, was made in [22] where the *transport capacity* of a network was introduced as the sum of products of bits and the distances over which they are carried, expressed in *bit-meter* (i.e., the network transports one bit-meter when one bit has been transported a distance of one meter towards its destination). The main results found under a certain model of communication, were that the transport capacity of a network of n nodes located in a region of unit area is $O(\sqrt{n})$ bit-meters/s. This implied that the per-user capacity scales as $O(\sqrt{n}/n) = O(1/\sqrt{n})$ (i.e., the achievable throughput per user in this point-to-point coding model tends to zero as the user density increases). [23] computed sharp information scaling laws under some conditions to establish the optimality of multihop operation in some situations, and a strategy of multistage relaying with interference in others. An information-theoretic construction scheme for obtaining an achievable rate region in communication networks, which are of arbitrary size and topology, and communication over a general discrete memory less vector channel were proposed in [24]. By using the proposed scheme, inner bounds for the multicast and all-cast capacities were derived. It was shown, further, that the proposed scheme achieves the transport capacity of $\Theta(n)$ bit-meters/s in a specific wireless network of n nodes located in a unit area region.

For general wireless ad hoc networks, without making simplifying assumptions, an upper bound can be computed in [19] by use of the *max-flow min-cut* theorem [12]. The cut-set bound found in [20] is a relaxed and easy to compute and could be applied to the investigation of the broadcast channels, the multiple-access channels, and the relay channels.

In the context of network information theory, the relay channel was first introduced in [25]. However, the capacity was established only for relatively degenerated channels. The upper and lower bounds to the capacity for the physically degraded Gaussian parallel relay channel was computed in [26]. The parallel relay channel consisted of a source node communicating with a destination and two relays assisting the data transmission. It was shown that the upper and the lower bounds coincide, resulting in reliably achievable rates, when considering one of the following two scenarios. The first scenario is when a natural staggered block coding scheme is used (i.e., the source transmits in the first half symbol period, and in the second half symbol period, the relays decode the received observations and then transmits identical corresponding code words (with high probability). The second scenario is when the relays view independent observations of the channel input. Then each relay acts as a simple transponder, amplifying both signal and noise. [20] investigated the *relay traffic pattern* composed of one source/destination pair, and a certain number of nodes assisting this transmission.

[12] noted that all the relay channel rates, established in the special cases, were proved to be achievable using information-theoretic coding strategies (e.g., *block Markov superposition encoding*, *regular block Markov encoding*, and *windowed decoding*). A comparison of these coding strategies can be found in [27]. These coding strategies are too complex for easy and practical implementation, as they apply to large sequences satisfying some mathematical assumptions [12]. Even if too theoretical for a practical implementation, the theory could help to determine the optimality and perhaps suggest some means of improving the communication rates.

A number of solutions exist for making use of the advantages of the relay channel in practice, namely, [28, 29]. [29] proposed an energy-efficient transmission protocol that offered diversity gains over single- and multihop transmission.

The FP6 IST project WINNER enhanced existing capacity relay channel concepts for a hot-spot area scenario and used them as a basis to develop a relay concept and integrate it in a ubiquitous multihop communication system concept (i.e., the WINNER system concept). As an initial study of the benefits of relaying to a broadband communication system, it was shown that the capacity that can be made available at the fixed relay station (FRS) (i.e., when the whole capacity of the access point is transferred to the area that is covered by one of the FRSs), amounts, depending on the FRS receive antenna gain, to values between 2.7 Mbits for 0-dBi gain and 15.87 Mbits for 30-dBi gain. This is shown in Figure 3.1 and Figure 3.2. The gap between the two curves in Figure 3.2 denotes the capacity that has to be invested into the extension of the coverage range by means of relaying.

In this case, a FRS serves another FRS according to the needs, besides serving the MTs that can be roaming in its local environment.

This investigation was performed for the purpose of the design of a MAC-frame based relaying concept [12]. The transmit capacity for the terminals directly associated to the AP (FRSs and MTs) is allocated by the AP. An FRS appears to the AP like a MT but sets up a *sub frame* (SF) structure, which is embedded into the MAC frame structure of the serving AP. The SF structure has available only the capacity

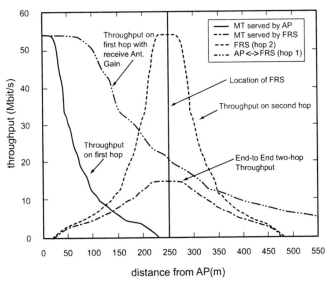

Figure 3.1 Throughput for separate hops and end-to-end for mobile terminals (MTs) served by FRS (16 dBi FRS receive antenna gain) [12].

assigned by the AP to the FRS. This is shown in Figure 3.3 for a HIPERLAN/2 (H2) MAC frame structure.

The SF is generated and controlled by the FRS and it is structured the same as the MAC frame used at the AP. It enables the communication with legacy H2 terminals without any modifications. The functions introduced to the H2 MAC frame to enable relaying in the time domain are shown in Figure 3.3. The capacity assigned in the MAC frame to the FRS to be used to establish a SF, is placed in the uplink (UL) frame part of the AP. When the FRS is transmitting in the downlink (DL), the data is addressed properly to its RMT and the AP will discard this data accordingly.

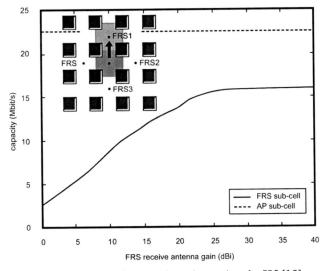

Figure 3.2 Capacity of the AP in single-hop mode and capacity of a FRS [12].

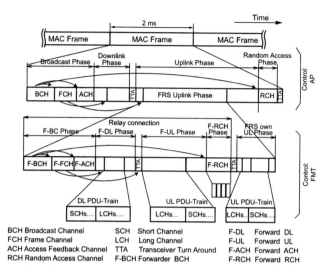

Figure 3.3 Standard-conformant enhancements of the H2 MAC frame [12].

The same applies for the data transmitted from the RMT to the FRS. The capacity to exchange the data between the AP and the FRS has to be reserved in both the UL and the DL directions upon request by the FRS [30]. A very similar operation is possible by using the hybrid coordinator access in IEEE802.11e [31].

Relaying can consume part of the capacity of an AP, since the relayed data has to go twice over the radio channel [12]. It was shown in [32] that for relay-based deployment concepts the MTs served at different relays that belong to the same AP can be served at the same time, whereby the capacity loss introduced by 2-hop communications can be compensated to a great extent [12]. This capacity loss can even be turned into a substantial gain, if directional antennas are used at the FRS.

3.1.1.1 Capacity of WSNs

In WSNs, the network capacity should be considered with the energy consumption (which determines the life time of the sensors) [10]. Taking into account the energy considerations, the network capacity (i.e., the maximum achievable data traffic inside the WSN under supposed ideal conditions; that is, optimal routing, perfect transmission medium, no noise, no interferences, and so on) should be considered as a trade-off with the energy consumption. The energy consumption can be modeled with considerations of the *modulation* techniques.

A WSN consists of a large number of small sensor nodes with sensing, data processing, as well as communication, and networking capabilities. These networks are characterized by a dense node deployment, which suffers from unstable sensor nodes, frequent topology changes, and severe power, computation, and memory constraints.

One of the largest energy consumers in the WSN node is the *radio part* used for the transmission of the information in the network. Two possible WSN architectures are the *flat sensor* network and the *clustered sensor* network [10]. A higher network capacity can be obtained when the sensor networks are divided into clusters, although in a non-energy-efficient fashion [33]. However, this capacity has a limited value and saturates if the number of clusters is above a certain value. The clustering method had been studied with and without the distributed coordination function and it was found that there is a critical value of the number of cluster heads beyond which the network capacity of the two models approaches the same [34].

The concept of *broadcast capacity* for mobile networks had been introduced in [35]. It corresponds to the maximum rate, at which the network may generate messages intended for distribution to the entire network, subjected to coverage and delay constraints. The constraints are completely different for WSNs, therefore, the capacity achievement must be redefined for this particular context.

In a WSN network a multihop strategy could be employed, where a message is only transmitted to nearby nodes, which then relays or retransmits the message towards its destination. The advantage of such a strategy is that as the transmit power increases quadratically with the distance, much lower transmit powers have to be used, which increases the overall lifetime of the node, and hence the capacity of the network [34].

Energy-efficient communications are then desirable for WSN, leading to low power consumption.

Since transmission techniques are directly related to power consumption, energy-efficient modulations and channel coding must be used in a WSN in order to maximize the lifetime of batteries. Another important aspect to be considered in a WSN is the media access, inherent to this type of networks. Since energy efficiency is a key point, the design of MAC protocols focused on the energy optimization can also lead to efficient utilization of limited power supply [34].

Sensor electronics design must take into account this important limitation, in order to minimize the power consumption as much as possible. In conventional wireless systems such as cellular networks and *wireless local area networks* (WLANs) most of the power consumption is made by the transmission electronics; on the contrary, in wireless sensor nodes, the power consumption is mainly due to the receiver electronics, therefore the proper design of receiving algorithms and receiver architecture has the potential to boost up the network capacity.

In [36], the performances of a WSN had been evaluated by the *sensing capacity* in a target detection system. The *sensing capacity* corresponds to the maximum ratio of target positions to sensors, for which inference of targets within a certain distortion is achievable and characterizes the ability of a sensor network to successfully distinguish among a discrete set of targets. The application itself enters to the capacity analysis [34].

Many WSN applications addressed do not require nodes mobility [34]. On the other hand, mobility can increase the complexity of data organizations and network protocols.

The introduction of mobility in WSN, on the contrary, can sometimes solve (at least partially) certain problems, such as energy consumption and improve characteristics such as the network capacity [37, 38].

In this way, mobility can used as an advantage; with proper mobility management mechanisms, better energy efficiency, improved coverage, enhanced target tracking, and superior channel capacity can be achieved [39].

Furthermore, the network performance can benefit in terms of scalability, energy efficiency, and packet delay [34].

3.1.2 Diversity

A fundamental characteristic of wireless communications is the time variations of the wireless channel. To mitigate the destructive effects of the wireless channel time variations, *diversity* is employed, where the basic idea is to improve the system performance by creating several independent signal paths between the transmitter and the receiver [15]. The wireless transmission medium is subject to random fluctuations of signal level in space, time and frequency caused by multipath nature of propagation and cochannel interference. Diversity is a way of combating channel fading and improving the link reliability.

The classical way of achieving diversity is by supplying the receiver with multiple (ideally) independently fading replicas of the data signal in time, frequency, or space. Appropriate combining at the receiver realizes a diversity gain, thereby improving the link reliability. Due to the scarcity of resources such as bandwidth and time, it is often desirable to achieve diversity through *spatial* techniques (i.e., using additional antenna) in a communication system. *Spatial diversity* techniques have been extensively studied in the context of multiple-antenna systems [40–43].

Despite the advantages of spatial diversity, it is hard to deploy a large number of antennas at a single network entity (e.g., relay) [12]. The *virtual antenna array* (VAA) structure is a pragmatic approach to communication network design, in which communicating entities use mutually the resources in a symbiotic relationship [12]. Cooperative diversity is achieved when the nodes in the network cooperate to form a VAA realizing spatial diversity in a distributed fashion [15].

Relays (mobile or fixed) cooperating in this manner may even emulate a MIMO system: a MIMO-VAA [44]. Especially, in scenarios where the number of antennas at the mobile terminals is the limiting factor, a MIMO-VAA architecture could be very helpful [12]. In a MIMO-VAA systems, the antenna elements are relatively further apart. Therefore, the space constraint that leads to mutual correlation among the antenna elements in conventional MIMO systems and, which may affect the performance of the MIMO systems is not a concern, anymore. [12] explains a simple diversity architecture that lays the foundation to generalized MIMO virtual channels that were investigated in relation to the WINNER radio system concept [14, 45, 46].

Various problems associated with VAA deployment (increased system load and interference, orthogonality constraint) occur only when using the *same* air interface for the main link and relaying link [12]. By employing a *different* interface for intra-VAA communication and transmission from the VAA to the final destination, these problems can be avoided. The traffic load created by the VAA is then shifted from the main access network to the relaying network. By using a different air interface for the relaying links, some of the core drawbacks of VAA could be avoided. However, due to the nature of this technique, it is no longer comparable to conventional cooperative diversity schemes since power normalization is impossible when using different wireless standards simultaneously [12].

Achieving diversity in a distributed fashion has been in the focus of research, especially with advances in the area of antenna techniques and the deployment benefits of multihop wireless networks [4, 49, 50]. Protocols for cooperation for channels with a single relay terminal have been analyzed in [28, 29]; [47–50]. It had been shown that for channels with multiple relays, cooperative diversity with appropriately designed codes can achieve a full spatial diversity gain [51]. Half-duplex relay channels were studied in detail in [52, 53]. In [52], a new protocol had been identified for the half-duplex single relay channel, which by maximizing the degree of collision and broadcast could achieve the diversity-multiplexing trade-off for *amplify-and-forward* based strategies (see Chapter 2 of this book). *Diversity-multiplexing* tradeoffs had been derived for a variety of half-duplex relay channels in [53], and a new *decode-and-forward* based protocol that achieves the diversity-multiplexing trade-off (for certain multiplexing gains) for the single relay channel had been identified. For channels with multiple relays, it had been shown that the diversity-multiplexing curve achievable by a class of slotted *amplify-and-forward* protocols dominate the diversity-multiplexing curves achievable by all other known *amplify-and-forward* protocols [54]. Finally, an array of works exists that studied and analyzed ways to achieve distributed beamforming in wireless networks and that many cooperative diversity schemes can be cast into the framework of network coding [55–57]. As a new strategy for information transmission in networks proposed in [55], network coding allows messages from different sources (or to different sinks) to mix in the intermediate nodes. Performance gains in terms of, e.g., network flow, robustness, or energy efficiency [58], are obtained. Network coding was originally proposed for error-free computer networks, however, the principles of network coding can be applied also to implement cooperative communications.

[59] in a follow up of the innovations achieved within the FP6 IST projects WINNER and WINNER II [4] proposed to combine wireless diversity and the capability of *max-flow achieving* of network coding. It was shown that the designed *non-binary* network codes can substantially decrease the outage probability/*frame error rate* (FER), for multiple-user cooperative communications. For the relay-based networks, [59] proposed a novel *network-coding-based* protocol with an efficient decoding approach at the receiver.

3.1.3   Scheduling, Resource Allocation, and Power Control

Packet scheduling, power control, resource allocation, and routing are commonly used techniques to enhance the performance in terms of throughput and data rate in wireless networks [60]. The traditional protocol layered approach that designs and optimizes the operations of each network protocol layer independently is not sufficient anymore in the context of next generation communication systems (and, in particular wireless ad hoc networks) to provide their optimal design and performance [61]. In *mesh* networks, the notion of UL and DL as in cellular networks disappears. Thus, the scheduling algorithms have to determine whether a given link is activated for transmitting or receiving (i.e., the *link-activation* problem). Reconfigurability is also a key design factor for next generation communication systems, and should be reflected into the resource management schemes (e.g., scheduling) [60].

One of the central tasks in *opportunistic* systems is the evaluation of the channel quality between collaborating radio nodes [13]. The channel quality measurement typically consists of interference/noise *power spectral density* (PSD) estimation in a given band, as well as the estimation of the *channel transfer function*. Based on the channel quality, radio resources are assigned accordingly in order to meet certain optimization criterions and constraints. Link adaptation algorithms help to design *decision-making* algorithms that decide if certain opportunities are exploitable or not. Figure 3.4 shows the overall relationship between channel estimation and opportunistic system performance.

The following example shows the importance of accurate channel estimation for the proper execution of strategies leading to an improved performance in opportunistic OFDM systems.

Consider the two-sided power spectral density N_0 of the interference/noise and using energy per data symbol E_S [13]. The instantaneous SNR γ can be defined as:

$$\gamma = \frac{|H|^2 E_S}{N_0} \tag{3.1}$$

where $|H|^2$ denotes the *channel transfer function* assuming flat fading conditions. The average SNR is given by:

$$\bar{\gamma} = \frac{E_S}{N_0} \tag{3.2}$$

The SNR estimate $\hat{\gamma}$ is given by means of the *estimated channel coefficient* $\hat{H} = H + W$ as:

$$\hat{\gamma} = \frac{|\hat{H}|^2 E_S}{N_0} = \frac{|H + W|^2 E_S}{N_0} \tag{3.3}$$

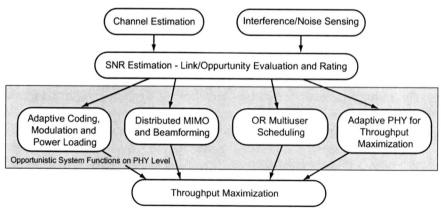

Figure 3.4 PHY layer functions that depend on accurate SNR and channel estimation in an opportunistic network [13].

The value of the SNR estimate in (3.3) is used for the following opportunistic system functions [13]:

1. Adaptive coding, modulation and power loading (i.e., link adaptation);
2. Decision making on the distribution of opportunities among users (i.e., multiuser scheduling);
3. Evaluation of the spatial distributed multi-antenna links for accurate combining and multiuser transmission (i.e., beamforming).

The accurate channel estimation can also used for the throughput maximization owing to the following reasons:

1. Generally, the accurate channel estimation increases the detection performance in case of coherent detection [13].
2. The *channel impulse response* (CIR) in frequency selective scenarios contains information on the scattering environment in terms of power delay profile and CIR length that can be used for PHY adaptation. For example, the CIR length knowledge can be used to adapt (minimize) the guard interval length of OFDM systems and hence increases the throughput.

Therefore, the accurate channel knowledge is one of the fundamentals of opportunistic behavior and is directly linked with the performance metrics of opportunistic systems [13]. Shows the statistics of the SNR estimation error in $\hat{\gamma}_{dB} - \gamma_{dB}$ in dB for different channel estimation schemes.

The channel estimation performance in terms of *mean squared channel estimation error* (MSE) is highly related to the CIR length L. The *frequency-domain channel length indicator* (FCLI) is an algorithm developed by [7] for joint channel and CIR length estimation. It can be seen in Figure 3.5 that the FCLI outperforms the conventional channel estimation methods in terms of SNR estimation error. Corrupted SNR estimates can potentially lead to wrong bit loading decisions among the subcarriers [13].

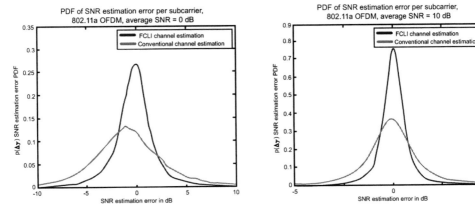

Figure 3.5   PDF of SNR estimation error under different average SNR (0 dB, 10 dB) using proposed FCLI algorithm and conventional channel estimation method [13].

FCLI requires extra signal processing capabilities, which puts requirements on the design of the opportunistic terminals (e.g., low-power opportunistic terminals will not be able to use FCLI for channel estimation). *Cyclic prefix* (CP) adaptation is one option for opportunistic MIMO-OFDM throughput enhancement by limiting the CP to the minimal necessary amount, without performing any complex signal processing at the receiver side [13]. CP-adaptation, however, requires extra signaling bandwidth and is sensitive to temporal channel length variations.

A relay system is a typical technique to exploit the space dimension opportunities (see Section 3.1.2). A fundamental issue to address is the subcarrier allocations for different links within the system in order to design an optimal scheme [13].

3.1.3.1   Power Control

Power control has been shown to be a very effective technique to combat channel fluctuation and interference, and thus improve the overall network performance. To design an appropriate framework for power control, it is necessary to acquire a good understanding of the channel characteristics [60]. Channel characteristics pertinent to power control include the *channel coherent time*, the *coherent bandwidth*, and the *delay spread*.

3.1.3.2   Scheduling

Current and next generation wireless networks make use of scheduling techniques to explore the multiuser diversity for achieving high system throughputs for wireless data applications [45, 60]. One particularly attractive strategy is to schedule the data transmission based on the relative channel quality of different users, while aiming for an acceptable balance between the system performance and fairness among users. A well-known algorithm that accomplishes this goal is the *proportional fair* (PF) scheduling, which is, for example, implemented in CDMA 1X EV-DO systems [62].

To efficiently utilize limited radio resources, existence of multiple users in the system and the time variability of channel quality for various users makes it possible to achieve performance gains by adopting suitable algorithms for packet scheduling [60]. Specifically, a scheduler decides to which user the next time slot is allocated according to a performance metric. At the beginning of each time slot, the scheduler updates the metric values of all ready users, and chooses the one with the maximum value to serve in the following time slot.

As a consequence, the effectiveness of the scheduling algorithm strongly depends on the accuracy of the performance metric, or to what extent the metric represents the benefit of the system gains from scheduling for the particular user. For example—PF scheduling allocates system resources according to the future channel quality for different users. Therefore, the accuracy of future channel-quality predictions is essential to achievement of the desirable performance of the PF scheduler [60].

In wireless communication systems, opportunistic scheduling algorithms are commonly used to take advantages of uncorrelated channel fading among users to achieve high data throughput, or multi-user diversity gain. As channels of different links in the network vary independently, a scheduler selects one or multiple channels with best channel conditions to transmit/receive data in the next time slot. Generally, systems using such scheduling algorithms provide higher performance than systems with other scheduling algorithms not taking advantage of the link quality, such as *Round Robin* and *random* schedulers [60].

Conventional opportunistic scheduling algorithms select only one user with the best channel condition for a transmission (service) at a time. For example, in cellular *time-division-multiplexing-access* (TDMA) systems, the packet scheduler at each *base station* (BS) collects the *channel state information* (CSI) from all terminals in the associated cell, and then chooses the terminal with the best channel quality to serve. Further details on this process can be found in [45]. Provided that the scheduler serves the user with the best channel condition in every slot, it had been shown [63] that the upper limit of the sum-rate of a cell containing K users is given by:

$$\log(1 + \rho \log K) \tag{3.4}$$

where K is the number of users and ρ is the expected SNR of the link of each user.

Comparing with the capacity that a single user could achieve with the same expected SNR:

$$\log(1 + \rho) \tag{3.5}$$

the opportunistic scheduler provides a performance gain increasing at a rate of *log log K* relative to the increase in the number of users.

In wireless mesh scenarios, the wireless routers are equipped with MIMO antennas to transmit and receive [60]. Although multiple antennas are available at a router, it is not preferable to use all antenna elements to transmit to one next-hop router because the sum rate of the MIMO channel will be saturated if the number of receiving antennas is less than the number of the transmitting antennas [64]. Thus, from a viewpoint of scheduling, it is desirable for the scheduler to choose to transmit to multiple neighboring routers simultaneously [60]. There are two potential advantages to transmit to multiple routers simultaneously, namely: a reduced complexity and the efficient use of resources. As a consequence, it is advantageous if scheduling algorithms for wireless mesh scenarios are capable of choosing and transmitting the different data to multiple users (i.e., next-hop routers) at the same time.

It is, however, challenging to achieve an optimal capacity with multi-user scheduling algorithms [60]. Simultaneous transmissions to multiple users cause interference between the different outgoing subchannels. To eliminate the interference, orthogonality of the transmission is required. A deliberate design of the coding schemes may solve this problem.

Dirty-paper-coding (DPC) [63] is a complicated coding scheme and had been shown to be able to guarantee the orthogonality and to achieve the theoretical upper limit of aggregated data throughput:

$$M \log(1 + \rho \log K) \tag{3.6}$$

given that there are K candidates for scheduling and M-antenna transmitters are used. As shown in (3.6), the MIMO gain scales as M and the scheduling gain scales as $loglogK$. A merit of DPC is that it can achieve this limit regardless of the number of the receiving antennas.

[65] proposed a *zero-forcing* scheduling algorithm, which is simpler than the DIP from a practical point of view. It had been shown that the proposed *zero-forcing* scheme could also achieve the performance limit in (3.6).

In addition to these challenges, tradeoffs between multiplexing and diversity [66], and the channel-hardening problem [65] exist. These issues make the grouping of antenna sets for simultaneous transmissions to various users challenging [60].

Pure opportunistic algorithms do not take the *fairness* issue into account. The key idea of fairness is not to let any user starve long. Fairness is a particularly important factor in the design of scheduling algorithms for wireless ad hoc scenarios. It is because all the nodes involved in the scheduling problem are wireless routers and are equally important. In a wireless mesh scenario, where a backhaul service must be provided in a geographic area [6], each wireless router is expected to serve as a source, intermediate (for data forwarding), and a sink node of data packets. As long as a router has a data packet to forward to the next-hop router, even if the associated outgoing link of the former router is not good, the packet should be transmitted (forwarded) within a certain amount of time, especially for delay sensitive traffic.

For scenarios where wireless routers are deployed at various locations to meet the traffic demands, the channels between any given pair of routers can demonstrate drastically different propagation loss [60]. The scheduling algorithms for such a scenario have to consider the radio propagation loss in order to provide fair transmission opportunities to the involved routers.

Given that nodes are all equipped with MIMO antennas, the design of desirable scheduling algorithms for a mesh scenario needs to address the following issues specifically related to the multiple antennas:

- Number of next-hop nodes that a given wireless router wants to transmit different signals to in the next time slot;
- The appropriate scheduling metrics for the MIMO schedulers given the scenario;
- Number of antennas and type, which should be chosen to ensure the best performance for each user;
- The determination of the set of antenna elements assigned to receive a signal at the receiving end to provide for desirable trade-offs between multiplexing and diversity gains.

In addition, from the system perspective the desirable characteristics of the schedulers for the ad hoc scenario networks should also be considered (e.g., in terms of scalability and link activation).

For example, in terms of *scalability*, in a large-scale multihop network, it is infeasible to use a central scheduler to schedule the packet transmissions on all nodes in the network. This is not practical because a prohibitive amount of control information needs to be exchanged between each node and the central scheduler, and the required computation complexity is excessively high.

If the communication nodes in the ad hoc network are *wireless routers*, which usually have continual power supply and sufficient memory, the *scheduling* as well as the *power control* algorithms do not have to be as power and memory-efficient as those used in hand-held devices. Furthermore, using a specialized control channel to support exchanges of control information with neighboring routers also may be practical [60].

If the nodes are widely deployed, cochannel interference among the nodes can be expected to be a performance limiting factor. Given the existence of interference, a reasonable level of *temporal correlation* of interference is required for accurate predictions of the channel quality, which often is used as an input parameter to the scheduler. [60] recommends to take the *interference temporal correlation* into account for the scheduler design for a wireless mesh backhaul scenario.

Finally, a proposed opportunistic scheduling algorithm should be in the optimal case applicable to any transmission mode [e.g., time-division duplex (TDD) and frequency-division duplex (FDD)].

On hand, *link activation* could be solved independently of the opportunistic scheduling. Such an action would require the identification of the duplexing status for every node in the network without any consideration of the pros and cons of being in such a duplexing state [60]. If the transmission or receiving decision is known, each node chooses the best link among those determined by the link-activation step to realize the multiuser diversity gain.

On the other hand, the outgoing and incoming links of a node have different channel qualities and receive different amount of interference. As a result, it is reasonable to expect potential gains by considering channel information of the outgoing/incoming links in the link activation decision (i.e., integrating the link activation process with the opportunistic scheduling).

Existing work in [67, 68] had shown that finding the optimal link activation is an *NP-complete* problem. Four types of approaches to link activation have been proposed in the literature, namely the following:

- *Centralized optimization*: This method assumes that there is a central controller having full knowledge about all links in the network. The method finds the optimal set of links to activate (for transmissions). The centralized link activation is based on the graph theory, in particular, the *matching and coloring* theory. A graph is composed of nodes and edges. The nodes in a graph correspond to the nodes in a communication network. Similarly, the edges in a graph represent the wireless links in the communication network. Methods based on the *conflict* graph theory find the matching of the graph (i.e., a set of links not sharing the same node [69, 70]). In each time slot, only links belonging to the same nonconflicting clique can be active simultaneously. Methods based on the *coloring* theory attempt to color the links in a network with the *minimum* number of colors [68] free of collision, then each color corresponds to a time slot. [68] also points out that to find a coloring method with the minimum number of colors for a graph is *NP-complete*. If the link activation problem is NP-complete, the centralized optimization approach could only be applied to *small-scale* networks due to its computational complexity. The centralized formulation was used to generate the optimal results as a benchmark to validate some heuristic algorithms for small-scale networks in [60].

- *Tree-structure activation*: This technique maps routers in the backhaul network into a tree [71] such that the gateway node to the Internet corresponds to the root of the tree and the wireless routers are mapped into branch and leaf nodes. All nodes at the same level of the tree have the same "duplexing status," that is, either transmitting or receiving in any time slot. The status for each node toggles in the next time slot. The main drawback of this approach is the poor reliability of the tree structure because once the root node fails, all other nodes have no alternate paths to communicate with the Internet. The same problem occurs when a parent node die. Another difficulty of the *tree-structure activation* is the mapping from the actual radio links into the tree structure. Efficient methods to perform such mapping must be identified.

- *IEEE 802.11 ad hoc mode*: The IEEE 802.11 *distributed coordination function* (DCF) with the *request-to-send* (RTS) and *clear-to-send* (CTS) mechanism [72] is commonly used for multihop wireless ad hoc networks. It is simple and fully distributed. However, the consequence of using RTS/CTS is two-folded [60]. On one hand, the RTS/CTS mechanism solves the hidden terminal problem. Its disadvantage is a decreased link utilization, and thus, an overall network throughput, because it allows at most one transmission in two partially overlapped areas, as represented by the transmission circles in Figure 3.6. For example—for scenarios where the wireless routers are equipped with MIMO antennas, [6] explored how the *carrier-sensing* (CSMA-CA) protocol without use of RTS/CTS can work efficiently using antennas with strong directivity.

- *Election-based scheduling*: The election-based scheduling algorithm is proposed in the IEEE 802.16 standard [28] as its scheduling scheme for the mesh mode. In the 802.16 mesh mode, a three-way mechanism involving request, grant, and confirmation messages is used to allocate bandwidth for transmission of a data subframe. The election-based scheduling algorithm is dealing with the scheduling of those handshake (control) messages. The exchange of handshake messages requires truly collision free transmission. As specified in the 802.16 standard, the recommended scheduling method for the exchange of control messages is called the *election-based* scheduling. All nodes run the same algorithm locally and obtain the same result about who will be transmitting in the next transmission opportunity in the neighborhood. As a result, it is fully distributed, fair and identical to all nodes [73].

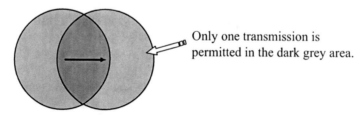

Figure 3.6 Low link utilization of CSMA-CA [60].

The first two approaches—the centralized optimization and the tree-structure algorithms—perform link activation independently and separated from the scheduling algorithms. Thus, they simply produce link activation results as inputs to the opportunistic scheduling algorithms in use.

On the other hand, the IEEE 802.11 and 802.16 approaches integrate link activation with the scheduling algorithms. However, since the determination of the transmitting links does not consider any performance metrics such as channel quality, and interference level, the link-activation steps do not yield any multiuser diversity gain.

[60] proposed a distributed scheduling algorithms for mesh wireless networks that used the 802.16 election-based scheduling as a reference to devise a timeframe structure for control information exchange.

3.1.4 Cross-Layer Techniques

Cross-layer techniques were introduced already in Chapter 2 in relation to cross-layer routing for wireless sensor networks.

In wireless mesh networks the spatial reuse of the spectral frequency and the broadcast, unstable, and error prone nature of the channel, make the layered approach suboptimum for the overall system performance. For example—a bad resource scheduling in the MAC layer can lead to interference that affects the performance of the PHY layer due to reduced SINR [60]. Local capacity optimization with opportunistic scheduling techniques that exploit the multiuser diversity may increase the overall outgoing throughput of the transceivers but they can also generate new bottlenecks in several routes in the network.

These are some of the reasons why the cross-layer design for improving the network performance has been a focus of much recent work. In a cross-layer paradigm, the joint optimization of control over two or more layers can yield significantly improved performance. Caution needs to be exercised, though, since the cross-layer design has the potential to destroy the modularity and make the overall system fragile. *QoS-routing* algorithms are a form of a cross-layer approach to the end-to-end path discovery and resource reservation problem in ad hoc networks where the behavior of the underlying layers is included in the routing decisions [60]. The FP6 IST project MEMBRANE [6] introduced a novel QoS routing framework that considers and harmonizes the flows of both the QoS and best-effort traffic in the network by introducing hard and soft bandwidth reservations. The MAC decisions are predicted during a new route discovery and included in the routing decision. Multipath routes are incorporated to provide QoS if this cannot be achieved by a single route and also to maximize the offered throughput to best-effort flows. Redundant route reservations are also included to minimize the delays in case of a primary route failure.

3.1.4.1 Routing

Routing was extensively described in Chapter 2 of this book. Here, a brief summary is given for clarity when explaining cross-layer techniques for performance enhancement.

The routing traditionally uses the knowledge of instantaneous connectivity of the network with emphasis on the state of network links. This is the so-called *to-pology-based* approach [74]. A different approach, called location-based or *posi-tion-based* routing, uses information related to the physical position of the nodes to help the task of routing. The *power/energy–aware* routing is an alternative to these approaches and uses the information related to the remaining battery lifetime of mobiles with the aim to produce paths that comprise nodes with a high remaining lifetime as well as to help them adjust their transmission power so as to keep the energy required to complete the routing task to a minimum.

If the power is not a critical issue (e.g., in wireless mesh backhaul networks), a topology-based approach can be a suitable choice for routing [60].

QoS Routing
A QoS routing technique determines a path that can guarantees the constraints requested by the incoming connection and rejects as few connections as possible. Specifically, the goal of QoS routing algorithms is three-fold, namely [60]:

1. The QoS constraints parameters of a single connection are usually specified in terms of minimum guaranteed bandwidth, maximum tolerable delay and/or jitter, and maximum tolerable loss rate.
2. Satisfying the QoS requirements for every admitted connection.
3. Achieving global efficiency in resource utilization.

The QoS metrics can be classified as *additive*, *concave*, and *multiplicative* met-rics [75]. If P is the path of hop of length h between a source-sink pair and L_i (m), $i=1, ...,h$ is the value of metric m over link L_iP, then an *additive* metric can be de-fined as:

$$A_m = \sum\nolimits_{i=1}^{h} L_i(m) \tag{3.7}$$

the *concave* as (3.8):

$$C_m = \min(L_i(m)) \tag{3.8}$$

and the *multiplicative* as (3.9):

$$M_m = \prod\nolimits_{i=1}^{h} L_i(m) \tag{3.9}$$

The *cost*, *delay*, and *jitter* are additive metrics, while the *bandwidth* is a con-cave metric since the available bandwidth on each link should be at least equal to the required value for the QoS [60]. The reliability or availability of a route based on some criteria such as *path-break probability* is an example of a multiplicative metric.

In some research focus is only on the bandwidth guaranteed paths, because they assume that all the other quality parameters can be controlled defining an equiva-lent flow bandwidth in a proper way [76, 77].

Existing QoS routing algorithms are the *min-hop algorithm* (MHA), which routes an incoming connection along the path, which reaches the destination node using the minimum number of feasible links. Using MHA can result in heavily loaded bottleneck links in the network, as it tends to overload some links leaving others underutilized [78]. The *widest shortest path algorithm* (WSP), proposed in [79], is an improvement of the MHA, as it attempts to balance the network traffic load. WSP still has the same drawbacks as MHA, since the path selection is performed among the *shortest feasible* paths, which are used until saturation before switching to other feasible paths. The *minimum interference routing algorithm* (MIRA), proposed in [80], explicitly takes into account the location of the ingress and egress (IE) routers. While choosing a path for an incoming request, however, MIRA does not take into account how the new connection will affect the future requests of the same ingress/egress pair (autointerference). The *QoS enabled Ad Hoc On Demand Distance Vector* (AODV) [81] is an extended version of the basic AODV routing protocol (see Chapter 2 of this book, Section 2.2.1.2). The main advantage of the QoS AODV is the simplicity of extension of the basic AODV protocol that can potentially enable QoS provisioning.

3.1.4.2 Cross-Layer Scheduling and Routing

A new link metric-based distributed joint routing and scheduling metric for ad hoc wireless networks was proposed in [82] in terms of the average consumed energy, delay, network lifetime, and communication overhead caused by the scheduling algorithm. In [83], a joint routing and scheduling scheme is proposed based on the *Bellman-Ford* algorithm, where the link distance consists of two terms. The first one takes into account the queue size, to encourage the usage of less congested links, while the second one is related to the power consumption, or the physical distance of the link, to minimize the energy consumption and also cause less interference to all other links in the network. [82] presents an algorithm that finds an optimal link scheduling and power control policy to support a given traffic rate on each link in the network. Then, a routing algorithm is guided by the computation of the minimum cost associated with the optimal scheduling and power control decisions that had already been taken. It had been shown that the optimum allocations do not necessarily route traffic over the minimum energy paths.

[84] studied the spatial multiplexing gain and diversity gain in MIMO links for routing in ad hoc networks. [85] showed the initial design and evaluation of two techniques for routing improvement using directional antennas in mobile ad hoc networks. The first one bridged the permanent network partitions by adaptively transmitting selected packets over a long distance, while the second one tried to repair routes in use, when an intermediate node moved out of the wireless transmission range along the route. The effects of the directional transmission on routing performance and more specifically the fundamental problem of "deafness" was analyzed in [86] where a new multipath routing scheme was introduced to overcome this problem.

3.1.4.3 Cross-Layer for WSNs

Narrowband Systems

The problem of minimizing the transmission energy in narrowband WSNs (e.g., Zigbee) had been considered in [87] by taking into account that every sensor may require a different bit rate and reliability according to its particular application. A cross-layer approach was proposed to tackle such a minimization in centralized networks for the total transmission energy consumption of the network: in the PHY layer, for each sensor the sink estimates the channel gain and adaptively selects a modulation scheme; in the MAC layer, each sensor is correspondingly assigned a number of time slots. The modulation level and the number of allocated time slots for every sensor are constrained to attain their applications bit rates in a globally energy-efficient manner. The SNR gap approximation was used jointly handle required bit rates, transmission energies, and symbol error rates.

Wideband and Multiband Systems

The use of multiband signaling in the 802.15.3 standard [88] allows that some of the cross-layer optimization techniques that had been proposed for ad hoc networks, and dealing with the development of efficient MAC protocols for or exploiting multiband capabilities, joint PHY – routing designs exploring the availability of several subcarriers and the general problem of radio resource allocation both in terms of channels and power, to be reconsidered for WSNs.

Although not specifically targeting the 802.15.3a standard the joint PHY-MAC design was considered in [89], where the authors considered the MAC protocol for multiband *ultra wideband* (UWB) technology. Specifically, the division of the spectra in several data bands and another one for control had been proposed. With such a design the number of available channels had been increased, while keeping the equalizers at moderate complexity levels.

A cross-layer optimization of the network throughput for multihop wireless networks and related to the joint PHY- and routing design was proposed in [90]. The authors split the throughput optimization problem into two sub-problems, namely the following: multihop flow routing at the network layer and power allocation at the PHY layer. The throughput is tied to the per-link data flow rates, which in turn depend on the link capacities and hence, the per-node radio power level. On the other hand, the power allocation problem is tied to interference as well as the link rate. Based on this solution, a CDMA/OFDM based solution was provided such that the power control and the routing are performed in a distributed manner.

The problem of optimum channel allocation for multiband multi-user UWB wireless systems had been considered in [91]. The optimization criteria involve the minimization of the power consumption under the constraints on the *packet error rate* (PER), the transmission rate, and related regulations. To deal with the NP problem, a sub-optimum algorithm based on the *greedy* approach had been proposed. Further results for the *multicarrier spread spectrum* (MC-SS) technology had been reported in [8] for the power allocation.

3.2   Enhancing the Performance of Relay Systems

The FP6 IST project WINNER [4] proposed a novel communication method based on *joint data packet encoding* and exploitation of a priori known information that enables reduced number of transmissions by relay nodes and enhances the aggregate system throughput [92].

3.2.1 Distributed Coding Strategies

The coding strategies range from *distributed Alamouti* coding and the use of *rate-compatible convolutional* codes to the application of *distributed turbo* coding. Some of these concepts have the power of overcoming the repetition-coded nature of the relay protocols, leading to benefits from additional coding gains at the cost of increased complexity. Coding as a general next generation radio interface procedure has been elaborately discussed in [45].

Distributed Alamouti coding [42, 93] assumes two amplify-and-forward relays to perform simple operations on the analog signals, namely delaying and conjugating as shown in Figure 3.7.

Decoding at the relay is not required and the procedure facilitates Alamouti coding for a single source and two relays.

A *hierarchically modulated Alamouti* code in a cascaded/windowed operation had been proposed in [94]. The scheme explicitly exploits the stronger source-relay links by using the concept of *multiclass information broadcasting* [95] to convey the information to cooperating relays, while simultaneously transmitting different information to the destination. The demonstrated advantages come at the cost of a strongly increased complexity. [12] investigated this scheme in more detail.

Other coding schemes proposed for relay communication are distributed block and convolutional coding and distributed turbo coding.

Use of *rate-compatible punctured convolutional codes* (RCPC) can yield a 1–2 dB gain over $R=1/4$ coded simple amplify-and-forward operation [96], while using one relay in connection with a strong turbo code, can outperform *conventional repetition* coding by 2 dB at a frame error rate of FER = 10^{-2} [97]. These coding gains increase correspondingly with the use of multiple relays [12].

3.2.2 Coded Bidirectional Relaying

The proposed coded bidirectional relay scheme assumes two-hop forwarding but extensions to cover more than one relay node are straightforward [92, 98]. The scenario assumes a two-hop cellular network with a BS, a *relay node* (RN) and a *user terminal* (UT) as shown in Figure 3.8.

In a cellular system, the traffic is normally communicated in both UL and DL between a BS and a UT. The classical rely scheme (see Figure 3.8, left) introduces an intermediate RN, extends the notion of DL and UL by use of *four orthogonal*

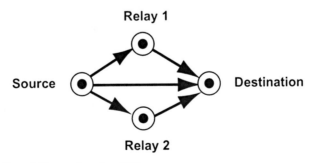

Figure 3.7 Distributed Alamouti coding [12].

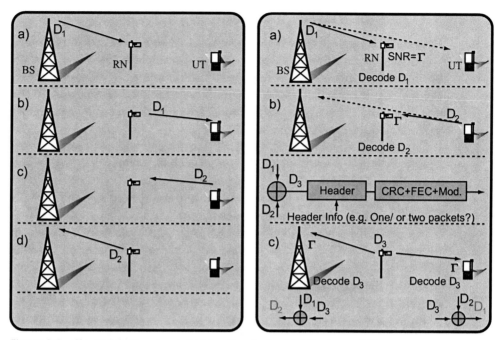

Figure 3.8 Classical (left) and coded bidirectional relay (right) scheme [92].

resources for the BS-to-RN, RN-to-UT, UT-to-RN, and RN-to-BS transmissions [92]. The order of the phases can also be varied. In Figure 3.8(a–d) represent different time instances.

A more efficient (energy and/or capacity) relaying solution assumes that the wireless medium is of a broadcast type, and a bidirectional topology. The basic idea is to jointly encode data received from the BS destined for some UT with data received from the same UT (destined for the BS) in the RN into a *jointly encoded data packet*. When each node (i.e., the BS and the UT) receives the *jointly encoded* data packet, each exploits a priori information of its originally transmitted (and stored) data to decode the jointly encoded data packet.

In phase (a) and (b) of Figure 3.8, transmissions take place from BS-to-RN and from UT-to-RN, and the relay node decodes the received data packets D_1 and D_2. In phase (c), the RN jointly encodes the data packets D_1 and D_2 with a bitwise *XOR* operation[1] into a common data packet D_3 (i.e., prior to modulation and *forward error correction* (FEC) encoding[2]. Subsequently, the RN multicasts D_3 to both the BS and the UT. Hence, instead of two transmissions from the RN, only one transmission is used, but the same amount of data is transferred with the same energy expenditure as for one (the worst) transmission. At decoding at the BS, a bitwise *XOR* operation of the data packet D_3 (after error correction and demodulation) and the a priori information D_1 is performed, which then yields D_2. The UT performs the

---

1.   N.B. the method for encoding is preferably based on XOR bitwise encoding, due to its simplicity, but other codes (with the desired invertability) may also be used such as an erasure code like Reed Solomon. Moreover, more than two packets may also participate in the bitwise XOR operation.

2.   N.B. by performing the FEC encoding after the XOR operation no extra complexity is added in the receiving nodes (i.e., BS and UT).

corresponding operation. If the data packets are of unequal length, zero-padding of the shorter data sequence is used. Moreover, when the link quality of the two links from the relay node differs, the relay node transmit power is preferably set according to the most stringent link requirement.

An example of a possible *code-frame* format is shown in Figure 3.9.

The packets from the two nodes are subjected to bitwise *XOR* in the RN. It is possible (and preferred) to replace the individual *cyclic redundancy check* (CRC) on D_1 and D_2 seen at the reception at the RN. Those CRCs are replaced with a common CRC over the bitwise *XORed* packet. It may be advantageous to let the RN append an extra header indicating the characteristics of the relayed bitwise *XORed* packet (this information may also be signaled out of band). The composite header may, for example, indicate if one or two packets are transmitted (e.g., if D_1 was in error, one may decide to only forward D_2). If the two packets are of unequal length, the composite header needs to indicate the length of the shortest packet and, which of the two packets, is the shortest. The extra header includes the necessary information to allow the receiver to identify and to exploit a previously sent packet, in order to extract the new information [92].

Several extensions of this basic scheme may be envisioned, the idea of bidirectional 2-hop relaying can be trivially extended to multiple hops. In addition, the direct signals BS-to-UT and UT-to-BS can be exploited (e.g., through *incremental redundancy* and *chase combining* strategies).

The proposed scheme was analyzed in [92] by means of the Shannon capacity bounds of the aggregate rate under an aggregate energy constraint. The proposed scheme is compared with a classical four-phase solution. A three-node system is considered, with nodes v_k, $k = \{1,2,3\}$. Nodes, v_1 and v_2 are nodes that have data packets of lengths L_1 and L_2 bits, respectively, queued for one another, whereas v_3 is a relaying node. Further, each node transmits a complex Gaussian signal with power P_k, $k = \{1,2,3\}$ over a flat channel with gain G_i, $i = \{1,2\}$, where G_1 and G_2 is the (reciprocal) gain between nodes v_1 and v_3 and nodes v_2 and v_3 respectively. At the reception, complex Gaussian noise (and interference) with variance σ_k^2 is added, however, to simplify analysis it was assumed that the noise level is equal at all nodes [92]. In the analysis, it was further assumed that the power and the data packet lengths are the variables that subject to optimization.

The transmission times for the proposed scheme are given by the number of bits transferred and divided by the channel rate, as follows:

$$t_1 = L_1/B \lg_2(1 + P_1 G_1/\sigma^2) \tag{3.10}$$

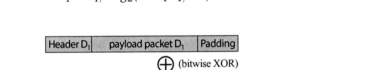

Figure 3.9   Encoding frame format [92].

$$t_2 = L_2 / B \lg_2(1 + P_2 G_2 / \sigma^2) \qquad (3.11)$$

$$t_3 = \max\{L_1, L_2\} / B \lg_2(1 + \min\{P_3 G_1, P_3 G_2\} / \sigma^2) \qquad (3.12)$$

where $t_k = T/3$, $k = \{1,2,3\}$, and T is the full frame duration. Here, it is assumed that any packet transmitted must fit into one time-slot, where the slot length is equal to one-third of the frame duration for the three phase case (and one-fourth in the four phase case). In order to compare the two schemes, an aggregate energy constraint is imposed. In the three-phase case, the constraint can be expressed as follows:

$$E = (P_1 + P_2 + P_3)T/3 \qquad (3.13)$$

The aggregate rate is the total number of transferred bits divided by the frame duration, namely:

$$R = (L_1 + L_2)/T \qquad (3.14)$$

After calculations, the optimum rate for the proposed scheme can be determined to be:

$$R_3^{(opt)} = 2/3 \cdot B \lg_2(1 + 3\Gamma_3^{(eff)}) \qquad (3.15)$$

where,

$$\Gamma_3^{(eff)} = \frac{\bar{\Gamma}_1' \cdot \bar{\Gamma}_2'}{\bar{\Gamma}_1' + 2\bar{\Gamma}_2'} \qquad (3.16)$$

$$\bar{\Gamma}_1' = \min\{\bar{\Gamma}_1, \bar{\Gamma}_2\} \qquad (3.17)$$

$$\bar{\Gamma}_2' = \max\{\bar{\Gamma}_1, \bar{\Gamma}_2\} \qquad (3.18)$$

$$\bar{\Gamma}_1 = \bar{P} \cdot G_1 / \sigma^2 \qquad (3.19)$$

$$\bar{\Gamma}_2 = \bar{P} \cdot G_2 / \sigma^2 \qquad (3.20)$$

$$\bar{P} = E/T \qquad (3.21)$$

The optimum rate for a classical four-phase relaying scheme can be derived in a similar way:

$$R_4^{(opt)} = 1/2 \cdot B \lg_2(1 + 2\Gamma_4^{(eff)}) \qquad (3.22)$$

where,

$$\Gamma_4^{(eff)} = \frac{\bar{\Gamma}_1 \cdot \bar{\Gamma}_2}{\bar{\Gamma}_1 + \bar{\Gamma}_2} \tag{3.23}$$

The relative throughput improvement using the proposed three-phased scheme as opposed to a classical four-phased relaying scheme is shown in Figure 3.10.

The relative throughput gain is given as a function of the experienced mean SNRs as defined in (3.19) and (3.20). The gain is upper bounded by a 33% gain improvement (or the ratio 4/3) at equal SNR and it is lower bounded to a unit value when the path losses differ significantly. The throughput gain in bi-directional multihopping (with large number of hops) can be shown to be upper limited to a factor of two [92]. Moreover, when optimizing the aggregate throughput under the current assumptions, the (end-to-end) rates in the UL and the DL directions are equal (i.e., $L_1 = L_2$).

It can be argued that one could achieve the same goal of reduced number of transmissions with a beamforming solution, with one beam to the BS and one to the UT, respectively, which in essence is a *spatial division multiplex access* (SDMA) solution. The obvious benefits for beamforming exist in general, but some disadvantages related to technical and cost aspects could also be identified with the proposed scheme [92].

A disadvantage of a beamforming-enabled relay node is that two concurrent beams will at least to some extent, interfere with each other. If at least one of the beams allows for adaptive beamforming (and then the other is of fixed beam type), the algorithm(s) for finding the antenna weights should consider the interference aspect. Some channel/direction realization may not even have a solution.

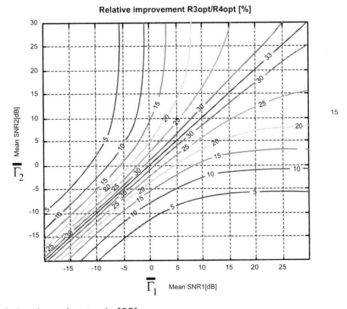

Figure 3.10 Relative throughput gain [92].

Further, because multiple beams require multiple amplifiers, one must ensure that intermodulation products due to back-coupling in the power amplifiers do not arise.

Supporting multiple concurrently transmitting beams, and disregarding whether adaptive or fixed beam solutions are considered, requires more hardware than the proposed scheme. More hardware could be expected to be more costly. For example—in the adaptive beamforming case, multiple antennas and the associated power amplifiers are required. In the fixed beam case, at least two antennas are needed as well as two different power amplifiers. In the proposed method, only one antenna and power amplifier are required.

3.2.3 Cooperative Cyclic Diversity

Cooperative relaying as an extension of *conventional* relaying was proposed by [12] as a viable strategy for future wireless networks. Namely, conventional relaying can serve as a means of providing coverage in areas where the direct communication and cooperative relaying are not viable; for the remaining areas, cooperative relaying is used to improve the network performance by lowering the transmit powers, reducing the effective interferences, or providing higher data rates [92].

Figure 3.11 summarizes the extension and fallback options for the joint use of conventional and cooperative relaying.

Figure 3.12 shows a single cooperating multiple-antenna relay (a); and a multiple cooperating multiple-antenna relays (b).

Each relay is equipped with L antennas and the relays are of the regenerative (digital) type (i.e., the decode-and-forward relaying strategy is considered). Since by design, the relay is expected to be cost-efficient and of low complexity, the relay may use a subset of these L antennas; in the limiting (worst) case scenario a relay uses only 1 antenna through selection combining of its L antennas. In this case, only one signal detection chain is required at each relay similar to the relays that do not use multiple antennas. Therefore, in terms of cost, this *selection diversity-based*

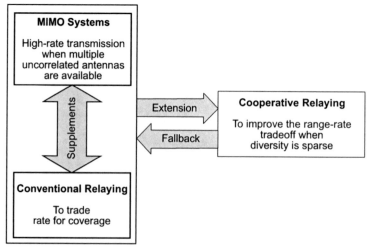

Figure 3.11 Cooperative relaying as extension and fallback option for MIMO systems and conventional relaying [92].

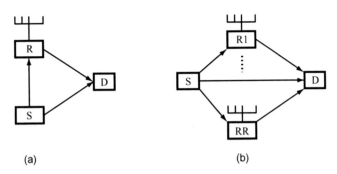

Figure 3.12   (a): A single cooperating multiantenna relay; and (b): multiple cooperating multi-antennas relays [99].

relay system incurs no significant penalty (other than the cost of the extra antennas and selection mechanism) [99]. If the relays can utilize signals from more than one antenna, then the classical generalized selective combining (GSC) scheme can be employed. For forwarding, however, each relay uses only one antenna.

A method that introduced *artificial frequency selectivity* and *spatial diversity* in a cooperative relaying wireless communication system was proposed in [92]. The *artificial frequency selectivity* was exploited in conjunction with *forward error correction coding* to provide a *coding diversity* gain.

The scenario assumed that each of the RNs consists of one or more antennas. The AP transmits to M RNs and the UT. The relay forwards the information received from a first node (e.g., AP) to a second node (e.g., UT) by using *cycle delay diversity* (CDD) [100, 101]. This can be done either with an *amplify and forward*, *decode and forward*, or a *hybrid* strategy [12].

The relay nodes demodulate and/or decode the signal (decode and forward based relaying assumed) received from the AP and forward the information to the UT using cycle delay diversity in two steps, namely the following:

- *Step one*: the AP transmits data, which is decoded by 1) the relay nodes where the information is stored 2) by the UT.
- *Step two*: each relay node encodes the data and applies different cyclic shift on different antennas and adds the cyclic prefix before transmitting the signals. The UT receives the combined signals decodes the data which may be combined with the data obtained from *Step 1*.

Figure 3.13 shows the structure of a transmitter in a cooperative relay system using CDD.

At the AP, the OFDM symbol S is subject to CDD. This is implemented by the *cyclic shifting* the OFDM symbol at each antenna. A different cyclic shift δ is applied on the different antennas. Following the cyclic shift, a guard interval is applied on each branch. The GI is implemented using the *cyclic prefix* method. Then the signals are upconverted from the baseband into the RF-band and transmitted. In the case when the AP is equipped with a single antenna, the CDD does not need to be applied to the OFDM symbol.

The receiver structure at the RN is shown in the top of Figure 3.13. For each receive antenna, the data is first down-converted from the RF-band into base-band

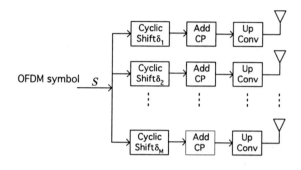

M antennas

Figure 3.13 Transmitter at an AP in a cooperative relay system using CDD [92].

and then the CP is removed. The data then undergoes a *fast fourier transform* (FFT) operation and is equalized. The data estimates from all the receive antennas may be combined using the *maximum ratio combining* (MRC) method. The coded output data is then stored in order to be processed and forwarded at the next time slot. There are following two possibilities:

- The coded output data is modulated and forwarded (*nonregenerative* relaying);
- The coded output data is decoded, re-encoded, modulated, and forwarded (*regenerative* relaying).

Forwarding is implemented using the CDD method. The transmitter in the RN is shown identical to the transmitter at the AP. The receiver at the UT is identical to the receiver in the RN.

An example of a CDD distributed relay is shown in Figure 3.14.

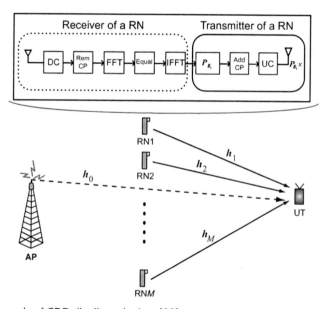

Figure 3.14 Example of CDD distributed relays [92].

There are M RN nodes and one AP in the example of Figure 3.14, where the AP and each relay node are equipped with one transmit and one receive antenna. Let x be the OFDM symbol of length N. A cyclic shift of length δ_0 and $\delta_m, m \in \{0, 1, \cdots, M\}$ is applied to x at the antenna of the AP and RN before transmission, respectively. Let h_m and H_m, respectively, denote the channel impulse response and the channel matrix from the mth transmit antenna to the receive antenna of the UT. Assuming that the CP is greater than the channel order, the received signal y_0 from the AP and y_1 from the relay nodes can be expressed as:

$$y_0 = H_0 P_{\delta_0} x, \quad y_1 = \sum_{m=1}^{M} H_m P_{\delta_m} x \tag{3.24}$$

where P_{δ_m} is a permutation matrix, which applies a cyclic shift of length δ_m to the data vector x. Since H_m is a circulant matrix, the channel matrix can be diagonalized as follows:

$$H_m = F^H D(Fh_m) F \tag{3.25}$$

where, $D(x)$ is a diagonal matrix with x on its main diagonal, and F is the unitary discrete Fourier transform matrix of size $N \times N$. The (n,m)th element of F is given by:

$$F(n,m) = \frac{1}{\sqrt{N}} \exp\left(-\frac{j2\pi(n-1)(m-1)}{N}\right) \tag{3.26}$$

P_k is a right circulant matrix with $e_{1+(1-k)\bmod N}$ as the first row [i.e., $P_k = \mathrm{circ}(e_{1+(1-k)\bmod N})$] where e_k is column vector where all elements are equal to zero except the element at position k which is equal to one. Since P_{δ_m} and H_m are circulant matrices then:

$$H_R = \sum_{m=1}^{M} H_m P_{\delta_m} \tag{3.27}$$

is also a circulant matrix [102]. Equation (3.27) can be decomposed as $H_R = F^H D(Fh_R) F$, where h_R can be called the effective channel impulse response from the relays, where

$$h_R = \sum_{m=1}^{M} h_m \circ F e_m \tag{3.28}$$

where $\circ$ denotes the Hadamard product.

The *discrete Fourier transform* (DFT) of the received signal in (3.24) yields:

$$Fy_0 = D(Fh_0) Fx, \quad Fy_1 = D(Fh_R) Fx \tag{3.29}$$

The signals can be combined using the MRC method. This approach does not require the explicit estimation of the channels from each antenna. The effective channel impulse response and the channel response from the AP can be estimated using a common time-frequency pilot pattern, which are not antenna specific. The same conclusion is reached when multiple transmit antennas are used in the AP and/or the RNs.

3.2.4 Fixed-Relay Assisted User Cooperation

User or terminal cooperative diversity requires that at least two users are in the network and willing to cooperate [92]. This might prove difficult, especially, if one is a *greedy* and *selfish* user. Furthermore, each user data have to be protected. To circumvent the problems of *user-dependent cooperative diversity*, [92] proposed fixed relay-assisted user cooperation.

In the proposed relay-assisted user cooperation scheme, the users are ignorant of the cooperation. They do not need to be able to decode any data of the other user. Privacy is not an issue because the user is not required to agree to the cooperation. The relay, however, may have to be given more processing power, but this implies that certain functions of the BS can be decentralized, which could translate into a reduced BS cost. Most importantly, this technique can be used in any network, because it does not require a major change to the UT.

The relays in this scheme need to have *multiuser detection* (MUD) capability. After performing MUD, the relays could enter into cooperation based on the level of information that the relays can exchange or based on predetermined mode. An example of the latter form of cooperation could be the *distributed space-time coding* [51].

A *generic relay-assisted cooperative* scheme is shown in Figure 3.15.

The analysis scenario assumes a basic wireless network with two users and two fixed relays [92]. A binary transmission and general asynchronous model are considered, which then is the specialized to a synchronous scenario.

3.2.4.1 Distributed Relay and MAC Model

Consider a multiuser system with K users. User k transmits a data sequence $b_k(i) \in \{\pm1\}$. This sequence has a length N. Let $s_k(t)$ be the unique signature wave-

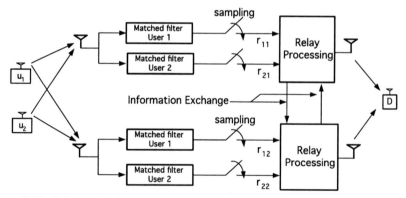

Figure 3.15 Relay-Assisted User Cooperation [92].

form with support on [0, *T*]. The received signal *y(t)* at relay *r* can be expressed as:

$$y^{(r)}(t) = \sum_{k=1}^{K}\sum_{i=1}^{N} \alpha_k^{(r)} b_k(i) s_k(t - iT - \tau_k^{(r)}) + n^{(r)}(t), \qquad (3.30)$$

where the superscript on $\tau_k^{(r)}$ explains the general case of the difference in the relative delays between the signals of users at the relays; *n(t)* is the *additive white Gaussian noise* (AWGN) with power spectral density N_0; and $\alpha_k^{(r)}(i)$ is an independent *wide-sense stationary* process that can modeled using any of the following: *Rayleigh, Ricean, Nakagami, lognormal,* or *composite* fading. This multiple-access model is fairly general since it takes into account the changing propagation conditions due to mobility and other channel impairment [92].

The system is considered in its simplest form, namely, the *K*-user synchronous case. This model is achieved by setting $\tau_1 = \tau_2, ..., \tau_K = 0$. Further, by setting *K*=2, a basis for employing both optimum and suboptimum detection strategies without the analysis can be established. This implies an MUD, which yields the *minimum achievable probability* of error or *asymptotic multiuser efficiency* and possibly optimum *near-far resistance* in the multi-user channels [103]. To achieve a scalable network with more than two users, there will be need for *suboptimum linear detection techniques* such as *single-user detection* (SUD), *decorrelating detector* (DD), or *minimum mean squared error* (MMSE) detectors [92]. In this case, the optimum detection has complexity that grows exponentially with *K*, which could become undesirable.

The performance evaluation in [92] employed an 8-PSK modulation scheme with the relay-assisted cooperative diversity scheme, which was then compared to a BPSK-based noncooperative scheme. The strategy to check the relays performing space-time coding based on erroneous detections was also examined. The relays request for retransmission each time when the decisions of the relays are different. Although, one could do better than this strategy it has been used only as a test case. The detection request is captured in the throughput curve given in

At high SNR (above 20 dB), the loss in retransmission is negligible. In these SNR regimes the error performance results in Figure 3.16 to Figure 3.18 show that

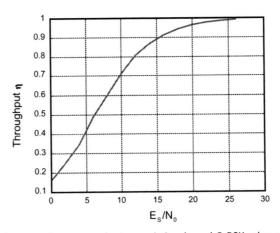

Figure 3.16 Throughput performance of retransmission-based 8-PSK relay-assisted user cooperative diversity scheme [92].

the scheme significantly outperforms the baseline, which is a 2-user noncooperative network. The two users communicate with the BS each employing BPSK modulation.

Figure 3.17 shows the system performance when the destination is equipped with two antennas. In this case, each antenna of the two (distributed) relays and the two antennas of the destination form a virtual 2 x 2 antenna scheme that could emulate conventional 2 x 2 Alamouti scheme (or MIMO channels). The single-hop (S-H) single user transmission is shown for comparison purposes. Figure 3.18 shows simulated and analytical results when 8-PSK constellation is employed in all the links, a reasonable agreement exists between these results.

This means that a crude transmission strategy such as the one described here will not be necessary at sufficiently high SNRs [92].

3.2.4.2   Realization of Fixed Relay-Enabled User Cooperation

In the proposed scheme shown in Figure 3.19, the signal processing at the relays can be classified into two types: *spatial division multiple access* (SDMA)- and MUD-based processing. In the multiantenna relay architecture the relays can use an optimum combining scheme to separate the users through the SDMA. For the single antenna relay architecture, a MUD capability will be required at the relays where it is assumed that the users employ some spreading (orthogonal) codes.

In general, after the relays detect the signal in the first hop, they could enter into cooperation using, for example, *distributed space-time coding* (DSTC).

The scheme using SDMA operates in the following manner. Each relay detects the signal of one user by directing a null towards the other. To engage the data of the two users in the DSTC, the signal of the nulled user is obtained by the other relay node by the means of exploiting the broadcast nature of the wireless channels. This exchange of user signals is accomplished in the second time slot when each

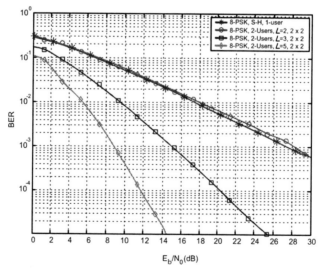

Figure 3.17   BER (BPSK) relay-enabled user cooperation for two receive antennas at destination [105].

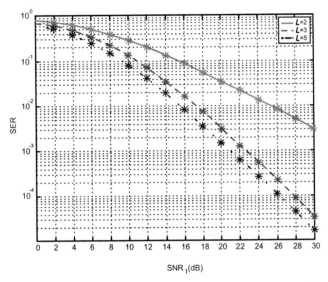

Figure 3.18 End-to-end performance for 8-PSK of relay-enabled user cooperation [105].

relay node transmits to the destination [see Figure 3.19(b)]. The three-slot protocol combines the two-hop relay network with two (virtual) transmit antenna scheme. The sum of the times allocated to the three slots is made equal to that of the reference scheme for fair comparison. Alternatively, to allow for the same bandwidth with the reference scheme, a higher modulation constellation could be employed in the links.

The second realization uses MUD at each relay to make the detections of the signals of the two users. In this realization, the system could operate in a mode we will refer to as a "predetermined" space-time coding, in which case, the relays do not exchange any instantaneous information as the cooperation is agreed upon a priori. One advantage of this is the removal of the *full-duplex* requirement on the relays [92] (i.e., the relays do not need to transmit and receive at the same time). In this mode, each relay assumes that the signals of the users it detects have the same reliability as those detected by the cooperating relay. They simply employ these signals to realize the distributed space-time coding. In the implementation of this scheme, however, *automatic repeat request* in conjunction with CRC might be

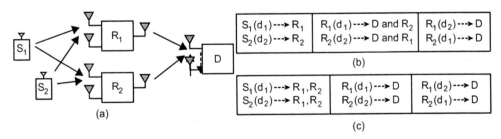

Figure 3.19 (a–c) Fixed relay-enabled user cooperation: Realization I and II: (a). Relay-enabled co-operative diversity scheme, (b). Protocol for realization I: SDMA is used to separate the users at the relay, Protocol for Realization II: MUD is used to detect the cooperating partners [104].

required to alleviate the problem of performing space-time coding with erroneous data.

Figure 3.20 shows results for the relay-enabled user cooperation where the relays use SDMA mode of detection.

Each user has single antenna and the destination has one antenna but it can implement an Alamouti-type receiver. The results are shown for various numbers of antennas (L) at the relay stations.

For L = 2, the scheme provides each of the two users in the network with an error rate that is not inferior to a single user network. Eight-PSK modulation format is adopted in both networks. This error performance advantage comes at the expense of 33% loss in spectral efficiency. The situation is different for large L. For instance, when L=3, the performance of each user in the new scheme is superior to that of the single user network using BPSK. This represents both diversity and spectral (multiplexing) gains. Considering the overall network, an improvement in spectral efficiency of 100% is achieved with the new scheme in addition to the large SNR gain at low error performance. The new scheme achieves an SNR gain as large as 7.5 dB (for L=3) and 9 dB (L=5) over the BPSK at a BER of 0.001. The margin of the gain is larger in low BER regime. In the uplinks, the base station is the destination. Therefore, the number of antennas could certainly be more than one. The discussions so far, therefore, represent pessimistic results. Figure 3.20 also shows the system performance when the destination uses two antennas. In this case, each antenna of the two (distributed) relays and the two antennas of the destination form a *virtual* 2x2 antenna scheme (see Section 3.1.2) that could emulate conventional 2x2 Alamouti scheme (or MIMO channels). There is a large performance improvement due to the additional antenna at the destination [104]. For example, for a BER of 0.0001, the virtual 2x2 of the second hop is superior its 2x1 counterparts by about 8 dB. Further performance gains can be obtained if the relays use their multiple antennas to transmit.

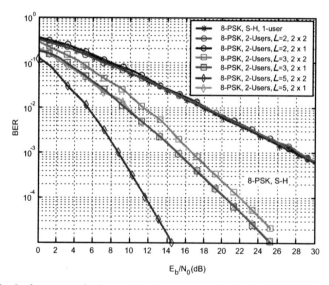

Figure 3.20 Performance of relay-enabled user cooperation in Rayleigh fading channels with one or two antennas at the destination (forming either virtual 2 x 1 or 2 x 2 in the second hop) [104].

3.2.5 Joint Routing and Resource Partitioning

The issue of *resource partitioning* for cellular networks enhanced with *fixed relay nodes* (FRNs) has been addressed in many literatures. However, most of these investigations were conducted with simple routing solutions (e.g., the *distance-based* or the *path loss-based* ones). Next generation relay-enhanced radio systems, however, can benefit from an optimized routing solution jointly with radio resource partitioning [105].

3.2.5.1 Resource Allocation/Scheduling in Multihop Relaying Networks

The multihop relay (n-hop, n≥2) scenario [104] is an extension of the 2-hop scenario that combines a linear deployment of RNs, from a logical viewpoint, along different branches; and *point-to-multipoint* (PMP) last-hop connections towards the UTs and around each RN. A novel resource request and allocation scheme, named, *connection based scheduling with differentiated service support* (CbS-DSS) was proposed in [92] and extended in [105]. The CbS-DSS mechanism is able to satisfy different QoS requirements for different service classes (e.g., non-real time and real-time traffic). Requests of different service classes are distinguished in order to allow the BS to assign resource allocation according to QoS constraints.

3.2.5.2 Routing and Resource Allocation for Relay Enabled User Cooperation Diversity Schemes

A radio resource management (RRM) scheme and its corresponding signaling structure to support fixed-relay cooperative communication in a next generation air interface (e.g., WINNER) was proposed in [105]. Centralized and distributed routing schemes were already discussed in Chapter 2 of this book.

From an architectural perspective, two basic routing strategies are envisioned for next generation radio systems, namely *network-centric* and *user-centric* [105]. Each of these strategies can be considered as a core for developing RRM for fixed relay cooperative relaying. The routing mechanism should recognize the cooperative routes as well as noncooperative routes.

The above RRM scheme is built on a network-centric routing and selects the optimal pair of active users. This cooperative route is then saved in the *candidate route database* in the routing module located in central radio resource management entity. A significant improvement in the achieved system throughput could be shown by system level simulations.

Let U be the set of users in the coverage area of the BS where there are two fixed relay nodes, namely FRN_1 and FRN_2. $U_1, U_2 \subset U$ are the set of users in the coverage area, which can communicate with FRN_1 and FRN_2, respectively. Therefore, $u_i, u_j \subset U_1 \cap U_2$ and a search for candidate pairs should be performed in $U_1 \cap U_2$ so that $r_i^c + r_j^c / r_i + r_j$ is maximized. This process is implemented in the *route discovery* module of the routing function of Figure 2.42 (see, Chapter 2 of this book). The scenario realization is shown in Figure 3.21. Cooperative pairs are then added to the *candidate route database* as alternative/optional routes. A cooperative route is then assigned by the routing algorithm in a case that this particular cooperative route results in a better achieved system performance. Considering only the best pair for each communication significantly reduces the system complexity [105]. The route for both users in the selected pair should be also the same.

3.2.5.3   System Model Description for Joint Resource Partitioning and Routing

Let consider an FRN enhanced cellular network consisting of 19 hexagonal cells with frequency reuse one [105]. In each cell, a BS is located at the centre and six FRNs are respectively placed on lines from the BS and six cell vertices.

A homogeneous air interface is adopted (i.e., BS-UT links, BS-FRN links, and FRN-UT links share the spectrum in each cell). It is assumed that the total available bandwidth of W Hz is partitioned into orthogonal channels, which are assigned to different links in each cell. In this way, there is no intracell interference and the system performance is mainly affected by the intercell cochannel interference.

3.2.5.4   Cell Partitioning Based Frequency Assignment

To facilitate effective control of inter-cell interference, a cell-partitioning based frequency assignment framework is considered [105]. With this framework, three adjacent cells constitute a virtual cluster, and the available bandwidth is divided into three orthogonal subsets F_1, F_2, and F_3. As shown in Figure 3.22, in each cluster the FRN-UT links in cell A, in cell B, and in cell C use subsets F_1, F_2, and F_3 respectively; while the BS-UT and the BS-FRN links are assigned the remaining bandwidth in each cell. In this way, the link quality of the two-hop users (usually edge-users) could be guaranteed due to the avoidance of strong intercell interference.

In each cell, a BS is located at the centre and six FRNs are respectively placed on lines from the BS and six cell vertices. Homogeneous air-interface is adopted, that is, BS-UT (mobile terminal) links, BS-FRN links, and FRN-UT links share the

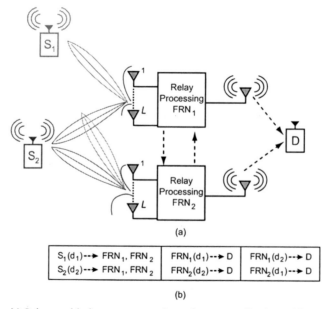

(a)

| $S_1(d_1) \dashrightarrow$ FRN$_1$, FRN$_2$ | FRN$_1(d_1) \dashrightarrow$ D | FRN$_1(d_2) \dashrightarrow$ D |
| $S_2(d_2) \dashrightarrow$ FRN$_1$, FRN$_2$ | FRN$_2(d_2) \dashrightarrow$ D | FRN$_2(d_1) \dashrightarrow$ D |

(b)

Figure 3.21   (a, b) Relay-enabled user cooperation schemes: realization without a full-duplex requirement [105].

spectrum in each cell. We assume that the total available bandwidth of W Hz is partitioned into orthogonal channels, which are assigned to different links in each cell. In this way, there is no intracell interference and the system performance is mainly affected by the intercell cochannel interference.

The orthogonal requirement in a virtual cluster imposes the following constraint

$$|F_1| + |F_2| + |F_3| \leq W \tag{3.31}$$

where $|\cdot|$ denotes the bandwidth. Such a constraint has in fact limited the number of UTs that can be assigned a two-hop route in each cell and, hence, the selection of an optimal set of two-hop UTs is an important problem.

3.2.5.5   Cell Capacity-Oriented Routing

The cell-partitioning constrained routing is approached with a cell capacity oriented solution [105].

Denote the number of UTs communicating over single-hop links as N_1, that over two-hop links as N_2, and $N = N_1 + N_2$ is the total number of UTs. Suppose that the routing controller is located at each BS and has the information of all SIR values on the BS-UT, BS-FRN, and FRN-UT links. The bandwidth of each channel is given by:

$$W_{CH} = \frac{W}{N_1 + 2N_2} = \frac{W}{N + N_2} \tag{3.32}$$

The channel bandwidth depends on the number of two-hop UTs, i.e., N_2. On the other hand, the cell-partitioning framework imposes a constraint on. N_2. In particular, this constraint can be reformulated as:

$$\frac{N_2}{N + N_2} W \leq \frac{1}{3} W \tag{3.33}$$

which is equivalent to:

$$N_2 \leq \frac{1}{2} N \tag{3.34}$$

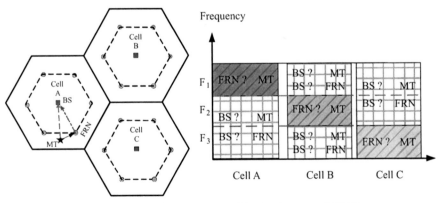

Figure 3.22   System architecture and spectrum partitioning framework [105].

Suppose that the routing controller is located at the BS and this controller has all the SIR values on the BS-UT, BS-FRN, and the FRN-UT links (e.g., denoted as $\text{SIR}_{\text{BM},i}$, $\text{SIR}_{\text{BF},i}$ and $\text{SIR}_{\text{FM},i}$ for the i-th UT. If a two-hop communication is of concern, the nearest FRN away from the i-th UT can be selected as the candidate FRN for the two-hop routing. In this way, the routing controller can calculate the single-hop and the two-hop spectral efficiency for the i-th UT as:

$$\eta_1(i) \underline{\Delta} \log_2(1 + \text{SIR}_{\text{BM},i}) \tag{3.35}$$

$$\eta_2(i) \underline{\Delta} \log_2(1 + \min(\text{SIR}_{\text{BF},i}, \text{SIR}_{\text{FM},i})) \tag{3.36}$$

Then, the achieved data rate of the i-th UT when communicates over single-hop link and two-hop link can be respectively expressed as:

$$R_1(i) = W_{\text{CH}} \eta_1(i) \tag{3.37}$$

$$R_2(i) = W_{\text{CH}} \eta_2(i) \tag{3.38}$$

On the basis of (3.37) and (3.38), the cell capacity can be expressed as a function of N_2.

$$C(N_2) = W_{\text{CH}} \left(\sum_{i=1}^{N_1} \eta_1(i) + \sum_{j=N_1+1}^{N} \eta_2(j) \right)$$
$$= \frac{W}{N + N_2} \left(\sum_{i=1}^{N-N_2} \eta_1(i) + \sum_{j=N-N_2+1}^{N} \eta_2(j) \right) \tag{3.39}$$

where $\sum_{n_1}^{n_2} (\cdot) = 0$ if $n_2 < n_1$. Then, the cell-capacity oriented solution can be formulated as:

$$\max_{\{m_1, m_2, \ldots m_{N_2}\}} C(N_2)$$
$$s.t. \, N_2 \leq \frac{N}{2} \tag{3.40}$$

where $\{m_1, m_2, \ldots m_{N_2}\}$ denotes the set of selected two-hop UTs, and the constraint comes from the cell-partitioning based spectrum assignment framework.

Equation (3.40) has a unique solution [105].

For notational convenience, the following notations were introduced:

$$\Delta\eta_0(i) = \eta_2(i) - \eta_1(i) \quad \text{for} \quad 1 \leq i \leq N \tag{3.41}$$

$$\kappa = \sum_{i=1}^{N} \eta_1(i) \tag{3.42}$$

$\{\Delta\eta(j)\}$ is obtained by sorting the elements in $\{\Delta\eta_0(i)\}$ in the descending order, where $\Delta\eta(j_1) \leq \Delta\eta(j_2)$ if $j_1 > j_2$.

The routing algorithm is supported by the following proposition:

Proposition: If and only if

$$\Delta\eta(j) > \frac{\kappa + \sum_{i=1}^{j-1}\Delta\eta(i)}{N+(j-1)} \tag{3.43}$$

the cell capacity increases when the j-th UT is assigned a two-hop link, indexed within $\{\Delta\eta(j)\}$.

Proof: In the following, x is a positive integer larger than two. According to the cell capacity formula in (3.39):

$$C(x) - C(x-1)$$

$$= \frac{W}{N+x}\left(\Delta\eta(x) - \frac{\sum_{i=1}^{x-1}\Delta\eta(i) + \kappa}{N+(x-1)}\right) \tag{3.44}$$

$$= \frac{W\left((N+x-1)\times\Delta\eta(x) - \sum_{i=1}^{x-1}\Delta\eta(i) - \kappa\right)}{(N+x)(N+x-1)}$$

If (3.44) is satisfied, then:

$$(N+x-1)\times\Delta\eta(x) - \sum_{i=1}^{x-1}\Delta\eta(i) - \kappa > 0 \tag{3.45}$$

$$C(x) - C(x-1) > 0 \tag{3.46}$$

Similarly, the capacity can be given as:

$$C(x-1) - C(x-2)$$

$$= \frac{W\left((N+x-2)\times\Delta\eta(x-1) - \sum_{i=0}^{x-2}\Delta\eta(i) - \kappa\right)}{(N+x-1)(N+x-2)} \tag{3.47}$$

Because the following holds:

$$\left((N+x-2)\times\Delta\eta(x-1) - \sum_{i=1}^{x-2}\Delta\eta(i) - \kappa\right) -$$

$$\left((N+x-1)\times\Delta\eta(x) - \sum_{i=1}^{x-1}\Delta\eta(i) - \kappa\right) \tag{3.48}$$

$$= (N+x-1)(\Delta\eta(x-1) - \Delta\eta(x))$$

and $\Delta\eta(j)$ is sorted in descending order. According to (3.46):

$$\left((N + x - 2) \times \Delta\eta(x-1) - \sum_{i=1}^{x-2} \Delta\eta(i) - \kappa\right) >$$

$$\left((N + x - 1) \times \Delta\eta(x) - \sum_{i=1}^{x-1} \Delta\eta(i) - \kappa\right) > 0 \tag{3.49}$$

Then:

$$C(x - 1) - C(x - 2) > 0 \tag{3.50}$$

$$C(x) > C(x - 1) > C(x - 2).... > C(0) \tag{3.51}$$

If (3.49) is not satisfied, then:

$$C(x) > C(x + 1) > C(x + 2).... > C(N) \tag{3.52}$$

Based on above proposition, the routing controller executes the following steps:

Step 1: Calculate $\Delta\eta_0(i)$, κ;

Step 2: Sort $\{\Delta\eta_0(i)\}$ to obtain $\{\Delta\eta(j)\}$;

Step 3: If $\Delta\eta(1) \leq \kappa/N$, let $N_2 = 0$ and exit; otherwise, let $j = 1$ and $\chi = \Delta\eta(1)$;

Step 4: Update $j = j + 1$. If $j = \lfloor N/2 \rfloor$, let $N_2 = \lfloor N/2 \rfloor$ and exit;

Step 5: If $\Delta\eta(j) \leq (\kappa + \chi)/(N + j - 1)$, let $N_2 = j - 1$ and exit; otherwise, update $\chi = \chi + \Delta\eta(j)$ and go to Step 4.

After executing above steps, the proper N_2 UTs are selected and assigned to communicate over the two-hop links, while the residual UTs are assigned single-hop links.

By jointly considering the routing and partitioning problems, the proposed algorithm can maximize the cell capacity under the cell-partition based framework.

The cell capacity of the proposed scheme is compared to a "simple" routing mechanism (UT is assigned a two-hop link if $\eta_2(i) > \eta_1(i)$). In particular, the DL of a multicell environment with 19 cells is simulated (see Section 3.2.5.3). In the simulations, $W = 25.6$ MHz and the cell radius is $R = 500$ meters. Figure 3.23 shows the improvement in cell capacity by use of the joint resource allocation and routing scheme.

With proper relay positioning and power control (with regarding to the ratio of the transmit power at FRN and that at BS) about 6%–15% benefit in cell capacity can be obtained in comparison to a simple route scheme [105].

3.3 Performance Enhancements for Wireless Mesh Networks

A fully distributed, cross-layer scheme, comprising PHY, MAC, and Network layer functionalities, which can guarantee QoS to a wide range of applications in wire-

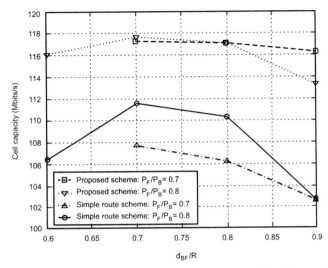

Figure 3.23   Improvement in cell capacity by use of joint resource allocation and routing [105].

less mesh networks was proposed in [106]. Due to the nature of distributed algorithms, distributed scheduling algorithms for mesh networks cannot guarantee 100% conflict-free decisions among neighboring nodes due to a limited information exchange. A generic approach to the design of distributed opportunistic mesh scheduling algorithms to be applied in a wireless mesh backhaul scenario [6] was initially integrated with a multiple-antenna PHY layer.

As a follow up, a methodology (based on a *soft resource reservation* scheme) for successfully merging the distributed opportunistic scheduling scheme with a novel multiconstrained routing algorithm with QoS provisioning was developed. The goal was to provide end-to-end QoS guarantees to applications that have multiple constraints, while at the same time the overall network performance can be optimized by avoiding creating bottlenecks that can negatively affect the network connectivity.

3.3.1   Scheduling Algorithms for Wireless Backhaul Networks

Backhaul networks transfer data between APs to the gateway nodes, which are connected to the wired Internet. Traditional backhaul networks are based on wired technologies such as *asynchronous digital subscriber line* (ADSL), T1, and optical fiber [107]. The FP6 IST project MEMBRANE [6] considered the realization of backhaul networks further and proposed the use of WMNs in order to achieve low cost and ease of deployment of the backhaul capabilities.

Figure 3.24 shows an example of a WMN backhaul scenario.

The backhaul traffic is collected from a number of sources (e.g., APs, BSs), which could be randomly located. Therefore, a key design challenge for the wireless backhaul networks is to provide very high network throughput while meeting the QoS. One of the approaches to overcoming this challenge is to use advanced scheduling algorithms to realize the throughput gain and ensure QoS.

3.3.1.1 Distributed Scheduling Framework with a Generic Utility Function

A single-link distributed framework for mesh network scheduling with TDD operations where nodes cannot transmit and receive simultaneously was proposed in [108]. The network nodes of the studied in [108] backhaul network were *wireless routers* (WRs), which are also shown in Figure 3.24. It must be noted, that the algorithm is applicable to general network topologies and layouts although, and that the hexagonal layout of nodes is adopted here, is only exemplary [106].

From a point view of graph theory, the hexagonal layout is a regular graph of degree *r*. In the assumption, each node can only transmit/receive to/from its one-hop neighboring nodes, whereas it can be interfered by all other concurrently active transmissions in the network. For the sake of discussion convenience, the "one" requirements are for various data flows. It is assumed that each node schedules one of the links associated with it at a time. The objective of the scheduling algorithm is to identify not only the *duplexing mode* (transmitting or receiving) but also the specific direction (to which neighbor) of the next communication in an *opportunistic* manner [106]. For example, if a node is receiving a great deal of interference, it may be more appropriate for the node to choose to transmit, provided that the intended receiver is expected to receive properly.

On the contrary, if a node finds that one of its incoming links of the highest profit among all of its associated links, then the node may prefer to receive from that link. In the described here scheduling algorithm [108], every directional link is assigned with a utility representing the benefit of transmitting on this link in the next time slot. The opportunistic approach, hence, is to choose a combination of concurrent links with the highest aggregated instantaneous utility.

The initial focus assumes the framework with a *generic utility* definition. It was shown in [109] that a collision-free method for utility exchange is feasible. Then it can be assumed that the utility values of both incoming and outgoing links are available to the node, and that the two ends of each link keep the same latest utility value to make the scheduling decisions.

The first stage of the framework is for each node to choose the link with the highest utility among all the incoming and outing links to activate for the next time

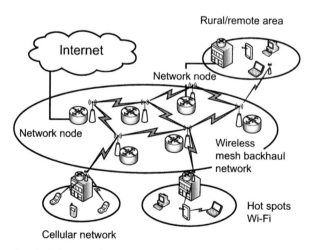

Figure 3.24 Scenario of a WMN-enabled backhaul network [106].

slot. Then in the ideal case, $N/2$ links with the highest utilities will be chosen to activate in an N-node mesh network [106].

The main difficulty in implementing this idea in a distributed way is the possibility that a node makes a decision conflicting with the neighbors in terms of duplexing mode. Figure 3.25 shows a node i, which identifies that it is most profitable to transmit to the neighboring node k and at the same time another neighbor node j of node i decides to send data to node i. Due to the TDD operation, node i cannot transmit and receive at the same time and hence it is necessary to solve the conflict. It is difficult to improve the scheduling on all the nodes in the network, in order to find a conflict-free solution that yields the best performance because fundamentally with a distributed algorithm, the nodes have no prior knowledge about the status of their neighbors at this decision making stage [106].

One suggestion is to exchange the initial decision made among the neighboring nodes and let the nodes with a collided destination give up the intended transmission (i.e., the dashed transmission line in Figure 3.25). The transmission (solid arrow) from node i must be more profitable than the incoming transmission (dashed arrow) to node i, because it is the reason why node i chooses the solid arrow to be the next communication rather than the dashed one.

A formal description of the described distributed scheduling framework comprises the following two control phases:

- Utility exchange and initial decision making;
- Initial decision exchange and final decision making.

The described scheduling is fully distributed without deadlock [106]. The nodes make scheduling decisions simultaneously, and do not need to wait for the decisions of the other nodes to make their own decision.

The scheduling framework exploits also *multiuser diversity*. Although in mesh networks, it is very likely that the fluctuation of the wireless links is weak, the multiuser diversity can be realized with other aspects such as the *differences in propagation loss* (with random node layout), the independent incoming and outgoing channel qualities, and the dynamic interference.

Compared to random schedulers, the scheduling framework described here generates smooth interference. Since the scheduling decisions are related to the *instant* utility, as long as the utility function is with strong time coherence, the link schedule shall generate interference with reasonably strong temporal correlation [106]. From this point of view, it is expected that a *tree-structure* method will not perform well in terms of maintaining the interference coherence because the different sets of links transmit in the even and odd time slots. In addition, smooth interference from either IEEE 802.11 or 802.16 scheduling algorithms cannot be expected because the links are activated based on random competition.

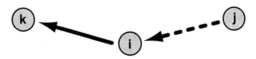

Figure 3.25   An example of scheduling decisions in conflict [106].

A feasible single-link schedule for mesh networks is a matching of the network graph [109], which is a set of independent links.

Then, the objective of the scheduling is to activate the independent link set, which aggregates the largest utility, assuming that an appropriate link utility is assigned to each link. Because of the possibility for conflicts in the scheduling decisions [108], one node can be involved into multiple transmissions. A solution that adds one more control exchange to drop some of the intended transmissions in order to guarantee 100% separated transmissions in the network would induce an efficiency loss in terms of the number of active links, compared to the centralized scheduling [110, 111], which is able to activate as many network nodes as possible by exhaustive searching for a maximal matching, which is NP-hard.

[106] proposed a new generic approach to the design of distributed scheduling algorithms, which only requires one round of control information exchange, towards the same objective. The efficiency of two specific algorithms was analyzed, which assumed that each node knows the link utilities of its one-round and two-round neighbors, respectively.

Let the *preferred link* of node i be defined as the link with the largest utility among all links associated with node i, denoted by $PL(i)$. The utility of link $PL(i)$ is denoted as A_i. The preferred node of node i is defined as the other end of the preferred link of node i, denoted by $PN(i)$.

The requesting node to node i is node j, denoted by $RN(i) = j$, if $PN(j) = i$ but $PN(i) \neq j$.

The preferred link of a requesting node to node i is called a requested link of node i, denoted by $RL(i)$. Let assume that every node has r neighbors and hence has $2r$ links in a regular network layout (see Figure 3.25). Among the $2r$ links, for every node there is always a preferred link $PL(i)$, which is the solid arrow starting from i labelled with A_i.

It must be noted that a preferred link of node i could be a requested link to node $PN(i)$, if node i and $PN(i)$ are not mutually preferred. For example, link A3 is the preferred link of node 3 and a requested link of node 1 as shown in Figure 3.26.

Nodes 3, 1, 0, 2, and 4 form a *utility* chain. The arrows always point from node i to $PN(i)$. The direction of data transmission depends on whether the preferred link is an outgoing or an incoming link. Node 3 is called the *end* node of the chain because there is no requesting node to node 3. Node 4 is called the *top* node because node 4 is not a requesting node to anyone. Node 1, 0, and 2 are in the middle of the chain.

For every node there are three kinds of links to activate, namely the *preferred* link, one of the *requested* links if there is any, and one of the *rest* links. For a scheduling algorithm aiming at maximizing the network aggregated utility, it is desirable to try to activate the *high-utility* links. For this reason, here it is only considered activating either the preferred link or the best requested link of a node. Along the utility chain from the end to the top, the utility values of the preferred/requested links are in ascending order. For example, in Figure 3.26, A3 < A1 < A0 < A2 = A4. Therefore, if nodes could see the entire chain, they could cooperate to activate the independent link set with the highest aggregate utility.

Due to limited utility exchanges between nodes, a node may not see the entire chain. The misunderstanding of the position on the utility chain causes the main

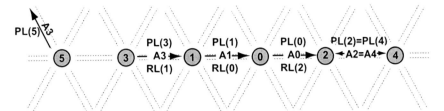

Figure 3.26 An example of utility chain [106].

difficulty in implementing a distributed algorithm for mesh networks [106]. That is, the conflicting decisions among neighboring nodes, which make the intended transmissions unsuccessful.

As an example, if the nodes in Figure 3.26 know one-round utility only, then node 1 knows that $PN(3) = 1$, $PN(1) = 0$, and $PN(0) = 2$ (i.e. node 1 is in the middle of a "partial" chain). At the same time, node 0 knows that $PN(1) = 0$, $PN(0) = 2$ and $PN(2) = 4$ (i.e. node 0 is also in the middle of a chain). Then, because node 1 and node 0 run the same distributed algorithm, they obey to the same scheduling rule. In this case, if the scheduling rule for a node in the middle of a chain is to select the preferred link, then node 1 will choose to activate $PL(1)$ (link libelled with $A1$), and node 0 will choose $PL(0)$ (link libelled with $A0$). Due to the TDD and single-link scheduling constraints, link $A1$ and $A0$ cannot be active simultaneously, which means that the intended transmissions will not be successful. Thus, to activate a middle link is dangerous because there is always a possibility for the other end of the intended transmission to choose a different link. As a counterexample, when a node is at the end/top of a chain, as long as the rule is defined as "to activate the end/top link," both ends of the end/top link will choose the same link, because for both of them, the link is absolutely an "end/top link."

Therefore, the approach is to activate a link, which is or is likely to be at the end or at the top of a utility chain, based on the utility information known to the node.

3.3.1.2 Distributed Scheduling Algorithms with Generic Link Selection Rules

Let assume that for each situation, the distributed scheduling determines, which link, the preferred or the requested link, to be activated. Further, the following specifics are assumed:

- The utilities of the different links are identical in space;
- The network layout is hexagonal, such that every node has $r = 6$ immediate/one-round neighbors around;
- Nodes can only talk to the immediate/one-hop neighbors.

In order to obtain the average number of successful transmissions, all the possible situations are enumerated, and the situation probability as well as the success probability are calculated for having one conflict-free link in the situation. The computation of the probability for every situation (position on the chain), requires the derivation of some key probabilities first [106].

Proposition Probabilities

In the following, the key probabilities are defined for the possible scenarios.

Equation (3.53) formulates the first proposition probability:

$$P_1 = \Pr\left\{PL(1) = PL(2) \mid A1 = \max\{N1\}\right\} \tag{3.53}$$

If we assume that there are two neighboring nodes, node 1 and node 2, then the links associated with node 1 compose set $N1$ and the links associated with node 2, excluding the two shared links with $N1$, compose set $N2$. That is, N1 contains $2r$ links, and N2 contains $2r-2$ links. A1 and A2 are utilities of $PL(1)$ and $PL(2)$, respectively. Then, if $PL(2) = PL(1)$, or $A2 = A1$, node 1 and node 2 are mutually preferred nodes. In a hexagonal layout where $r = 6$, the probability can be formulated as follows:

$$\begin{aligned} P_1 &= \frac{\Pr\left\{A1 = A2 = \max\{N1 \cup N2\}\right\}}{\Pr\left\{A1 = \max\{N1\}\right\}} \\ &= \frac{|N1|}{|N1| + |N2|} = \frac{2r}{4r-2} = 0.545 \end{aligned} \tag{3.54}$$

Equation (3.55) formulates P_2:

$$P_2 = \Pr\left\{PN(2) = 3 \mid PN(1) = 2\right\} \tag{3.55}$$

There are three neighboring nodes, 1, 2, and 3 on a chain. Link sets N_1, N_2, and N_3 denote the non-overlapping link sets associated with them respectively (i.e., $|N2| = |N3| = 2r - 2$ and $|N1| = 2r$). Given that node 2 is the preferred node of node 1, the probability that node 3 is the preferred node of node 2 is given by:

$$\begin{aligned} P_2 &= 1 - \Pr\left\{A3 < A2 \mid A2 > A1\right\} \\ &= 1 - \frac{\Pr\left\{A2 = \max\{N1 \cup N2 \cup N3\}\right\}}{\Pr\left\{A2 = \max\{N1 \cup N2\}\right\}} \\ &= 1 - \frac{|N1| + |N2|}{|N1| + |N2| + |N3|} \\ &= 1 - \frac{4r-2}{6r-4} = 0.3125 \end{aligned} \tag{3.56}$$

The third probability is defined as follows:

$$P_3 = \Pr\left\{PN(3) = 4 \mid PN(2) = 3, \quad PN(1) = 2\right\} \tag{3.57}$$

Similarly to the derivation of P_2, (3.57) can be reformulated as follows:

$$P_3 = \Pr\{A4 > A3 \mid A3 > A2 > A1\} = 0.238 \tag{3.58}$$

Equation (3.59) gives the fourth probability:

$$P_4 = \Pr\{|RN(1)| > 0\} \tag{3.59}$$

First, assume node 1 and node 2 are two neighboring nodes. If node 2 is a requesting node to node 1, $PL(2)$ should be either of the two links between node 1 and node 2 out of $2r$ links. That means that:

$$\Pr\{PL(2) = l(1,2) \ or \ l(2,1)\} = 2/2r = 1/r \tag{3.60}$$

Besides, the utility of $PL(2)$ should be smaller than $PL(1)$. The probability that $2 = RN(1)$ is:

$$\begin{aligned} P_4^0 &= \Pr\{A2 < A1, PL(2) = l(1,2) \ or \ l(2,1) \mid A1 = \max\{N1\}\} \\ &= \frac{\Pr\{A1 = \max\{N1 \cup N2\}\} \cdot \Pr\{PL(2) = l(1,2) \ or \ l(2,1)\}}{\Pr\{A1 = \max\{N1\}\}} \\ &= \frac{|N1|}{|N1|+|N2|} \cdot \frac{1}{r} = \frac{2r-2}{4r-2} \cdot \frac{1}{r} = 0.076 \end{aligned} \tag{3.61}$$

Because node 1 has $r-1$ neighbors except for the preferred node, the probability for node 1 to have at least one requesting node is:

$$P_4 = 1 - (1 - P_4^0)^{r-1} = 0.3265 \tag{3.62}$$

For scenario five:

$$P_5 = \Pr\{|RN(2)| > 0 \mid RN(1) = 2\} \tag{3.63}$$

Similar to scenario four, there are three nodes here. It is known that node 2 is a requesting node to node 1. Then the probability that node 3 is a requesting node to node 2 is:

$$\begin{aligned} P_5^0 &= \Pr\{A3 < A2, \quad PL(3) = l(2,3) \ or \ l(3,2) \mid A2 < A1\} \\ &= \frac{\Pr\{A1 > A2 > A3\}}{\Pr\{A1 > A2\}} \cdot \Pr\{PL(3) = l(2,3) \ or \ l(3,2)\} \\ &= \frac{1}{3} \cdot \frac{1}{r} = 0.056 \end{aligned} \tag{3.64}$$

Then the probability for node 2 to have at least one requesting node is:

$$P_5 = 1 - (1 - P_5^0)^{r-1} = 0.2503 \tag{3.65}$$

For scenario six:

$$P_6 = \Pr\{|RN(3)| > 0 \,|\, RN(2) = 3, \quad RN(1) = 2\} \tag{3.66}$$

and it can be calculated that:

$$P_6 = 1 - (1 - P_6^0)^{r-1} = 0.1931 \tag{3.67}$$

Efficiency of One-Round Scheduling
To derive the scheduling efficiency, all possible situations must be listed and a rule must be set for every situation to choose one link to activate. Then the success probability for every situation is calculated together with an average over all situations in order to obtain the *overall success* probability as the *efficiency* [106].

Knowing the link utility of the one-hop neighbors (node 1's and node 2's), node 0 could be in the following situations as shown in Table 3.1.

The situation probabilities and decisions are given per situation in Table 3.2.

The approach is to activate the end and the top links of a chain.

Following this approach, the link $PL(0)$ is activated in situation 2 because this link is at the end of the chain. Also, in situations 3 and 4, the link $PL(0)$, is activated because this link is at the top of the chain. In situation 1, $PL(0)$ is given up and $RN(0)$ activated, because this link is likely to be the end of a chain.

The situation probability and the success probability for all situations is derived using the probabilities calculated in A. There are two conditions for a transmission to be successful, namely the following:

1) The other end of the intended transmission chooses the same link to activate;
2) No other neighboring node decides to transmit to either end of the intended transmission.

By averaging the success probability subject to the situation probabilities (see Table 3.2), the overall success probability for one node to have one successful transmission/reception using the one-round scheduling is

$$P_S^1 = \sum_{i=1}^{4} P^1(i) \cdot P_S^1(i) = 0.5912 \tag{3.68}$$

Table 3.1 Possible Situations for One-Round Scheduling [106]

Given $PN(0) = 2$	$PN(2) \neq 0$	$PN(2) = 0$
$|RN(0)| > 0$	Situation 1	Situation 3
$|RN(0)| = 0$	Situation 2	Situation 4

Table 3.2   Situations and Situation Probabilities for
One-Round [106]

i	Situation	Decision	$P^1(i)$	$P_S^1(i)$
1	1 --> 0 --> 2 --> 4	1-->0	0.1139	0.7109
2	1 0 --> 2 --> 4	0-->2	0.3411	0.2477
3	1 --> 0 <--> 2	0-->2	0.1779	0.5821
4	1 0 <--> 2	0-->2	0.3671	0.8776

Therefore, the average number of conflict-free active links L in the network containing N nodes is:

$$L_{active}(1) = 0.5 \cdot N \cdot P_s^1 \tag{3.69}$$

Compared to the perfect scheduling algorithms, which could activate at most $N/2$ links without collision, the efficiency of the *one-round* scheduling is 59.12%.

Efficiency of Two-Round Scheduling
With two-round information, every node knows the link utilities of its one-hop and two-hop neighbors [106]. The utility exchange is more complicated and the utility exchange for the two-hop-link utilities may experience larger delays relative to the one-round scheduling.

Because the network nodes in backhaul networks are stationary, the delay in the two-round utility exchange would most probably not degrade the scheduling performance noticeably [106].

One possible solution to the two-round control exchange can be found in IEEE 802.16 specifications.

Similar to the one-round scheduling, the probabilities are calculated with node 0 as the reference node of interest. The situations and corresponding probabilities are given in Table 3.3.

The derivation method is the same as the derivation for the one-round scheduling algorithm. The success probability is given in the last column of Table 3.3 or:

$$P_S^2 = \sum_{i=1}^{8} P^2(i) \cdot P_S^2(i) = 0.7311 \tag{3.70}$$

Then the efficiency of our two-round scheduling is 73.11%.

Figure 3.27 verifies the obtained theoretical efficiency with simulation results. In the simulations performed in [6], the hexagonal layout network consists of

Table 3.3   Situations and Situation Probabilities for Two-Round
Scheduling [106]

i	Situation	$P^2(i)$	Decision	$P_S^2(i)$
1	3 --> 1 --> 0 -->2<-->4	0.0173	idle	0
2	3-->1-->0-->2-->4-->6	0.0043	idle	0
3	3 1-->0 -->2<-->4	0.07	1-->0	1
4	3 1-->0 -->2 -->4-->6	0.0221	1-->0	1
5	1 0 -->2<-->4	0.2354	idle	0
6	1 0 -->2 -->4-->6	0.1057	0-->2	0.8895
7	1 0<-->2	0.4087	0-->2	1
8	1-->0<-->2	0.1363	0-->2	1

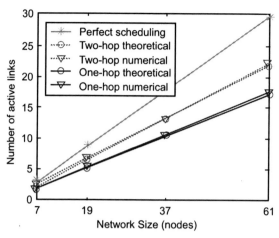

Figure 3.27   Number of active links versus the network size [106].

7, 19, 37, and 61 wireless routers. Independent Rayleigh-fading channels between different sender/receiver pairs were implemented. A low Doppler frequency of 20 Hz, corresponding to a 4.32-kmph velocity for a 5-GHz carrier frequency, was assumed to approximate the stationary topology. The SIR is based on the presently measured interference as the link utility for the next time slot. The Shannon capacity [112] based on the instantaneous SIR was used to obtain an upper bound of the network throughput [106].

Figure 3.27 includes also the maximum number of concurrent links for every network size to show the upper bound, which is $N/2$ for an N-node network with a TDD constraint. In Figure 3.27, both the one-round and two-round numerical curves match the theoretical curve. The greatest error between the numerical and theoretical results is 2%, appearing in the one-round case for the 61-node network. The two-round scheduling algorithm can activate 25% more conflict-free concurrent transmissions over the simple one-round algorithm.

To summarize, the proposed *one-round* and *two-round* scheduling algorithms have the following advantages [106]:

- They are fully distributed. As long as the utility exchange is accurate and prompt, the scheduling decisions are made simultaneously by each node without the need for a central controller.
- The scheduling algorithms achieve multiuser diversity gain in terms of time-varying wireless channel quality and space-varying interference, because the algorithms activate the highly demanded links, namely the preferred and requested links, with high priority.
- The proposed algorithms provide higher spatial link-utilization than the IEEE 802.11 with RTS/CTS [72], where at most only one link could be active in a two-hop neighborhood.

3.3.1.3   Distributed Opportunistic Scheduling with Utility Definition

In order to deliver a good backhaul service, the scheduling algorithm should be able to cooperate with the routing algorithms [106]. In wireless networks, the uncertainty in link capacity due to the randomness of the lower-layer protocols and the wireless channels degrades the performance of the routing protocols (see Section 3.1.4). Furthermore, it is difficult to guarantee the system performance if an opportunistic MAC layer is deployed, because the opportunistic approaches usually introduce more fluctuating instantaneous performance at individual nodes.

[106] proposed a utility function as a scheduling metric, which achieves an opportunistic gain and also supports QoS as committed by the routing algorithm in use.

The proposed co-operation between the scheduling and routing algorithms is in a "request-enforce" manner. The routing protocol would determine the routes for the packet flows based on the total costs of the paths. The cost of a link is usually computed using the long-time average capacity of that link. Thus, for routing purposes, the estimation of the future link capacity is crucial in order to maintain an effective routing table for various source and destination node pairs. As a result, it is desirable for the routing layer to specify a target throughput allocation among the links on each node along a route and then to request the scheduling algorithm to enforce such throughput allocation.

Rather than achieving the precise target throughput for each link, the objective of the algorithm described here is to achieve the relative target throughput for each link scaled by a per-node (not per-link) proportionality constant. Thus, achieving the relative target by the proposed scheduler effectively yields the actual throughput target.

The derivation of a new utility function has a focus on the individual node as described for the generic distributed algorithms (see Section 3.3.1.1 and Section 3.3.1.2). The incoming and the outgoing links are equally treated as competitors. For an arbitrary node i with r neighboring nodes, it has $2r$ candidate links to schedule for transmission in every time slot, assuming that each link constantly has traffic that is ready for transmission. Figure 3.28 shows how the routing algorithm periodically estimates the throughput demand on each link associated with each node in the next time duration (e.g., thousands of time slots), and provides the scheduler with a target throughput allocation $\mathbf{a}_i = (a_i(1), a_i(2), \cdots, a_i(2r))$ to achieve the desired QoS.

The goal is to define an appropriate utility with which the allocation of the scheduler of the long-run throughput for all links $\mathbf{R}_i = (R_i(1), R_i(2), \cdots, R_i(2r))$ is proportional to the target allocation $\mathbf{a}_i$ for $\mathbf{R}_i^* = c_i \cdot \mathbf{a}_i$, where c_i is a positive proportionality constant for that node i.

In order to define the utility, an optimization problem for each node i with an objective function $f(R_i^*)$ is defined first. The optimal solution R_i^* for the optimization problem is proportional to the target throughput allocation $\mathbf{a}_i$. If a scheduler

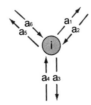

Figure 3.28   Target throughput allocation among links for one node when $r = 3$ [106].

chooses a link for the transmission according to this utility, the allocated throughputs by this scheduler converge to the optimal solution R_i^* to the defined optimization $f(R_i^*)$ in the long run.

Lemma 1: If the optimization problem for each node i is to maximize the objective function:

$$\max_{R_i} f(\mathbf{R}_i) = \sum_{k=1}^{2r} a_i(k) \cdot \log R_i(k),$$

$$\text{s.t.} \sum_{k=1}^{2r} R_i(k) \leq C \tag{3.71}$$

Then the optimal solution $\mathbf{R}_i^* = (R_i^*(1), R_i^*(2), \cdots, R_i^*(2r))$ is directly proportional to $\mathbf{a}_i = (a_i(1), a_i(2), \cdots, a_i(2r))$ element by element.

Proof: This is a classic constrained optimization problem, which could be solved by *Lagrange* multipliers. Specifically, the new Lagrange problem with multiplier λ_i for node i is:

$$L = \sum_{k=1}^{2r} a_i(k) \log R_i(k) - \lambda_i \cdot \left(\sum_{i=k}^{2r} R_i(k) - C \right) \tag{3.72}$$

The first order necessary optimality condition for (3.72) is:

$$\begin{cases} \nabla L = 0 \\ \lambda_i \cdot \left(\sum_{k=1}^{2r} R_i^*(k) - C \right) = 0 \end{cases} \tag{3.73}$$

The constraint is binding, $\lambda_i \neq 0$ and

$$\sum_{k=1}^{2r} R_i^*(k) - C = 0$$

then (3.73) becomes:

$$\nabla L = 0 \Leftrightarrow \frac{a_i(k)}{R_i^*(k)} = \lambda_i \quad k = 1, 2, \cdots, 2r \tag{3.74}$$

or equivalently:

$$R_i^*(k) = \lambda_i^{-1} a_i(k) \quad k = 1, 2, \cdots, 2r \tag{3.75}$$

λ_i is the same for all links $k = 1, 2, \ldots, 2r$ associated with node i. That means that the final throughput allocation in the long run is proportional to the target throughput allocation $\mathbf{R}_i^* = \lambda_i^{-1} \cdot \mathbf{a}_i$ with the with the per-node proportionality constant equal to λ_i^{-1}.

Lemma 2: If node i uses the following scheduling utility (or metric) for all links k from 1 to $2r$:

$$M_i(k) = a_i(k) \frac{\rho_i(k)}{R_i(k)} \tag{3.76}$$

the scheduler maximizes the objective function iteratively where $\rho_i(k)$ is the instantaneous supportable data rate for link k and $R_i(k)$ is the long-time average of $\rho_i(k)$.

Proof: Since the objective function in (3.76) is convex, the sufficient and necessary *optimality* condition for this convex optimization problem [113] is:

$$\nabla f|_{R_i^*} \otimes \Delta R_i \leq 0 \tag{3.77}$$

where R_i^* is the optimal long-run throughput vector, and R_i is any of the non-optimal average throughput vector and $\Delta R_i = R_i - R_i^*$.

Expanding (3.77), where $\otimes$ is the inner product, gives:

$$\sum_{k=1}^{2r} a_i(k) \frac{R_i(k) - R_i^*(k)}{R_i^*(k)} \leq 0 \tag{3.78}$$

where R_i and R_i^* are the time average of $\rho_i(k)$ and $\rho_i^*(k)$, respectively. (3.78) can be expressed as:

$$\sum_{k=1}^{2r} a_i(k) \frac{E\{\rho_i(k)\}}{R_i^*(k)} - \sum_{k=1}^{2r} a_i(k) \frac{E\{\rho_i^*(k)\}}{R_i^*(k)} \leq 0 \tag{3.79}$$

Then the instantaneous data rate vector $\rho_i^* = (\rho_i^*(1), \rho_i^*(2), \cdots \rho_i^*(2r),)$ in each time slot will maximize the negative term of (3.79) in the following:

$$\max_{\rho_i} \sum_{k=1}^{2r} a_i(k) \frac{\rho_i(k)}{R_i(k)} \tag{3.80}$$

Since ρ_i in the first summation term of (3.79) is not optimal as in (3.80), the second summation term in (3.79) must be greater than the first term in magnitude. (3.78) is satisfied. For the scheduler choosing one best link for transmission at a time, (3.80) becomes:

$$\max_{k} a_i(k) \frac{\rho_i(k)}{R_i(k)} \tag{3.81}$$

Then, the scheduler choosing the link with the largest metric to transmit in each time slot maximizes the objective function in (3.71).

The scheduling metric in (3.81) is parallel to a technique derived for the scheduling of TCP traffic in [114].

Based on *Lemmas* 1 and 2, it can be concluded that the scheduling algorithm with a metric or utility defined in (3.80) maximizes (3.71), which leads to an actual throughput allocation according to the target allocation. With other words, the scheduler with utility function defined by (3.80), allocates the throughput directly proportional to the routing demands in the long run [106]. Combining the framework and the utility definition, a *novel integrated, distributed, and opportunistic* scheduling algorithm for wireless mesh networks can be defined [106].

The performance of the proposed scheduling algorithm was evaluated by simulations and in terms of the throughput allocation convergence, the interference temporal correlation, and the average network throughput.

A hexagonal layout network consisting of 37 wireless routers and independent Rayleigh-fading channels between the different sender/receiver pairs were implemented using the Jakes model in MATLAB [6].

Due to the stationary topology for the backhaul networks, a low Doppler frequency of 5 to 20 Hz was assumed, which corresponds to 1.08- to 4.32-kmph velocity for a 5-GHz carrier frequency and 1-msec time slot.

The convergence of the algorithm in terms of throughput allocation was assessed by a random generation of the target throughputs in the simulation, and the actual time-averaged throughput of each link was observed slot by slot [106]. The actual throughput gained by each link is recorded and compared relatively to the predefined throughputs. The comparison is examined by the ratio I (3.82) at the centrally positioned node of the network to eliminate the edging effect:

$$\beta(k,t) = \frac{R(k,t)}{a(k)} \quad k = 1, 2, \cdots, 2r \tag{3.82}$$

In (3.82), the numerator is the *actual long-run* throughput of the candidate links up to time slot t, while the denominator is the *target* throughput allocation, which is a pre-defined constant (by the routing algorithm). It is important to note the similarity among the elements of the vector β: the more similar to each other the elements, the better the convergence [106].

Let consider the ideal case where all the links associated to one node have the same ratio:

$$\beta^*(t) = \frac{R(k,t)}{a(k)} \quad k = 1, 2, \cdots, 2r \tag{3.83}$$

Then, $R(t) = \beta^*(t) \cdot \mathbf{a}$, which is the original objective of the scheduling based on the throughput allocation. Thus, the similarity of the ratios is assessed by taking the normalized mean deviation of β for every time slot:

$$\sigma(t) = \frac{1}{2r} \sum_{k=1}^{2r} \left| \frac{\beta(k,t) - \mu_R(t)}{\mu_R(t)} \right| \tag{3.84}$$

where t is the time-slot index, $\mu_R(t)$ is the ergodic mean of $\{\beta_i(k,t), k = 1, 2, \cdots, 2r\}$ and $\beta_i(k,t)$ is the ratio for link k in time slot t. Figure 3.29 proposed algorithm converges to the target throughput allocation among links, showing that the normalized mean deviation of the ratio drops rapidly from the original high value (1.82) to near zero (0.15).

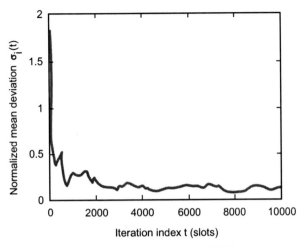

Figure 3.29   Normalized mean deviation of the ratio vector [106].

For the temporal correlation of interference, the interference at the central node was measured at every time slot and its autocorrelation coefficient was computed based on statistics from running the simulation for 10,000 time slots as follows:

$$\xi_I(\tau) = \mathrm{E}\left\{ \frac{|\{(I(t) - \mu_I)(I(t - \tau) - \mu_I)\}|}{\sigma_I^2} \right\} \tag{3.85}$$

where $I(t)$ is the interference encountered by the central node at time t, and μ_I and σ_I^2 denote the *ergodic mean* and the *variance* of the random process $I(t)$, respectively. Theoretically, the coefficient $\xi_I(\tau)$ is in the range of $[0, 1]$, where $\xi_I(\tau) = 0$ implies a temporary independent process and $\xi_I(\tau) = 1$ indicates the strongest temporal correlation. This is shown in Figure 3.30.

Figure 3.30 shows that the proposed scheduling framework is able to maintain a reasonable *interference temporal* correlation for all studied utilities. The SIR

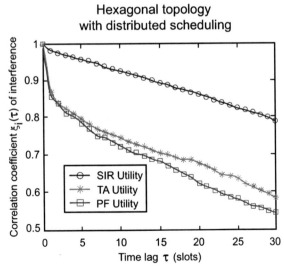

Figure 3.30   Temporal correlation of interference for distributed scheduling [106].

utility, which chooses the link with the highest SIR for transmission, yields higher temporal correlation than the other two methods (*throughput allocation*—TA and *proportional fair*—PF). The TA utility generates slightly more correlated interference than the commonly used PF utility. This is due to the fact that the PF utility is a special case to the TA utility with an identical throughput allocation to all the links. The more the allocations among links differ; the more likely it is that the same link would be scheduled in successive time slots, thus generating more temporally correlated interference [106]. Another important observation is that if the target throughput is proportional to the observed actual link capacity in the past, which is likely to be the case for *adaptive routing* algorithms, then the interference temporal correlation generated by the TA utility would be as strong as that provided by the SIR utility. This is because of the fact that the target allocation a_i and the actual throughput R_i in (3.80) will be cancelled and the scheduling becomes primarily determined by the (opportunistic) SIR of each link.

Figure 3.31 shows that a tree structure generates oscillating interference irrelevant to which utility is used. Because both the distributed power control and the scheduling algorithms rely on temporally coherent interference to predict the future channel quality [106], the oscillating interference can drastically reduce their potential performance gain. Figure 3.32 shows the aggregated instantaneous network throughput as a measurement of the opportunistic gain achievement.

The proposed distributed and opportunistic algorithm provides an enhancement up to 19% in network throughput compared to the *tree* method. The throughputs for both the distributed and tree scheduling increases as the channel fluctuates faster because the tree method is also an opportunistic approach and takes advantage of the channel fluctuation to some extent.

Due to the static tree mapping, many links are not included in the tree, therefore, the network throughput by the tree method is uniformly below that of the distributed algorithm. As a result, the distributed scheduling algorithm is superior to the tree method in terms of both interference temporal correlation and an op-

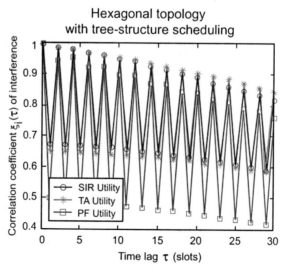

Figure 3.31 Temporal correlation of interference for *tree-structure* scheduling [106].

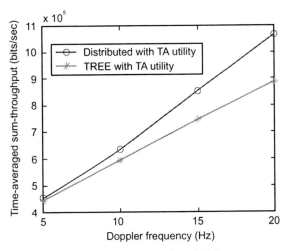

Figure 3.32 Network throughput for hexagonal topology [106].

portunistic gain achievement. Furthermore, the proposed TA utility in (3.80) can help routing algorithms to guarantee QoS [106].

3.3.1.4 Proportional Fair Scheduler

The use of *proportional fair scheduling* (PFS) has a great potential for enhancing the throughput while exploiting the fairness in fading wireless environments via *multi-user diversity* and *game-theoretic* equilibrium.

The calculation of the throughput is one of the fundamental problems for a wireless network integrating the PFS algorithm. Due to the NP-completeness of the PFS problem, analytical work on understanding the performance of the PFS is burdensome. Most of the existing research on PFS prior to the FP6 IST program, was related to single-hop wireless networks. [6, 4, 7] were some of the FP6 IST projects that focused on the integration of the multiple-antenna technology with scheduling as a means to increase the throughput in next generation wireless networks via spatial reuse and/or spatial division multiplexing.

Proportional Fair Scheduling Criteria
For the mesh network shown in Figure 3.33, the link scheduling is performed for node 0, which has radio link connections to its neighboring nodes m_1, m_2, ..., and m_N. The traffic is collected at node 0 before forwarding it to the wired Internet.

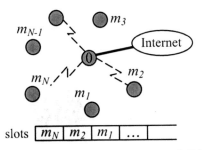

Figure 3.33 Slot-based scheduling in a wireless backhaul network [106].

The time is divided into small scheduling intervals called *slots*. In each slot only one neighboring node is chosen to transmit. In a next scheduling slot, the system will estimate the rates by estimating the SNR, by use of a pilot signal broadcasted periodically, with a very short delay. The selection of the node to schedule is based on a balance between the *current possible rates* and *fairness*. The PFS [115, 116] performs this by comparing the ratio of the feasible rate for each node to its average throughput tracked by an exponential moving average, which is defined as the preference metric. The node with the maximum preference metric will be selected for transmission at the next scheduling slot. This is described mathematically as follows.

The end of a slot n is called *time n*. In the next time slot $n+1$, the instantaneous data rate of node j will be $R_j [n + 1]$. Its k-point moving average throughput up to time n is denoted by $r_{j;k} [n]$ and the preference metric by $M_{j;k}[n + 1] = R_j [n + 1]/ r_{j;k} [n]$. Node $i = \arg \max_j M_{j;k}[n + 1] = \arg \max_j R_j[n+1]/r_{j;k} [n]$ is chosen to transmit is chosen to transmit in the next time slot $n+1$. The moving average throughput of node j up to time $n+1$ of is updated by:

$$r_{j;k}[n + 1] = \left(1 - \frac{1}{k}\right) r_{j;k}[n] + I_j[n + 1] \times \frac{R_j[n + 1]}{k} \qquad (3.86)$$

where the indicator function of the event that node j is scheduled to transmit in time slot $n+1$ is given by:

$$I_j[n + 1] = \begin{cases} 1, & \text{node } j \text{ scheduled in time slot } n + 1 \\ 0, & \text{else} \end{cases} \qquad (3.87)$$

By introducing a *user utility function*: $U_j = Log[r_j]$ [115] had proven that the sum of the user utility (user satisfaction indicator) is maximized under the PFS criteria. For a time-varying fading environment, when the node number is large, there will always be high probability that some nodes are in the good channel status [106]. On the other hand, the PFS provides some sense of fairness, in the sense that the nodes that are frequently in the bad channel status have low throughput, which in turn tends to increase their probability of being scheduled. It is the *logarithm utility maximization* characteristics, the *multi-user diversity* gains, and the *possibility to schedule* bad-channel-condition nodes that make the PF scheduler superior to the traditional ones such as *round-robin* (RR) and the opportunistic scheduler.

The PFS criteria described above have a fairly low implementation complexity because each scheduling slot requires only N addition operations and $2N+1$ multiplication operations for given N nodes, however, in a network-wide context, the proportional fairness is an NP-hard problem for a close-optimal solution [117].

For the communication between one node with t transmit antennas and another one with r receive antennas, the exact distribution of the capacity in a Rayleigh fading environment can be found, in principle, however, results have been found too complex to be of practical use. A certain limit theorem exists [118] that shows that the distribution of the standardized capacity is asymptotically Gaussian as $r \to \infty$, $t \to \infty$ and $r/t \to y$ for some constant y.

By assuming identical fading on all $r \times t$ paths, [36] found similar properties, namely, that the channel capacity in a Rayleigh or Ricean fading environment can be accurately modeled by a Gaussian; and that the *variance* of the channel capacity is not sensitive to the number of antennas and is mainly influenced by the SNR.

This approach was used by [6] to analyze PFS for the single-antenna case when there is one transmit antenna and one receive antenna per node; and for the multiantenna case where each node in the scenario has multiple transmit antennas and multiple receive antenna.

With Gaussian approximation, for single input single output (SISO) transmission, the feasible rate over a Rayleigh fading environment follows a Gaussian distribution, with the mean and variance, respectively, as follows [106, 119, 120]:

$$E[R] = \int_0^\infty \log(1 + SNR \times \lambda) \times e^{-\lambda} d\lambda \qquad (3.88)$$

$$\sigma_R^2 = \int_0^\infty (\log(1 + SNR \times \lambda))^2 \times e^{-\lambda} d\lambda$$
$$-\left(\int_0^\infty \log(1 + SNR \times \lambda) \times e^{-\lambda} d\lambda\right)^2 \qquad (3.89)$$

The mean and standard deviation given by (3.88) and (3.89), respectively, are the normalized value in [bps/Hz], the SNR denotes the received signal to noise ratio.

Figure 3.34 plots the $E[R]$ versus σ_R^2 in for various SNR up to 60 dB.

Let $P(A)$ and $P(B)$ be the positions of the node A and B, respectively. All nodes are stationary and all antennas of all nodes have the same transmission range, denoted by d_0.

Two node A and B can communicate with each other when their Euclidean norm is less than d (i.e., $|P(A) - P(B)| \le d_0$) or when the scheduling policy allows them to do so. In the context of UL communication, a receive antenna i of an AP n is called a *scheduling entity*, denoted by $RE(n, i)$. The transmit antenna j within range is called $RE(n, i)$'s *scheduled entity*, denoted by $TE_{RE(n, i)}(m, j)$, given the transmit antenna belongs to node m. A scheduling entity $RE(n, i)$ has $TEN(n, i)$ scheduled entities.

Let define $R_{(m, j, n, i)}$ to be the feasible rate for the communication of $TE_{RE(n,i)}(m,j)$, and $RE(n, i)$.

Each scheduled entity corresponds to one and only one scheduling entity [106]. Different scheduled entities of the same node must correspond to different scheduling entities. A scheduling entity grants the channel access to its scheduled entities in

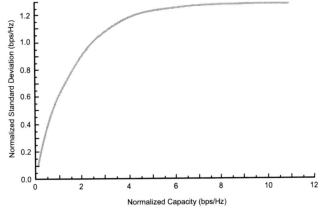

Figure 3.34   Feasible rate of the mean versus standard deviation [106].

the PFS-manner described above. The scheduled entities of a node could be simultaneously scheduled by different scheduling entities.

Assume the *same power* for the communication between a scheduled entity and a scheduling entity. Each node has one receive antenna and then $TEN(n, i)$, $TE_{RE(n, i)}(m, j)$ and $RE(n, i)$ can be reformulated as $TEN(n)$, $TE_{RE(n)}(m, j)$ and $RE(n)$, respectively, for the sake of notation simplicity. Also assume SISO transmission for the communication between the two nodes. The channels are statistically symmetric and independent of each other. Then, the *communication capacity* between node m and n, denoted by $R_{m,n}$, is the *channel capacity* between the scheduled entity of node m and the scheduling entity of node n, which follows the Gaussian distribution characterized by (3.88) and (3.89).

The analysis can be performed for two cases [106]: (*i*) Single-antenna case when there is one transmit antenna and one receive antenna per node; and (*ii*) multi-antenna case, where each node in the scenario has multiple transmit antennas and multiple receive antenna.

PFS for Single and Multiple Antenna Scenarios
For the single-antenna case, for the network of N+1 nodes, the *mean throughput* of node *j* can be written as follows:

$$E[r_j] = \frac{E[R_{j,0}]}{N} + \frac{\sigma_{R_{j,0}}}{N} \left[N \int_{-\infty}^{\infty} y f_{(0,1)}(y)(F_{(0,1)}(y))^{N-1} dy \right] \tag{3.90}$$

where $R_{j,0}$ is the feasible rate between node 0 and its neighboring node j, $f_{(0,1)}(.)$ and $F_{(0,1)}(.)$ denote the *pdf* and *cdf* of the standard normal distribution with zero mean, unit variance, respectively.

Equation (3.90) indicates that *the mean throughput of a PFS node depends solely on its own channel statistics*. Although (3.90) shows a linear relationship between the mean throughput and the first-order statistics of the channel capacity, one should not assume that the mean throughput is linearly proportional to the feasible rate [106], due to the fact that $E[R_{j,0}]/\sigma_{R_{j,0}}$ is not a constant value as can be verified by (3.88) and (3.89). It also suggests that one can expect higher *mean throughput* of PFS under a fading environment (larger $\sigma_{R_{j,0}}$). On the other hand, when the channel is always in a good status (i.e., very large value of $E[R_{j,0}]/\sigma_{R_{j,0}}$ or $\sigma_{R_{j,0}} \approx 0$), the PFS value will exhibit the same performance as the simple *round-Robin* scheduling.

A few observations can be made from (3.90). First, as $E[R_j]/N$ is the *mean throughput* of node *j* when using *round-Robin* scheduling, the throughput gain of node *j*'s for PFS over round-Robin scheduling is:

$$\begin{aligned} G_{PFS}(j) &= \frac{E\lfloor r_j \rfloor}{E_{RR}\lfloor r_j \rfloor} = \frac{E\lfloor r_j \rfloor}{E[R_j]/N} \\ &= 1 + \frac{\sigma_{R_{j,0}}}{E[R_{j,0}]} \left[N \int_{-\infty}^{\infty} y f_{(0,1)}(y)(F_{(0,1)}(y))^{N-1} dy \right] \end{aligned} \tag{3.91}$$

Figure 3.35 shows the throughput gain versus the SINR for a different number N of the neighboring nodes (curves up to down are for N=1000, 500, 200, 70, 20, 5, and 1, respectively).

In Figure 3.35 the $G_{PFS} \in (1.4, 2.0)$ for a typical network scenario where the SINR $\in (6 \text{ dB}, 28 \text{ dB})$ and $N \in 2$–200. The G_{PFS} is in fact the multi-user diversity gain, and Figure 3.35 shows that it is dependent on the SINR and the number of nodes. This means that a node with good channel conditions (i.e., with large SINR), will not benefit much from the PFS as the PFS principle requires that it should provide compensation for the nodes with bad channel conditions [106].

Figure 3.36 shows a comparison of the PFS gain between (3.91) and a linear rate model described in [121] for different values of N and some typical values of the SNR. The PFS gain in the linear rate model is unrealistic as it solely depends on N. Figure 3.36 shows that the bad-channel nodes will benefit more from use of PFS.

Figure 3.37 shows the difference between the node throughput when employing the PFS model described here and a model assuming a logarithm rate model [121].

The difference in node throughput between the one defined in (3.91) to the one assuming a logarithm rate model is about 72.5% when $N = 10$, and increases with N to about 80.6% for $N = 300$. Both the linear rate model and the logarithm rate model over-estimate the PFS throughput [106].

In a multiantenna case, the nodes operate in duplex mode, and are multi-antenna capable. For the scenario adopted here, each node has multiple transmit antennas and only one receive antenna [106].

For the *multiantenna mesh* networks with N nodes, the mean throughputs of node j can be written as follows:

$$E[ri_j] = \frac{1}{TEN(j)} \sum_{\substack{\forall n \neq j, \\ |P(n)-P(j)| < d_0}}^{N} \left(E[R_{n,j}] + \sigma_{R_{n,i}} \left[TEN(j) \int_{-\infty}^{\infty} y f_{(0,1)}(y) (F_{(0,1)}(y))^{TEN(j)-1} dy \right] \right)$$

(3.92)

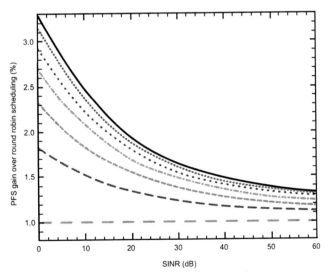

Figure 3.35   Throughput gain G_{PFS} versus the SINR for various N values [106].

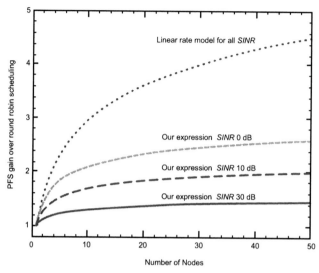

Figure 3.36 PFS gain compared to a linear rate model [106].

$$E[ro_j] = \sum_{\substack{\forall n \neq j, \\ \|P(n)-P(j)\| < d_0}}^{N}$$

$$\left(\frac{E[R_{j,n}]}{TEN(n)} + \frac{\sigma_{R_{j,n}}}{TEN(n)} \left[TEN(n) \int_{-\infty}^{\infty} y f_{(0,1)}(y)(F_{(0,1)}(y))^{TEN(n)-1} dy \right] \right) \tag{3.93}$$

where ri_j and ro_j are the *mean input* throughput and *mean output* throughput of node j, respectively.

Equation (3.92) and (3.93) indicate a difference between the *mean input* and *mean output* throughput in the general cases; however, it can be verified that the ri_j

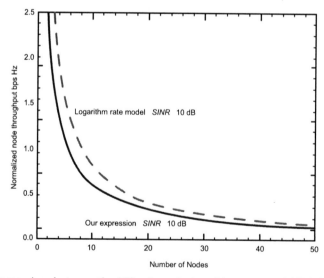

Figure 3.37 Comparison between the PFS gain and a logarithm rate model [106].

and the ro_j are equal, that is, $\sum E[ri_j] = \sum E[ro_j]$. Further, $E[ri_j] = E[ro_j]$ when the number of nodes within ranges are the same for all the nodes [106]. Then, for *fully mesh networks* with N nodes, the *mean throughput* of node j is:

$$E[r_j] = E[ri_j] = E[ro_j] =$$
$$\frac{1}{N-1} \sum_{n=1,n\neq j}^{N} \left(E[R_{n,j}] + \sigma_{R_{n,i}} \left[(N-1) \int_{-\infty}^{\infty} y f_{(0,1)}(y)(F_{(0,1)}(y))^{N-2} dy \right] \right) \qquad (3.94)$$

Equation (3.90) and (3.94) are the closed-form expressions for the mean throughput for the single-and multiple-antenna scenarios, respectively. The performance to the network based on use of the PFS proposed here, is further investigated through simulations [106].

For the single-antenna case, the feasible rates over the Rayleigh fading channels are statistically independent Gaussian distribution for different nodes and the initial moving average throughputs of the nodes are randomized. The system parameters are: 20-MHz bandwidth, $k = 500$. For simplicity, it was assumed that For simplicity, the notation $n_j[\mu, \sigma]$ is assumed to indicate that the feasible rate of node j has a mean value μ and a standard deviation σ (in Mbps).

In a first scenario, the data rates for all nodes are Gaussian distributions. Figures 3.38 and 3.39 show the accuracy of the analytical expressions for the throughput of an 11-nodes network (e.g., node 0 has 10 neighboring nodes) for some typical values of the SINR.

Figure 3.40 shows the results for the normalized throughput (bps/Hz) versus the number of neighboring nodes for various values of the SINR.

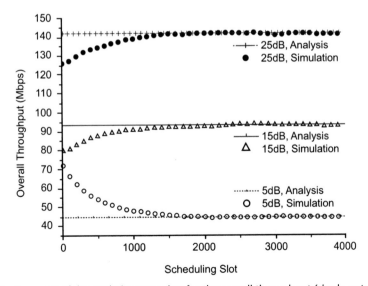

Figure 3.38 Accuracy of the analytic expression for the overall throughput (single-antenna, Gaussian distribution) [106].

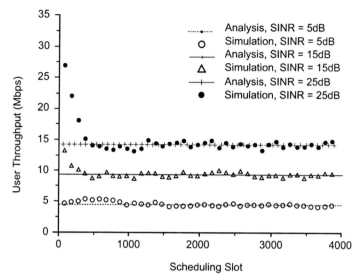

Figure 3.39 Accuracy of the analytic expression for the node throughput (single-antenna, Gaussian distribution) [106].

In a second, more generalized scenario for the single-antenna case, for any j, node j has randomized data rate mean $E[R_j]$ and a proportional standard deviation σ_{R_j}. The feasible rates (in Mbps) of the nodes are randomized as follows:

$$n_1[80,40], n_2[40,20], n_3[26,13], n_4[80,40], n_5[35,17.5],$$
$$n_6[28,14], n_7[10,5], n_8[38,19], n_9[18,9], n_{10}[45,22.5],$$

where a 0.5 proportional factor is assumed. Figure 3.41 shows the results for the mean throughput of node j under the above assumptions.

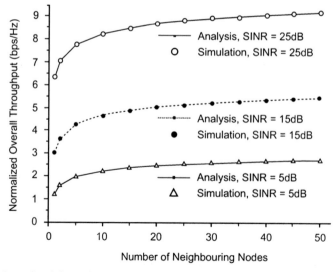

Figure 3.40 Normalized throughput for the single antenna case, Gaussian distribution [106].

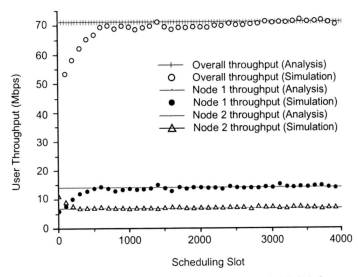

Figure 3.41 Results for the mean throughput of node j (Equation 3-90) [106].

The simulation results in the two scenarios above (Figure 3.38 to Figure 3.41) indicate a high accuracy of the closed-form analytical formulae for the PFS obtained when assuming a *proportional* relationship between the mean and the standard deviation of the moving average throughput [106].

In a third, more realistic scenario, the feasible rates over the Rayleigh fading channels are statically independent Gaussian distribution and the initial throughputs of the nodes are randomized. The feasible rates (in Mbps) of the nodes are randomized as follows:

$$n_1[43.3, 18.9], n_2[87.3, 24.0], n_3[24.1, 13.7], n_4[34.9, 16.9], n_5[129, 25.3],$$
$$n_6[91.8, 24.2], n_7[70.6, 22.7], n_8[37.4, 17.6], n_9[150, 25.5], n_{10}[27.7, 14.9],$$
$$n_{11}[45.9, 19.4], n_{12}[78.0, 23.4], n_{13}[34.2, 16.8], n_{14}[102, 24.6], n_{15}[86.9, 23.9],$$
$$n_{16}[49.9, 20.1], n_{17}[25.5, 14.2], n_{18}[32.5, 16.3], n_{19}[118, 25.1], n_{20}[79.4, 23.5],$$
$$n_{21}[141, 25.4], n_{22}[27.7, 14.9], n_{23}[34.9, 16.9], n_{24}[46.3, 19.5], n_{25}[37.8, 17.6],$$
$$n_{26}[86.0, 23.9], n_{27}[70.6, 22.7], n_{28}[77.2, 23.3], n_{29}[127, 25.2], n_{30}[108, 24.8],$$
$$n_{31}[86.0, 23.9], n_{32}[60.0, 21.6], n_{33}[133, 25.3], n_{34}[27.7, 14.9], n_{35}[86.0, 23.9],$$
$$n_{36}[60.0, 21.6], n_{37}[47.1, 19.6], n_{38}[90.5, 24.1], n_{39}[158.9, 25.5], n_{40}[182, 25.6],$$
$$n_{41}[64.2, 22.1], n_{42}[36.7, 17.4], n_{43}[24.4, 13.8], n_{43}[34.2, 16.8], n_{45}[42.1, 18.6],$$
$$n_{46}[88.2, 24.0], n_{47}[65.5, 22.2], n_{48}[90.5, 24.1], n_{49}[71.9, 22.9], n_{50}[91.8, 24.2]$$

The given traffic pattern reflects a real network scenario where the different nodes may experience channel characteristics of great difference. Figure 3.42 shows that (3.90) provides high accuracy in terms of the normalized overall throughput and the node throughput (bps/Hz) even in cases that exhibit Rayleigh fading characteristics of a real wireless environment.

For the simulations in the multiantenna mesh network scenario it was assumed that when the SNR at the reference distance d_r is PL_r, the SNR at distance d is $PL_d =$

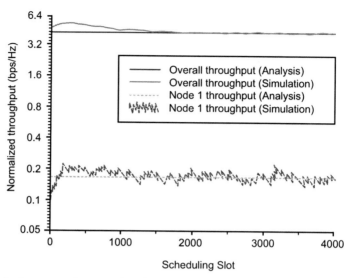

Figure 3.42 Performance of a network with real traffic pattern and a single antenna [106].

$PL_r \times (d_r/d)^\alpha$ given the path loss exponent α. The mean throughput in the multi-antenna mesh case can be evaluated by (3.92) and (3.93).

Figure 3.43 shows a typical network scenario frequently seen on the highway. The simulations assume the SNR at the reference point $PL_r = 20$ dB; the reference point distance $d_r = 50$ meter, range $d_0 = 110$ meter, the path loss exponent $\alpha = 2.5$, and the *node distance* = 100 meters.

Figure 3.44 shows the mean throughputs of *node* 3, [i.e., the node at position (3, 1)], obtained analytically and through simulations.

Figure 3.45 shows a typical hexagonal mesh deployment scenario.

The SNR at the reference point $PL_r = 20$ dB, reference point distance $d_r = 50$ meters, the range $d_0 = 110$ meters, the path loss exponent $\alpha = 2.5$, and the node distance = 100 meter.

Figure 3.46 shows the mean throughputs of node 16 (i.e., the node at position $(2; \sqrt{3})$ obtained analytically and through simulations.

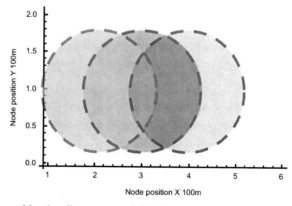

Figure 3.43 Node position in a linear network [106].

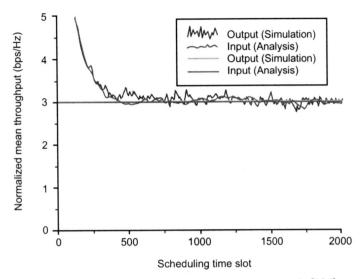

Figure 3.44   Results for the mean throughput: in a linear network scenario [106].

Figure 3.47 shows a random mesh scenario. The SNR at the reference point PL_r = 20 dB, reference point distance d_r = 50 meters, the range d_0 = 110 meters, the path loss exponent α = 2.5, and the node distance = 100 meter. Figure 3.48 the mean throughput of node 6, [i.e., the node at position (3.5; 2)], obtained analytically and through simulations.

In summary, the PFS algorithm can improve the performance in various Rayleigh fading scenarios.

Summary
The presented PFS methodology is a simple and mathematically elegant solution applicable to both practical and theoretical experiments. The theoretical results

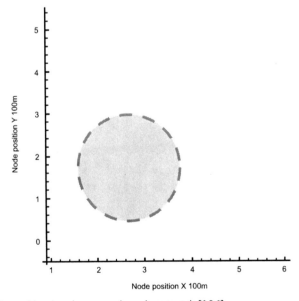

Figure 3.45   Node position in a hexagonal mesh network [106].

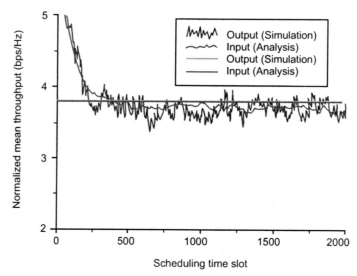

Figure 3.46 Mean throughput of node 16 in a hexagonal mesh network [106].

and findings provide guidelines and an analytical support for the system design, simulation-based modeling, and the performance analysis of the PFS algorithm in the context of a cross-layer design.

The derived mean and variance of the PFS throughput shows that it may be possible to model it as a simple distribution (e.g., Gaussian) and still obtain realistic results.

3.3.2 Cross-Layer Framework for Wireless Mesh Networks

The FP6 IST project MEMBRANE developed a fully distributed, cross-layer framework, comprising MAC and network layer functionalities, which can guarantee QoS to a wide range of applications in a wireless mesh network. The cross-layer framework described here incorporates the novel opportunistic scheduling algorithms described in Section 3.3.1 and a novel QoS routing algorithm (see Chapter 2 of this book and [60, 61]).

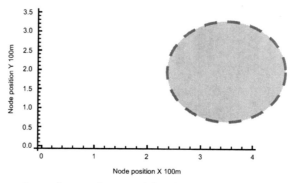

Figure 3.47 A layout of a random mesh network [106].

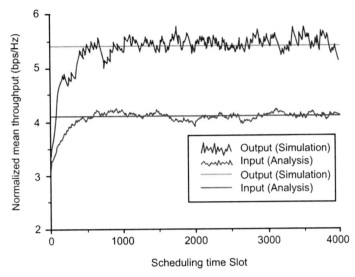

Figure 3.48   Mean throughput in a random mesh network [106].

In wireless mesh networks the long term channel prediction and resource reservation is an extremely difficult task. In order to overcome this challenge, a *soft resource reservation* scheme that is triggered by the routing algorithm can be introduced, while the MAC scheduler is used to guarantee the required resources [106].

The distributed opportunistic scheduling algorithm proposed for the cross-layer framework brings in the advantages of multiuser diversity gain while it maintains a strong temporal correlation for the interference, without which the channel quality and the interference cannot be tracked and predicted with reasonable accuracy [106]. A *multiobjective utility* function is introduced to jointly optimize the multiple constraints required by the overlaying application. This approach guarantees end-to-end QoS for multiple constraints, while at the same time allows for the optimization of the overall network performance by avoiding creating bottlenecks that can negatively affect the network connectivity.

In order to find the optimum route that satisfies multiple QoS constrains, [106] defined a *dissatisfaction ratio R*, for each of the QoS requirements, as the ratio between the expected metric values and the value defined by the QoS requirements. For example, in a wireless mesh network, each wireless mesh router independently generates data sessions/flows. Each QoS flow with *flow index q* has to fulfill a set of QoS constraints that includes the *end-to-end* (ETE) *packet delay* D_q^r, the throughput T_q^r, and the *packet error rate* (PER) E_q^r. This set can be denoted as (D_q^r, T_q^r, E_q^r). A route Ω_{st}^k from a source wireless mesh router with index s to a destination Internet gateway (IGW) indexed t within the route set Ω_{st}^k is concatenated by a set of links $\{(v_i, v_j)\}$ for all $v_i, v_j \in V_R \cup V_G$, where V_R is the set of wireless mesh routers and V_G is the set of IGWs.

The route from s to t can be formulated as:

$$\Omega_{st}^k = \left\{(v_i, v_j) \mid \forall v_i, v_j \in V_R \cup V_G\right\} \tag{3.95}$$

where the total m candidate routes exist.

Then, the ETE packet delay dissatisfaction ratio $R_k{}^D$ for route $\Omega_{st}{}^k$ is defined as the actual delay measurement over the delay requirement. On the other hand, the throughput dissatisfaction ratio is formulated as the ratio between the throughput requirement $T_q{}^r$ and the actual and actual bottleneck link throughput; the minimum of all link throughputs along route $\Omega_{st}{}^k$. Finally, the dissatisfaction ratio for the PER, $R_k{}^E$, is defined as the multiplication of all one-hop PER over the PER requirement $E_q{}^r$ since this is a multiplicative constraint. Since a session has to fulfil the set of QoS requirements, a source-to-gateway route will be feasible if and only if all defined ratios of relevant constraints are less than one. In order to efficiently cope with the above-mentioned coexisting QoS flows with different relevant requirements in terms of delay, throughput, and PER, the *indication* function $I_D{}^q$, $I_T{}^q$, $I_E{}^q$ is introduced for each QoS constraint, where:

$$I_a^q = \begin{cases} 1 & \text{if requirement } a \text{ is required for flow } q \\ 0 & \text{elsewhere} \end{cases} \tag{3.96}$$

In order to find the optimal route for different sessions, another set of *resource reservation factors* is introduced as β_D, β_T, and β_E, respectively for the delay, throughput, and PER. These are the factors that provide the *cross-layer resource allocation interfaces* between the network and MAC layers.

Equation (3.97) introduces a multiobjective function composed from the utility functions in order to find the optimal route for session q in a heuristic way and by taking into account the multiple QoS constraints simultaneously [106].

$$\min_{\forall \Omega_{st}^k \in \Omega_{st}} \max\left[U_k^D, U_k^T, U_k^E\right]$$
$$= \min_{\forall \Omega_{st}^k \in \Omega_{st}} \max\left[I_D^q R_k^D, I_T^q R_k^T, I_E^q R_k^E\right] \tag{3.97}$$

subject to:

$$\begin{cases} R_k^D(q) = \dfrac{\sum_{(i,j)\in\Omega_{st}^k} D_{ij}^a}{(1-\beta_D)D_q^r} \leq 1 \\[4mm] R_k^T(q) = \dfrac{(1+\beta_T)T_q^r}{\min_{(i,j)\in\Omega_{st}^k} T_{ij}^a} \leq 1 \\[4mm] R_k^E(q) = \dfrac{1-\prod_{(i,j)\in\Omega_{st}^k}(1-E_{ij}^a)}{(1-\beta_E)E_q^r} \leq 1 \end{cases} \tag{3.98}$$

Equation (3.98) is a *distributed optimization* function that runs on each node, where a "min-max" operator takes the minimum dissatisfactory ratio among the set of all possible routes Ω_{st} from a source wireless mesh router to a certain IGW.

A distributed opportunistic scheduling metric with long-term throughput satisfaction can be defined as:

$$U_{ij} = a_{ij}^r \frac{\rho_{ij}}{C_{ij}} \tag{3.99}$$

where a_{ij}^r is the QoS throughput requirement for link (v_i, v_j) from routing perspective, ρ_{ij}^r and C_{ij}^r are the instantaneous data rate and link throughput in the long run, respectively. Equation (3.99) is the algorithm interface for an interface for cross-layer QoS routing and distributed scheduling framework with resource reservation and allocation [106].

Let B_{ij}^r denote the amount of *bandwidth resources* that need to be reserved for session q in the link (v_i, v_j). If B_{ij} is the set of resource reservation requests related to link (v_i, v_j) for different QoS flows, and the routing demand a_{ij}^r can be related to B_{ij} in the network layer as:

$$a_{ij}^r = \sum\nolimits_{\forall q, \, if \, (v_i, v_j) \in P_q} B_{ij}^q \leq C_{ij} \tag{3.100}$$

where P_q is the final route chosen for the QoS session q, which makes sure that the total bandwidth requests do not exceed the channel capacity of link (v_i, v_j).

The cross-layer performance is assessed by a *slotted, time-driven* simulation platform [106]. A number of wireless mesh routers and IGWs are randomly and independently deployed on a rectangular two-dimensional space. Sessions are generated according to a Poisson process. Each session has to fulfil three QoS constraints, i.e., ETE packet delay, throughput, and end-to-end PER. The ETE packet delay consists of the queuing delay, the transmission delay, and the circuit processing delay. The circuit processing delay could be neglected. The Jake's Model is used for the wireless channel representation, while the required PER is derived based on the SINR curves for the used adaptive modulation and coding scheme. Each WMR is equipped with directional antennas, while accurate positioning was assumed.

At given time t, the receiving SINR γ_{ij} for the link (v_i, v_j) is calculated as:

$$\gamma_{ij} = \frac{P_{ij}^{TX} C_{ij} d_{ij}^{-\alpha}}{\sum_k P_{kj}^{TX} C_{kj} d_{kj}^{-\alpha} + N_0} \tag{3.101}$$

where P_{ij}^{TX}, C_{ij}, and $d_{ij}^{-\alpha}$ are the transmission power, the channel gain (the antenna gain has been also included here), and the path loss between link (v_i, v_j), respectively. The typical value for the path loss coefficient α is 3.5. N_0 is the single-sided power spectrum density for *the* additive white Gaussian noise. A retransmission scheme is assumed in the case of packet loss.

Each wireless mesh router is equipped with directional antennas and assumes accurate positioning.

The simulation topology for fifteen wireless mesh routers is shown in Figure 3.49.

The performance of the proposed cross-layer framework is investigated in terms of the *gateway throughput* and the probability of *successful end-to-end packet reception* as a function of the *packet interarrival time*. The probability of success is defined as the ratio of the packets that arrive at the gateway and have fulfilled the delay QoS requirements throughout the whole route. The modified *round robin*

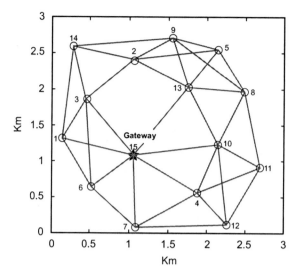

Figure 3.49 Simulation topology with 15 wireless mesh routers, of which one is the IGW [106].

scheduler proposed in Section 3.3.1.1 and the AODV routing protocol (see Chapter 2 of this book) are used as benchmarks and for four different combinations [106].

The opportunistic scheduler can guarantee high throughput even for small inter-arrival rates when the offered network traffic is getting high. This is shown in Figure 3.50 in a comparison of the scheme proposed here (i.e., IQoSR) with resource reservation and a distributed AODV-based scheme.

On the other hand the resource reservation scheme provides a constant throughput because the channel resources are reserved (at MAC level) independently of the offered traffic. The combination of opportunistic scheduling and the QoS routing scheme can satisfy the packet QoS requirements, while at the same time guarantees high throughput rate at the gateway side.

Figure 3.51 shows that the outage probability is less than 0.1 even for high traffic conditions.

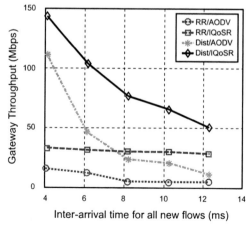

Figure 3.50 The effect of network traffic on the network throughput [106].

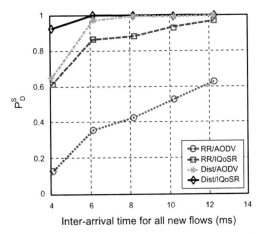

Figure 3.51   The effect of network traffic on probability of successful end-to-end packet reception in terms of end-to-end packet delay [106].

Figure 3.52 shows that the PER probability of the distributed IQoSR is worse compared to the three other schemes. Still, the dominant effect for the end-to-end packet delay is the queuing delay in each node's buffer, which gives a better overall delay performance.

Figure 3.53 shows the effect of the resource reservation factor β on different system performance parameters.

By increasing the resource reservation factor β, more bandwidth resources are reserved from the routing protocol in order to further secure the QoS constraints from the possible channel or interference fluctuations. However, this may result in a smaller admission control rate for the new incoming flows.

A higher β results in lower end-to-end delays and PER values. This is shown in Figures 3.54 and 3.55.

Although the PER curves with respect to the traffic load in Figure 3.55 for both the Dist/IQoSR and Dist/AODV combinations are always better than the other two using round robin scheduler, it also shows the increase of PER values and decrease

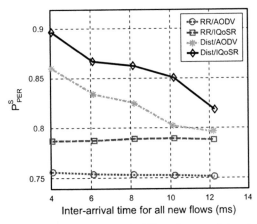

Figure 3.52   The effect of network traffic on probability of successful end-to-end packet reception in terms of PER [106].

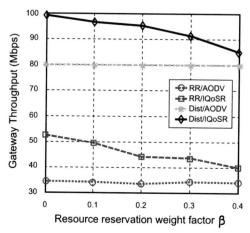

Figure 3.53   The effect of the resource reservation factor β on network throughput [106].

of satisfactory ratios. This is because the distributed opportunistic scheduler loses multiuser diversity gain when the traffic load is relatively low. In other words, this scheduler is not smart enough to stop the scheduling packet transmissions when channel quality is poor, thus resulting in lower receiving SINRs as shown in Figure 3.56.

In other words, this scheduler is not smart enough to stop the scheduling packet transmissions when channel quality is poor, thus lower receiving SINR trend could also be expected. Nevertheless, the PER satisfactory ratio for the Dist/IQoSR combination is always higher, which means that the algorithm can successfully meet the PER requirements.

In summary, the joint optimization of multiple constraints with the overall network performance can avoid creating bottlenecks that can negatively affect the network connectivity and consequently increases the capacity of the network. Further,

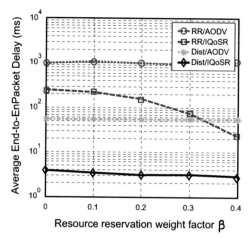

Figure 3.54   The effect of the resource reservation factor β on probability of successful end-to-end packet reception in terms of end-to-end packet delay.

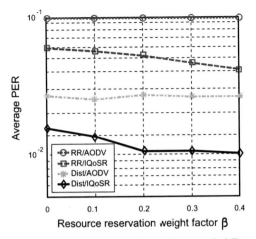

Figure 3.55   The effect of the resource reservation factor β on probability of successful end-to-end packet reception in terms of PER.

a fully distributed approach can comply with the scalability requirements and does not necessary have to be complex.

Further performance enhancements can be achieved by use of optimized and distributed power control algorithms (e.g., by use of non-cooperative game theory approaches) [106].

3.4   Capacity Enhancements in WPANs

For effective peer-to-peer connectivity in the presence of two independent piconets (clusters), each device in one piconet should be able to communicate with any other device in the other piconet [122]. The *high data rate* (HDR) WPAN standard [123] provides a parent-child communication model for the purpose. One limitation in the parent-child model to enable all the devices in one piconet to communicate with the devices in the other piconet result in scheduling issues. The FP6 IST project

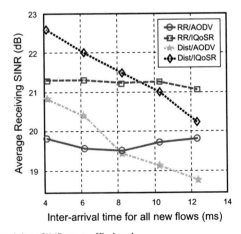

Figure 3.56   Average receiving SINR vs. traffic load

MAGNET Beyond analyzed the inter-PAN communication problem in the scope of the optimized low data rate (LDR) and HDR) WPAN-based air interfaces that were in the core of the proposed within the project concept of global PAN communications [8].

WPANs span a limited operating space consisting of low-cost and low-power devices. There are different standards for WPANs depending on the type of air interface supported (e.g., Bluetooth, UWB, and the data rate (low with extended range in IEEE 802.15.4 and high with limited range in IEEE 802.15.3). HDR WPANs operate in a communication system called a *piconet*. Each piconet has a central coordinator, which performs the administrative functions. The central coordinator is called the *piconet coordinator* (PNC). The PNC sends periodic beacons to provide timing information for the member devices. To communicate with devices in a piconet and access its resources, a device has to associate with the PNC by sending an association request command. The PNC then allocates a *Device ID* (DEVID) to the device upon a successful association and informs the other member devices about the newly associated device and its advertised capabilities. The communication takes place in a superframe, which consists of a beacon, a *contention access period* (CAP) and a *channel time allocation period* (CTAP) as shown in Figure 3.57.

The CAP is CSMA/CA-based and it can be used for sending association requests and other commands if allowed by the PNC. Small amounts of asynchronous data can also be sent in the CAP by the member devices. If the devices require channel time on a periodic basis, they can request the PNC for the channel time in the CTAP, which is TDMA-based. The *channel time allocation* (CTA) request can be for either a subrate allocation (CTAs in alternate superframes) or a super rate allocation (multiple CTAs in the same superframe) depending on the type of traffic the device wants to send and its constraints (e.g., frame interarrival time). The communication in a piconet is single-hop, therefore, to extend the range of the piconet the IEEE 802.15.3 [123] standard allows the formation of *child piconets*, which are dependent on the channel time from the established parent piconet. The devices in the *parent piconet* and the parent PNC can communicate with the child PNC and vice versa. The limitation in this extension is that the devices in the child piconet cannot communicate with the devices in the parent piconet and the parent PNC. Since the devices communicate with each other through a single-hop link, in case of bad channel conditions, the devices have to reduce their data rates. If the CTA rate

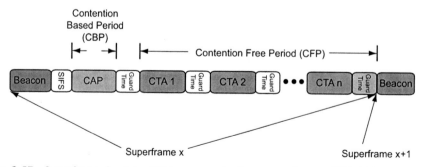

Figure 3.57 Superframe for inter-PAN communication according to IEEE 802.15.3 standard [122].

allocation is the same, then by lowering the data rates, the duration of each CTA has to be increased, which reduces the number of devices that can be supported by the superframe. The reason is the increased duration of the CTAs for devices with more than one CTA per superframe, which shall occupy more time slots and, hence, decrease the capacity for other devices [122]. If most of the devices are suffering due to bad channel conditions, then the PNC has an option to change the channel. If two devices are at the two extreme ends of the piconet, then they have to transmit at a relatively higher transmission power to keep the required SNR. Transmitting at a higher power results in increased energy use, which is not beneficial for energy constrained devices [122].

3.4.1 Parent-Child Communication Model and Its Limitations

The HDR WPAN standard, 802.15.3 defines the transmitter data rates (∂_T) of 11, 22, 33, 44, and 55 Mbps [123]. The beacon and the MAC headers of the frames are sent at the base rate of 22 Mbps and the rest of the payload at any of the desired values of ∂_T. Since the CTAP is TDMA-based, it is not possible to achieve the defined ∂_T. Therefore, the throughput achieved at the MAC layer is always much less than the transmitter data rate at the PHY layer. If a device wants to send small amounts of asynchronous data in a single CTA, then it can transmit and achieve the defined data rates. For isochronous transmission, the requirement is to allocate more than one CTA per superframe (depending on the tolerable inter arrival delay) for the device. The achievable actual data rate (∂_A) is always less than the ∂_T and depends on certain factors such as number of CTAs allocated to the device, number of frames sent in each CTA and the time duration of each CTA (which depends on the required data rate). The number of devices in a piconet influences the decision of the PNC to allocate a particular number of CTAs to a device to ensure fair allocation. Theoretically, there can be 256 devices supported by the PNC in a piconet. Since some of the Device IDs (DEVIDs) are reserved for special purposes, the maximum number of devices that a single PNC can support as allowed by the 802.15.3 standard, is 243. The practical number of devices that a single PNC can support is, however, much lower than 243 if multimedia transfers are taking place between most of the devices. The increased number of devices also imposes additional processing overhead on the PNC.

Though the administration of the *child piconets* is done autonomously by a child PNC, the channel time is provided by the parent PNC from its transmitted superframe through a private CTA. that the time period in the superframe of a child piconet after the private CTA is reserved till the start of a subsequent private CTA in another superframe of the parent piconet. This is to keep synchronized with the time allocated by the parent PNC to the child PNC.

Figure 3.58 shows an issue related to the child superframe about the allocation of super rate CTAs for isochronous streams with strict delay constraints. If the reserved time after the private CTA allocated to the child PNC exceeds the maximum tolerable delay for most of the real time applications, then it is not possible for the child piconet to support them. If the reserved time is decreased by increasing the time allocation of the private CTA, it can disturb the CTA allocations in the parent piconet.

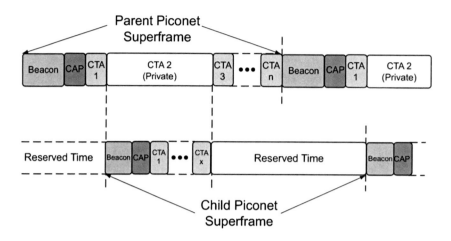

Figure 3.58 Child superframe time allocation [122].

3.4.1.1 Scheduling Problems

Considering the above limitations, it is difficult in a parent-child relationship to maintain QoS for certain multimedia applications, and especially voice. This can be a particular problem if the formation of a child piconet within a child piconet is considered [122]. This problem is shown in Figure 3.59.

If η_{PS} is the duration of the parent piconet superframe and η_{C1S} is the duration of the superframe for the level 1 child piconet, then

$$\eta_{C1S} = \eta_{PCTA1} + \eta_{RSVD} = \eta_{PS}$$

$$\eta_{PCTA1} = k_1 \, \eta_{PS} \qquad (0 < k_1 < 1) \qquad (3.102)$$

where η_{PCTA1} is the duration of a private CTA allocated to the level 1 child PNC and η_{RSVD} is duration of the reserved time in η_{C1S} for η_{PS} till the start of next successive private CTA for the level 1 child PNC. Similarly for the level 2 child piconet which is formed within the level 1 child piconet as follows:

$$\eta_{C2S} = k_2 \, \eta_{PCTA1} + \eta_{RSVD1} = \eta_{PS}$$

$$\eta_{PCTA2} = k_2 \, \eta_{PCTA1} \qquad (0 < k_2 < 1) \qquad (3.103)$$

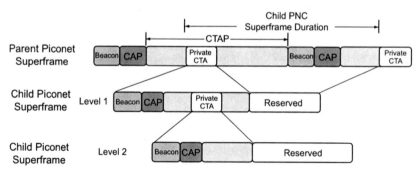

Figure 3.59 Time allocation for hierarchial child piconets [122].

where η_{C2S} is the duration of level 2 child piconet superframe and η_{PCTA2} is the duration of a private CTA allocated to the level 2 child PNC by the level 1 child PNC. The value of k_2 is less than 1 to indicate that $\eta_{PCTA2} < \eta_{PCTA1}$. The number of super rate CTAs allocated to a device which is sending a real-time traffic depends on the maximum tolerable delay and jitter for that particular traffic type and the required data rate ∂_R. Since the super rate CTAs are evenly spread throughout the superframe, the duration of a private CTA allocated to a child PNC is a significant factor to determine if the parent and child piconet can support a particular real time traffic type with specific requirements of maximum tolerable delay and jitter. If X_{MTD} denotes the value of maximum tolerable delay and jitter for a particular real time traffic type, then the superframe can be split into logical partitions to make time allocations easier. The smallest partition size is taken to be equivalent of the strictest requirement for delay and jitter, which according to [124] is for voice (<10 ms). If $X_{MTD(\min)}$ denotes the minimum compulsory logical partition size for the superframe, then:

$$X_{MTD} = \text{n } X_{MTD(\min)} \tag{3.104}$$

where n can be any positive integer. The value of n shall be set to 1 to indicate voice applications. The number of logical partitions can be found out by:

$$\text{Number of Partitions } (N_p) = \frac{\eta_{PS} - (\tau_{PNC} + \eta_{CAP})}{X_{MTD(\min)}} \tag{3.105}$$

where τ_{PNC} is the beacon overhead and η_{CAP} is the CAP duration. The expression $\eta_{PS} - (\tau_{PNC} + \eta_{CAP})$ gives us the CTAP duration, which is divided by $X_{MTD(\min)}$ into a number of partitions. If the value of $X_{MTD(\min)}$ is taken to be 8 ms, then the superframe is split into 8 partitions, each of approximately 8 ms. Once the superframe is partitioned, the time can be allocated much more easily for real-time applications keeping the boundaries of the logical partitions into consideration. The time allocation for a private CTA should be done very carefully because it can have a significant effect on the isochronous streams with super rate CTA allocations [122].

If $\eta_{RSVD} > X_{MTD(min)}$ and $\eta_{PCTA1} > X_{MTD(min)}$, then, both the parent piconet and the child piconet, cannot support voice applications as required. If $\eta_{RSVD} < X_{MTD(min)}$ and $\eta_{PCTA1} > X_{MTD(min)}$, then the child piconet can support voice applications but the parent piconet cannot. In order for both the parent piconet and child piconet to support voice applications (these applications have the strictest upper limit on the tolerable delay and jitter), the following two conditions must be true:

$$\eta_{RSVD} < X_{MTD(\min)} \text{ and } \eta_{PCTA1} < X_{MTD(\min)} \tag{3.106}$$

It can be shown that the above two conditions cannot be true at the same time. Since $\eta_{RSVD} < X_{MTD(\min)}$ and $\eta_{PS} = \eta_{C1S}$, with $\eta_{PCTA1} + \eta_{RSVD} = \eta_{PS}$. If η_{RSVD} is assumed to be equal to $X_{MTD(\min)}$, then $\eta_{PCTA1} = \eta_{PS} - X_{MTD(\min)}$. This means that $\eta_{PCTA1} = N_p - 1$ and thus takes the major portion of the parent superframe. Therefore, the parent piconet cannot support voice applications. The same theory can be applied to other traffic types as well. This shows that when the level 1 child piconet cannot support voice applications, there is no possibility for a level 2 child piconet or above to support multimedia applications [122]. The increase or decrease in ∂_T determines the length of the CTA required to send a particular type of traffic. With higher values of ∂_T, the overhead per CTA increases but the capacity of superframe also increases due to the reduced size of CTAs required by devices.

IEEE 802.15.3a [123] defines an alternate PHY layer based on UWB to achieve much higher data rates using the same MAC layer of 802.15.3. Higher data rates (in Gbps) are proposed in IEEE 802.15.3c [125] for the 60-GHz frequency band. Although, by using much higher data rates, the capacity of the superframe is increased and much smaller CTA durations can be used using the frame aggregation, the spacing of the super rate CTAs depending on the factor $X_{MTD(min)}$ does not change [122].

3.4.1.2   Inter-PAN Communication Model

The inter-PAN communication process has to address the following issues:

- Seamless merging of two or more piconets;
- Seamless splitting of two or more piconets;
- User's PAN device identity is not lost during the merging and splitting process;
- All devices in each and every piconet are able to communicate with one another directly, provided that they are in the transmission range of each other;
- The modifications should take into consideration the MAC reserved fields that are available in the IEEE 802.15.3 standard;
- The scheduling issues proposed in case of the parent-child model should be resolved so that the QoS for real time applications is not affected.

When two piconets merge for the purpose of inter-PAN communication, the intra-PAN piconet association and communication should not be disrupted. When the piconet splits, the inter-PAN communication should end tidily and not abruptly, and intra-PAN communication should not be disrupted. Each piconet that merge and split must be able to maintain its current association with its own piconet PNC. All devices in the inter-PAN communication should be able to communicate with one another directly; however, channel access can be monitored by the PNC. The transmission range of the devices limits the range, in which a device can communicate between piconets.

All modifications to support inter-PAN communication proposed in [122] considered the reserved fields in the IEEE802.15.3 MAC layer only. The proposed modification could be appended onto the child piconet that was already part of the standard.

The process of inter-PAN communication is first initiated by the discovery of an existing piconet through the *MLME-SCAN.request primitive* as shown in Figure 3.60.

The passive scanning request is carried out by the PNC or a device in a piconet. The PNC may allocate a CTA such that there is an unallocated channel in the

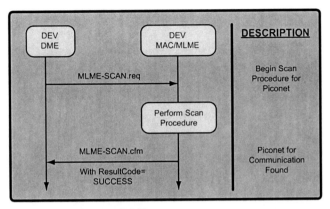

Figure 3.60   Piconet scan initialization [122].

CTAP, which provides quiet time for the PNC or a device to scan channels for other 802.15.3 piconets. When the PNC carries out the piconet scanning, it goes into a *silent* mode, where it shall cease the piconet beacon transmission for one or more beacon intervals. It is not allowed to suspend the beacon transmission, however, for more than twice *aMinChannelScan* [123]. If the device is doing the scanning, then the PNC will make the request to the device using the *MLME-REMOTE-SCAN. request*.

When the desired piconet for the communication is found, the DME of the PNC will initiate the *MLME-SYNC.request* and receive an *MLME-SYNC.confirm*. Once completed, the PNC can begin associating itself with the new piconet using the *MLME-ASSOCIATE.request* primitive. Since the PNC is about to associate with a new piconet it is referred as a device to the other PNC for descriptive purposes. The association process between the device (PNC of an established piconet) and PNC is shown in Figure 3.61.

The inter-PAN device that has associated with another piconet is referred to as *interdevice*. As soon as the device (PNC) of one piconet associates itself with the PNC of the different piconet, it can request the formation of a dependant child piconet. This process is triggered by the MAC layer using the *MLME-START-DE-PENDANT.request* primitive. The device shall send a *channel time request* command to request a pseudostatic private CTA. In a private CTA, the SrcID and the DestID are identical. The device shall set the SrcID and TrgtID fields in the Channel Time Request command to the DEVID of the originating device, the Stream Index field to zero and the PM CTRq Type field to ACTIVE. The PNC will then recognise that this is a request for a child piconet. If the PNC rejects the formation of a child PNC for any reason such as insufficient channel time or unable to allocate a pseudo-static CTA, it shall send a Channel Time Response command with the Reason Code field set to *request denied*. In this case, inter-PAN communication is not possible and the device should dissociate from the current piconet and return to its own piconet. If the device receives a private CTA from the PNC, the device DME configures the child PNC parameters using the *MLME-START-DEPENDANT.request* and confirm primitives.

Before the child PNC can begin transmitting its own beacon, it should return to its existing piconet channel and initiate moving the current channel to the newly

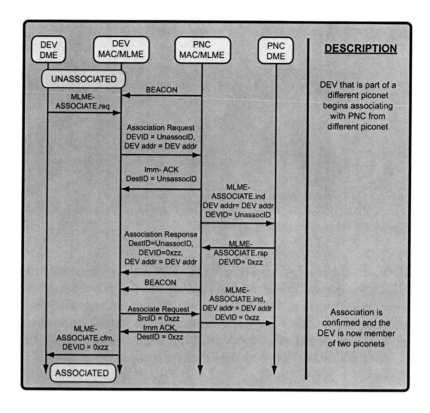

Figure 3.61   Association procedure [122].

allocated child piconet channel. The PNC will broadcast the *Piconet Parameter Change* Information Element with the change type set to CHANNEL in its current channel via its beacon for *NbrOfChangeBeacons* consecutive beacons. The Piconet Parameter Change IE shall contain the channel index of the new channel to which the PNC will be moving the piconet, and the *Change Beacon Number* field that contains the beacon number of the first beacon with a beacon number equal to Change Beacon Number field in the previous Piconet Parameter Change IEs. The device receiving this message shall change from their current channel to the new channel before the first expected beacon on the new channel. The devices shall not transmit on the new channel until a beacon has been correctly received on the new channel. To enable every device in the child piconet and parent piconet to communicate with one another, all members of the child and parent piconet should associate with one another.

A new command frame called *Inter-PAN Association Request* is created for this purpose and the process is shown in Figure 3.62.

The command frame is sent by either the child or parent PNC or both PNC to its members. This new command frame has a type value '011' which indicates that it is a *command* frame. The PNID is set to PNID of the originating piconet. The SrcID is set to the PNC's DEVID and the DestID is either set to BcastID if the PNC requires all its members to Inter-PAN Associate or to individual DEVID if requires only a specific device to associate. The ACK policy bit is set to '01' Immediate

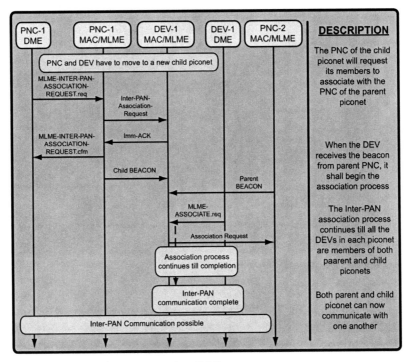

Figure 3.62   Inter-PAN association procedure [122].

Acknowledgement (Imm-ACK). The Inter-PAN Association Request MAC Frame payload will have the following fields:

- Inter PAN BSID (6-32 Octets);
- Inter PAN PNC Address (8 Octets).

The Inter PAN BSID and Inter PAN PNC Address are set to the target piconet address that the PNC requires its devices to associate. Upon receiving this command, the device(s) will begin listening for the specific beacon with the PAN BSID and PNC Address and begin the process of associating with the new piconet. The association process is similar to that described in the standard. Once the association process is successful, the devices that are member of both piconets can now communicate with one another using similar protocols defined in the standard. If the piconet to which a device is associating does not support inter-PAN communication, a new *reason code* is created within the *association response* message called *Inter-PAN communication not supported*. The reason code will use one of the reserved fields that are available in the *association response* fields.

This process of disassociation is an extension to the way, in which a dependant piconet ends its relationship with the PNC. Since devices in each piconet are potentially associated to more than one piconet, modifications are necessary so that both piconets split seamlessly. Either the child or parent PNC should send a new command frame, *Piconet-Splitting-Request* to the PNC of the other inter-PAN requesting to split from one another. This new command frame has a type value '011' which indicates that it is a Command frame. This process is described in Figure 3.63.

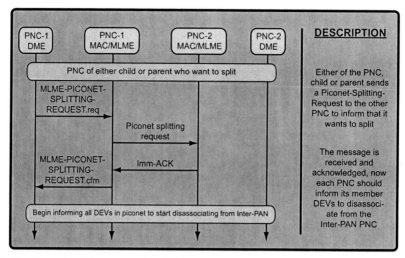

Figure 3.63 Piconet splitting procedure [122].

The PNID is set to PNID of the originating piconet. The SrcID is set to the PNC's DEVID of the originating piconet and the DestID is set to PNC's DEVID of the destination piconet. The ACK policy bit is set to '01' Immediate Acknowledgement (Imm-ACK). The MAC Frame payload is empty. Upon receiving Piconet Splitting Request command frame, both PNCs should begin informing their devices to disassociate themselves from the inter-PAN associated piconets. The new command frame is called *Force-Inter-PAN-Disassociation-Request* as shown in Figure 3.64.

The term '*forced*' is used because it is the PNC requesting its devices to dissociate instead of the devices. This new command frame has a type value '011' which indicates that it is a Command frame. The PNID is set to PNID of the originating piconet. The SrcID is set to the PNC's DEVID of the originating piconet and the DestID is set to BcastID of its member piconet. The ACK policy bit is set to '01' Immediate Acknowledgement (Imm-ACK).

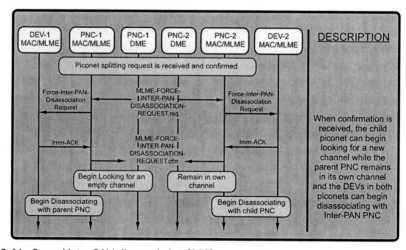

Figure 3.64 Forced inter-PAN disassociation [122].

The MAC Command frame will have the following fields:

- Inter-PAN BSID (6-32 Octets);
- Inter-PAN PNC Address (8 Octets);
- Mass Forced Disassociation (1 bit).

The Inter-PAN BSID and Inter-PAN PNC Addresses are set to the required piconet address that the PNC requires its devices to dissociate from. Both these fields are variable in size depending on the number of piconets that the PNC is requesting its members to dissociate from. The Mass Forced Disassociation bit is normally set if the PNC requires its devices to dissociate from every single inter-PAN that they are currently associated with.

When devices receive the Force-Inter-PAN-Disassociation Request message, they should initiate the Disassociation process with the given piconet addresses as defined in the standard. This is shown in Figure 3.65.

A new Reason Code in the Disassociation Request called *Inter-PAN Split* is created for the new piconet splitting procedure. The Reason Code will use one of the reserved fields that are available in the Disassociation Request fields. The parent piconet will remain in its own channel once the piconet splitting request is initiated while all child piconets would have to shut down or move to a different channel. If the child piconet decides to maintain its piconet, it shall begin scanning for a new channel to move its network. The scanning process is similar to that described in the standard.

3.4.2 Parent-Child Scheduling Solution

Some scheduling algorithms were proposed for HDR WPANs [126–128] with the focus mainly on supporting VBR MPEG streams without detailing the superframe utilization efficiency or about the parameters to support QoS.

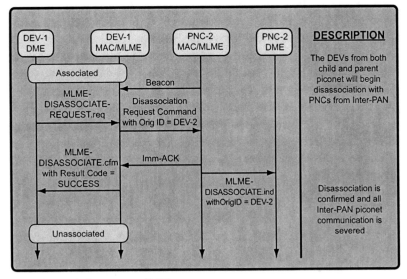

Figure 3.65 Disassociation process [122].

When one of the PNCs associates with the parent PNC as the child PNC, it sends a channel time request command to the parent PNC for channel time. The source DEVID and the destination DEVID are the same in the channel time request command so that the parent PNC can determine that it is a request for the private CTA from the child PNC. Upon reception of the channel time request command, if there is enough capacity in the superframe of the parent PNC, it shall accept the request of the child PNC and send a channel time response command. If no device in either the parent piconet or the child piconet is supporting any real time traffic with a particular value of $X_{MTD(min)}$, then the parent PNC can allocate a single private CTA to the child PNC. However, if either a device in the parent piconet or the child piconet or in both the piconets intends to request channel time for real time traffic with a certain value of $X_{MTD(min)}$, then if the parent PNC allocates a single private CTA to the child PNC, then the QoS for the device in the parent piconet and child piconet can be affected as explained. Since the upper limit on the tolerable delay and jitter for voice applications are the strictest, the CTAP of the superframe is partitioned into equal sized slots called Medium Access Slots (MASs).

The concept of dividing the superframe into MASs has been defined in [129], however, an appropriate size was not specified. The size of the MAS is defined in the proposed here solution to be 8 ms, which allows for QoS support of voice applications [122]. If the maximum size of the superframe is considered (i.e., 65,535 µs), there has to be at least 8 CTAs per superframe to support the voice applications. Therefore, the value of Np is 8. Because the QoS requirements of video are more relaxed than those for voice [123], the CTA rate factor for video traffic can be in factors of 2 per superframe according to the throughput requirements and the available capacity in the superframe. The proposed structure of the superframe when inter-PAN communication is considered is shown in Figure 3.66.

The parent and child PNCs send their beacon in the *beacon period* (BP). The BP can be extended in the presence of multiple piconets and more than two beacons can be sent in it. A single CAP is shared between the parent and child piconets for simplicity so that the inter PAN association requests by the devices from either the parent PNC or the child PNC can be sent in it. When the parent PNC receives a request for a private CTA from the child PNC, it checks the requested CTA duration η_{CTA-R} and compares it with the available time in all of the 8 MAS durations (η_{MAS}). If $\eta_{CTA-R} < \eta_{(MAS-A)i}$ (where $\eta_{(MAS-A)}$ is the available time in a MAS and the index i indicates the MAS number and $1 \le i \le 8$), then the parent PNC can accept the channel time request from the child PNC. If there are devices in the child piconet which intend to request time for voice traffic, the parent PNC shall allocate 8 private CTAs to the child PNC spread evenly throughput the 8 MASs in the superframe. In this way, the QoS can be supported for devices in both the parent and child piconets subject to available capacity. For video traffic, the parent or child PNC shall allocate

Figure 3.66 Superframe sharing in Inter-PAN communication [122].

CTAs to requesting devices in factors of 2 depending on the available capacity and throughput requirement specified in the request. The number of child piconets that can be supported depends on the *superframe capacity*.

3.4.3   Capacity Analysis

In order to undertake the capacity analysis for inter-PAN communication, a simulation model was developed in an OPNET Modeller ver 14.0 [122]. OPNET is a discrete-event simulator, which offers a modular approach to develop and simulate wireless network protocols. Because the HDR MAC requires TDMA for the isochronous stream allocations, all the timing related functions were implemented through self-timed interrupts. Upon a self-timed interrupt, the simulation kernel goes to the appropriate state and executes the required functions at that particular state.

3.4.3.1   Simulation Model

The inter-PAN communication was modeled by two-state-based process models. One model was developed according to [123] for the device with the necessary functionalities to communicate with the PNC and to send traffic to other peer devices. The second model was developed to have the functionalities of the PNC.

The state based process model for an IEEE 802.15.3 device is shown in Figure 3.67. In order to support the necessary modulation schemes as defined in [123], changes were made to the pipelines stages of OPNET so that the modulation scheme could change dynamically with the change in the transmission data rate [122]. The pipeline stages in OPNET refer to the PHY layer used for a particular wireless com-

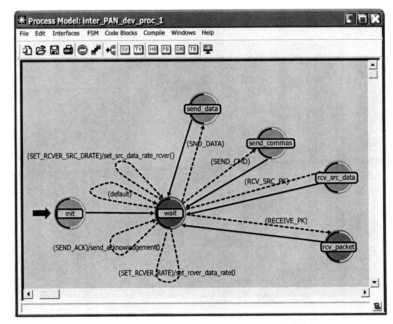

Figure 3.67   State-based process model of a PAN device in OPNET [122].

munication model. Because OPNET does not support the odd bit constellations of QAM (i.e., 32 QAM, which is required for the data rate of 33 Mbps), the bit error curve for 32 QAM was imported from MATLAB. OPNET uses bit error curves (BER versus E_b/N_o) to emulate a particular modulation scheme.

Upon initialization, a device waits for a beacon from its PNC. After the reception of the beacon, the device associates with the PNC. The piconets are identified on the basis of their *piconet identifiers* (PNIDs). Therefore, if a device receives a beacon or any other frame, which has a different PNID, it shall reject that frame. In case of inter-PAN communication, when one of the PNC becomes the child PNC, it shall contain the parent piconet IE and, therefore, the devices in the child piconet shall know about the PNID of the parent PNC as well. The time to send the channel time request command is kept configurable for a device. Furthermore, the number of frames to be sent per CTA, the size of the MPDU, type of traffic, transmission data rate and the CTA rate factor are also kept user configurable so that they can be changed from simulation to simulation. Also these factors are very important in determining the superframe capacity, the CTA overhead, the CTA utilization, and the superframe utilization [122].

The PNC contains all the relevant functionalities for inter-PAN communication in accordance with [1]. The state-based model for the PNC is shown in Figure 3.68.

The flexibility provided in the HDR IEEE 802.15.3 MAC standard is used to implement the additional functions for the inter-PAN communication model. The PNC sends periodic beacons after an interval of 65535 μ s as the maximum superframe duration is considered for all the analysis. Upon the reception of a *channel time request* command from a member device, the PNC switches to the appropriate state to execute the functions required to check the superframe capacity and update the list of structures containing the timing information for individual member devices. When a PNC receives a frame, which either has a different PNID than its own

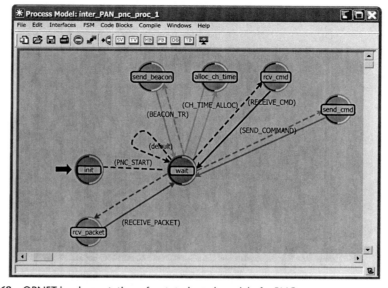

Figure 3.68 OPNET implementation of a state-based model of a PNC.

or when it receives a beacon from another PNC, it detects the presence of another piconet. After the identification of another piconet, one of the PNCs becomes the parent PNC and the other becomes a child PNC.

3.4.3.2   Inter-PAN PNC Selection Criteria

When a PNC receives a beacon from another PNC with a different PNID, it includes the PNC capabilities IE in its subsequent beacon frame [122]. The criteria for the PNC selection are given in [123] for PNC capable devices. For inter-PAN communication, where the capacity is a major issue, three extra parameters are defined and used in the simulation model apart from the ones mentioned in [123]. The three parameters are the number of supported child piconets, the number of active devices, the type of traffic communicated by the devices, and their CTA durations, and the PNID. The PNC, which already has dependant child piconets is given the preference. If none of the PNCs are already supporting child piconets, then the PNC with the higher number of active devices communicating in its piconet, is given the preference. If the number of devices is the same, then the PNC with more superframe utilization is given the preference. If none of the above are applicable, then the PNC with the higher PNID becomes the parent PNC. Both of the PNCs perform this comparison, therefore, the child PNC sends an association request to the parent PNC. The child PNC also informs its member devices and starts including the parent piconet IE in its beacon. The child piconet calculates the utilized total time in its superframe along with extra time (500 μs more per superframe in the simulation model) that it requires and sends the request to the parent PNC. If the child piconet has devices, which are communicating *voice* or *video* traffic, the parent PNC shall allocate eight private CTAs to the child PNC. If there is no device with voice or video traffic in the child piconet, and no device in the child piconet is capable or has any intention to send such traffic in the future, then the parent PNC shall allocate a single private CTA to the child PNC [122].

3.4.3.3   Capacity Analysis of HDR WPANs

Different multimedia codecs encode data at different rates with different data rate requirements. As a result, the size of the MSDU received from the higher layers to the MAC layer varies. If the size of the MSDU is larger than the largest MPDU size supported by the MAC layer, the MSDU has to be fragmented into smaller MPDUs. To simplify the fragmentation process, the MSDU is divided into equal size fragments (MPDUs). Through the DEV capabilities field in the DEV association IE, each device indicates its preferred fragment size and supported data rates to all the member devices in the piconet. If a specific application requires a data rate of x Mbps, then the MAC layer has to at least support a data rate of $(x + Layer \, Overhead)$ Mbps. The layer overhead can be calculated by considering the preamble added at the network layer and MAC layer as shown in Figure 3.69. Figure 3.69 shows that $\partial_A > \partial_E$ by an amount equal to the *Layer Overhead* in order to support the application.

To determine the efficiency of utilization of a CTA, it is mandatory to consider the particular ACK scheme, which is to be used. The ACK scheme in use uses a

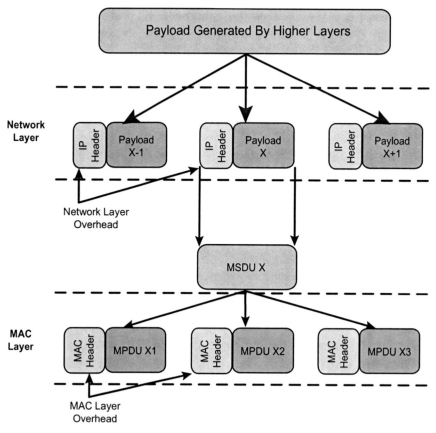

Figure 3.69 Overhead added at the network and MAC layer [122].

certain *interframe space* (IFS) duration between successive frames, which has an impact on the CTA overhead [122]. The impact of the IFS on the TA overhead is shown in Figure 3.70.

For the consideration of the voice and video traffic, the *delayed acknowledgement* (Del-ACK) scheme is used in the capacity analysis. When using the Del-ACK, either the *short interframe space* (SIFS) or the *minimum interframe space* (MIFS) can be used between successive frames. The CTA overhead when the SIFS is used between successive frames is given by:

$$\tau_{DACK(SIFS)} = \sum_{n=1}^{x_d+b}(\eta_{(SIFS)n} + b\eta_{DACK} + \eta_{GT}) \qquad (3.107)$$

where x_d is the number of frames sent in the CTA, η_{SIFS}, η_{DACK}, and η_{GT} are the duration of the SIFS, the time to send the Dly-ACK frame, and the guard time, respectively. The parameter b is set to 1 if there is a Dly-ACK frame in the CTA, otherwise, it is set to 0.

The total time allocated to each device (η_D) in the superframe can be given by:

$$\eta_D = \sum_{i=1}^{x}\eta_{(CTA)i} \qquad (3.108)$$

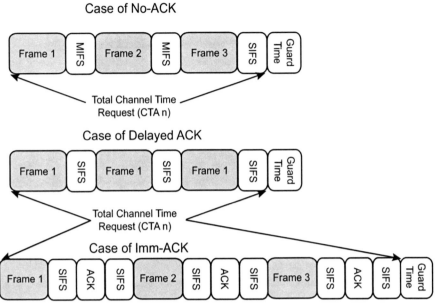

Figure 3.70   CTA structure in case of different ACK schemes [122].

where η_{CTA} is the duration of a single CTA allocated to the device and x is the total number of CTAs allocated to the device in the superframe. The actual data rate (∂_A) of the device can be given by:

$$\partial_A = \sum_{i=1}^{x} \eta_{(CTA)i} \tag{3.109}$$

The effective data rate (∂_E), at which the actual payload is delivered is given by:

$$\partial_E = \frac{\sum_{i=1}^{x} [\eta_{CTA} - \tau_{CTA}]_i}{\eta_S} \tag{3.110}$$

Where τ_{CTA} is the CTA overhead for each CTA, which includes the IFS durations, the ACKs, and the GT. Since the 802.15.3 MAC is a TDMA MAC, the following relation holds:

$$\partial_E < \partial_A < \partial_T \tag{3.111}$$

In order to make sure that the applications running on the devices are running smoothly during the communication, it should be made sure that the following condition is kept:

$$\partial_E \geq \partial_R \tag{3.112}$$

The capacity of the 802.15.3 superframe is calculated for voice and video traffic because these types of traffic consume the most of the networks resources. For voice, the G.711 codec is considered and for video, H.264 is considered [122].

Table 3.4 summarizes the parameters used in the capacity analysis for the voice traffic.

Table 3.4 shows that the IP overhead considered for the voice is 40 Octets and the number of the frames per second is 50. Therefore an additional 16 Kbps is required to send the IP layer overhead apart from the 64-Kbps data rate, which is for voice payload. The *MSDU size < MPDU size,* therefore, an MPDU size of 256 Octets is chosen to send the MSDU, which is 200 Octets in size. The MPDU size of 256 Octets also includes the *frame check sequence* (FCS), which is 4 Octets in length. While sending the MSDU, an additional overhead of MAC header has to be taken into account, also. Therefore, the total data rate that needs to be supported is 106.4 Kbps. The time required for the device per second can be calculated by:

$$\partial_R \times \frac{1}{\partial_T} \tag{3.113}$$

Since the superframe size considered is 65535 μs, there are 1/65535 μs super-frames in one second. Therefore, the time required per superframe for a device with a required data rate of 106.4 Kbps is:

$$\partial_R \times \frac{1}{\partial_T} \times \eta_S \tag{3.114}$$

The maximum tolerable delay and jitter for voice applications should be < 10 ms. Therefore, the available channel time (i.e., CTAP) is divided by 8 to limit the delay and jitter and maintain it under 10 ms. Figure 3.71 shows the PNC overhead and the ($\tau_{PNC} + \eta_{CAP}$) sum for 100 devices.

The variation in the overhead with increase in the number of devices is because of the additional information put into the CTA IE by the PNC. The PNC sends the beacon at a base rate of 22 Mbps; therefore, the overhead is the same for the devices operating at different transmitter data rates. When a device has been allocated a CTA by the PNC, then, depending on the number of frames sent in the CTA and the transmitter data rate, the superframe overhead (τ_S) increases or decreases. Although η_{MIFS}, η_{SIFS} and η_{GT} remain the same with the increase or decrease in the transmit-ted data rate, the time required to send the MPDUs (η_{MPDU}) within a CTA increases or decreases, respectively. Since with the increase in transmitted data rates, the ratio of (η_{MPDU}) to τ_{CTA} decreases, the overhead increases. This is shown in Figure 3.72, where the percentage superframe overhead is plotted against data rates of 22, 33, 44, and 55 Mbps with different number of frames sent in a CTA.

Table 3.4 Parameters for Voice Traffic [122]

S. No.	Parameters	Value
1.	Frame duration	20 ms
2.	Number of frames/ second	50
3.	Size of each frame	160 Octets
4.	Network (IP) layer overhead	40 Octets
5.	Fragment (MPDU) size considered	256 Octets
6.	Data rate for G.711	64 Kbps

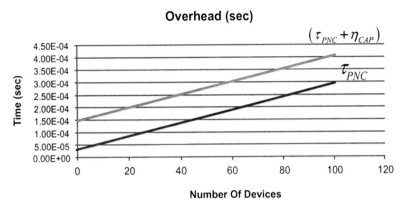

Figure 3.71   PNC overhead for various number of devices using voice [122].

As the number of frames per CTA increase, the ratio of (η_{MPDU}) to τ_{CTA} also increases and hence the overhead decreases.

The superframe overhead apart from the transmitter data rate, also depends on the number of MPDUs sent in the CTA and is calculated using the following (3.115):

$$\tau_s = \tau_{PNC} + (\tau_{CTA} \times N_{CS} \times N_D)$$

(3.115)

Where τ_S is the superframe overhead, N_{CS} denotes the number of CTAs per superframe and N_D denotes the number of supported devices. The CTA overhead τ_{CTA} is calculated by:

$$\tau_{CTA} = \eta_{CTA} - (\eta_{MPDU} \times N_{MC})$$

(3.116)

where η_{MPDU} is the time required to send an MPDU and N_{MC} denotes the number of MPDUs per CTA. Increasing ∂_T also has an advantage. The CTA duration (η_{CTA}) decreases with the increase in ∂_T and as a result, the superframe capacity increases.

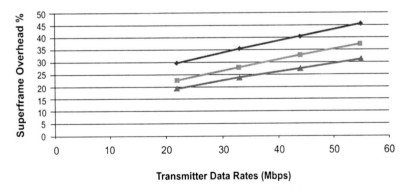

Figure 3.72   Overhead compared to the transmitted data rate [122].

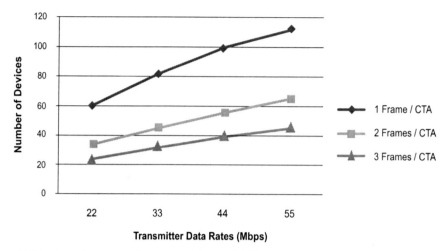

Figure 3.73 Superframe capacity versus data rate [122].

Figure 3.73 shows the superframe capacity plotted against ∂_T for different number of frames per CTA.

Figure 3.74 shows that the CTA overhead increases with the increase in the transmitter data rate. The reason is that with the increase in ∂_T, the time required to send the MPDU decreases but the IFS remains constant. Furthermore the time required to send the MAC header remains the same as it is always sent at the base rate of 22 Mbps. The CTA overhead decreases when the number of frames per CTA is increased.

The requirements for video traffic are more resource intensive than those for voice, therefore, the capacity for video traffic is analyzed in order to find an *upper limit* of data rate. For video traffic, 4 different levels of H.264 were considered for mobile content (3G video), Internet/ *standard definition* (SD), *high definition* (HD), and *full high definition* (FHD). Each level has different data rate requirements and

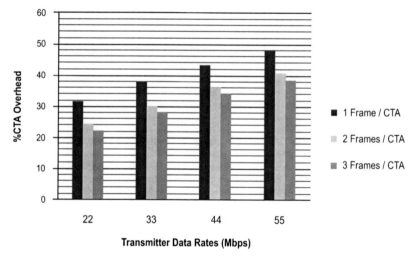

Figure 3.74 Percentage CTA overhead (MPDU size = 256 Octets) [122].

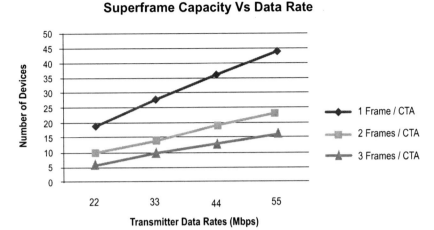

Figure 3.75   Superframe capacity against data rate for video (MPDU size = 1024 octets) [122].

a different number of frames sent per second. When sending mobile content at a resolution of 176 by 144 and a frame rate of 24 fps, the data rate required is about 160 Kbps [122]. The average size of each frame comes up to 834 Octets. If the IP overhead is considered, the frame size becomes 874 Octets. The nearest fragment size of 1024 Octets is used to efficiently carry an MSDU size of 874 Octets. If the MAC layer overhead is taken into account, ∂_R becomes ≈ 200 Kbps.

The maximum tolerable delay and jitter for video applications should be less than 100 ms, therfore, there is more flexibility in assigning super rate CTAs to the video applications depending on the required data rate. Figure 3.75 shows the superframe capacity when considering mobile content with an MPDU size of 1024 Octets.

Figure 3.76 shows that by increasing ∂_T, ∂_A only increases by 2%.

However there is a two fold increase in ∂_A by sending more frames in the CTA. Also ∂_A can be increased by increasing or decreasing the number of CTAs in the

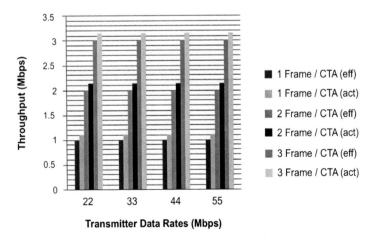

Figure 3.76   Video throughput (MPDU size = 1024 Octets) [122].

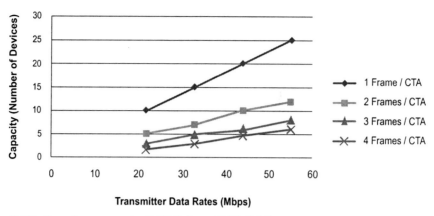

Figure 3.77   Superframe capacity (MPDU size = 2048) [122].

superframe. ∂_A does not change with the increase in ∂_T due to the TDMA MAC format. Since the number of bits sent per superframe remains the same for a device, ∂_A remains the same. For Internet/ Standard Definition (SD), HD and full HD, the data rate requirements are much higher than those for 3G mobile content. The *MSDU size* > *MPDU size* and therefore an MPDU size of 2048 is chosen. Data rates considered are 2, 6 and 8 Mbps.

Figure 3.77 shows the capacity of superframe when an MPDU size of 2048 is considered for up to 4 frames / CTA.

It can be seen that for lower values of ∂_T (e.g., 22 Mbps), only five devices can be supported for 2, 3, and 4 frames per CTA. Figure 3.78 shows that for the same 2, 3, and 4 frames per CTA, the ∂_A achieved when ∂_T is 22 Mbps, is up to 8 Mbps. Therefore, a practical limit of 8 Mbps can be set for the devices (when the full duration of 65535 μs is used) when the number of devices is low (i.e., 5–10) in the piconet. To achieve fairness among the higher number of devices, the upper limit should be further dropped.

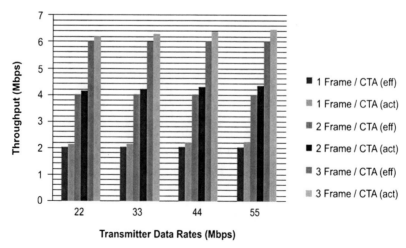

Figure 3.78   Actual data rate for video (MPDU size = 2048 Octets) [122].

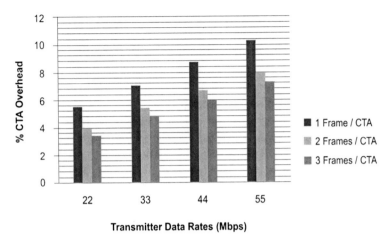

Figure 3.79 CTA overhead (MPDU Size = 2048 Octets) [122].

Figure 3.79 shows that the increase in CTA overhead is relatively less when compared with the use of smaller MPDU sizes.

The number of piconets that can take part in the inter-PAN communication system depends on the capacity of the superframe and therefore, a detailed capacity analysis is required. The proposed solution can maintain the QoS for isochronous flows. The proposed scheduling solution can be used to implement efficient scheduling algorithms along with admission control strategies applicable to HDR WPANs and an inter-PAN communication scenario based on a WPAN HDR air interface.

3.5 Conclusions

This Chapter described various methodologies that can be applied to enhance the capacity of different types of ad hoc networks.

Technology enablers for wireless ad hoc networks include but not limited to multihopping, various multiple antennas techniques, and novel MAC and routing algorithms. In wireless networks, the network capacity enhancement can be achieved as a tradeoff with the transmission delay.

The usage of diversity, advanced modulation and coding, and so forth, bring the advantages of coverage extension, interference reduction, and advantages over multipath. Furthermore, the introduction of adaptive antennas allows a considerable capacity increase through SDMA, which reuses the bandwidth by multiplexing the signals based on their spatial signatures. The introduction of the SDMA approach impacts heavily on the PHY and MAC layer architectures, which requires a modification of the basic strategies developed for traditional (e.g., cellular) wireless networks.

Multi-user diversity, provided by independent fading channels across different users, can be exploited through scheduling, which should be performed according to the knowledge of the CSI available at the scheduler.

Scheduling can be conveniently combined with linear pre-coding (e.g., beamforming) in multiantenna systems and in this context can be studied as a specific problem of power allocation or matrix modulation design.

In order to obtain a representative picture of the performance of a communication system as a whole, very extensive computations must be carried out to take all the factors affecting the system performance in to consideration, such as the geographical distribution of the nodes, the number of antennas and antenna patterns of each node, the effectiveness of signal processing algorithms involved, and so forth. To simplify this, the investigations (e.g., simulations) may be divided into several stages—first performing performance measuring of a single link between two nodes (link-level simulations), then building an abstraction model of the PHY layer based on the results of link-level simulations, and finally measuring the system performance with the help of system-level simulations taking into account the geographical parameters of the system deployment, mutual interference, where the performance of each link is obtained from the PHY abstraction model with much less computation efforts than using direct link-level modeling of the PHY layer.

The investigation of the throughput performance of the described in this chapter scheduling algorithms, can be extended by extensive simulation results regarding the throughput performance of the distributed schedulers with various network scenarios (e.g., single-antenna and multiple-antenna equipped network). Further benefits can be obtained by a new density control phase to enhance the distributed scheduling framework by optimally adjusting the link activation frequency as a means to tune the link density in the network. The latter would require a mathematical derivation of the updating mechanism for the transmission probability based on the steepest descent method.

Further work related to the described PFS can extend to a multicarrier, frequency-selective system.

A carefully designed cross-layer framework can guarantee long-term QoS, while at the same time significantly increase the overall network throughput as compared to the conventional methods. Exploiting the MIMO properties, such as multiplexing gain and interference cancellation, and the benefits of SDMA in a cross-layer context, can help not only to increase the overall network throughput but also to optimally distribute the data traffic over the network topology in a way that can guarantee the required long-term QoS throughput to the underlying applications.

Finally, further benefits for certain ad hoc scenarios can be obtained by use of cognitive radio and related techniques that increase the intelligence of the nodes.

References

[1]   Zhang, H., and J. Hou, "Capacity of Wireless Ad-hoc Networks under Ultra Wide Band with Power Constraint," Technical Report, July 2004, at http://hdl.handle.net/2142/10882.

[2]   FP6 IST Projects at http://cordis.europa.eu/ist/ct/proclu/p/mob-wireless.htm.

[3]   FP6 IST projects in Broadband for All, at http://cordis.europa.eu/ist/ct/proclu/p/broadband.htm.

[4]   FP6 IST Projects WINNER and WINNER II, at http://cordis.europa.eu/ist/ct/proclu/p/mob-wireless.htm.

[5]   FP6 IST Project FIREWORKS, at http://cordis.europa.eu/ist/ct/proclu/p/mob-wireless.htm.

[6]   FP6 IST Project MEMBRANE, at http://cordis.europa.eu/ist/ct/proclu/p/mob-wireless.htm.

[7]   FP6 IST Project ORACLE, at http://cordis.europa.eu/ist/ct/proclu/p/mob-wireless.htm.

[8]   FP6 IST Projects MAGNET and MAGNET Beyond, at www.ist-magnet.org.

[9]   FP6 IST Project E-SENSE, at http://cordis.europa.eu/ist/ct/proclu/p/mob-wireless.htm.

[10]   FP6 IST Project CRUISE, at http://cordis.europa.eu/ist/ct/proclu/p/broadband.htm.

[11]   FP6 IST Project PULSERS and PULSERS II, at http://cordis.europa.eu/ist/ct/proclu/p/mob-wireless.htm.

[12]   FP6 IST Project WINNER, Deliverable 3.1, "Description of identified new relay based radio network deployment concepts and first assessment by comparison against benchmarks of well known deployment concepts using enhanced radio interface technologies, " December 2004, http://cordis.europa.eu/ist/ct/proclu/p/mob-wireless.htm.

[13]   FP6 IST Project ORACLE, Deliverable 5.3, "Design and Evaluation of MAC Protocols and Algorithms for Space Opportunities Exploitation," November 2007, at http://cordis.europa.eu/ist/ct/proclu/p/mob-wireless.htm.

[14]   FP6 IST Project WINNER II, Deliverable D6.13.14, "WINNER II System Concept Description," November 2007, at http://cordis.europa.eu/ist/ct/proclu/p/mob-wireless.htm.

[15]   FP6 IST Project MEMBRANE, Deliverable 3.1, "Capacity Analysis of MIMO Multi-hop Interference Relay Networks," September 2006, http://cordis.europa.eu/ist/ct/proclu/p/broadband.htm.

[16]   FP6 IST Project MEMBRANE, Deliverable 4.1.2, "IA/MIMO-Enabled Techniques for Reconfigurable Transmission and Routing," July 2007, http://cordis.europa.eu/ist/ct/proclu/p/broadband.htm.

[17]   FP6 IST Project E-SENSE, Deliverable 3.3.1, "Efficient and Light Weight Wireless Sensor Communication Systems," May 2007, at http://cordis.europa.eu/ist/ct/proclu/p/broadband.htm.

[18]   FP6 IST Project E-SENSE, Deliverable 3.3.2, " Novel Cross-Layer Optimization," January 2006, http://cordis.europa.eu/ist/ct/proclu/p/broadband.htm.

[19]   Cover, T., and J. Thomas, *Elements of Information Theory*, Wiley, 1991.

[20]   Gastpar, M., and M. Vetterli, "On the Capacity of Large Gaussian Relay Networks," in *IEEE Transactions on Information Theory*, Vol. 51, No. 3, March 2005.

[21]   Ford, L., R., and D. R. Fulkerson, *Flows in Networks*, Princeton University Press, 1962.

[22]   Gupta, P., and P. R. Kumar, "The Capacity of Wireless Networks," in *IEEE Transactions on Information Theory*, Vol. 46, No. 2, March 2000, pp. 388–404.

[23]   Xie, L.-L., and P. R. Kumar, "A Network Information Theory for Wireless Communication: Scaling and Optimal Operation," in *IEEE Transactions on Information Theory*, April 2002.

[24]   Gupta, P., and P. R. Kumar, "Towards an Information Theory of Large Networks: an Achievable Rate Region," in *IEEE Transactions on Information Theory*, Vol. 49, No. 8, August 2003, pp. 1877–1894.

[25]   Van der Meulen, E. C., "A Survey on Multi-Way Channels in Information Theory," in *IEEE International Symposium on Information Theory*, Vol. 23, No. 1, July 2002, pp. 1–37.

[26]   Schein, B., and R. Gallager, "The Gaussian Parallel Relay Network," in *IEEE International Symposium on Information Theory*, June 2000, p. 22.

[27]   Kramer, G., and M. Gastpar, "Capacity Theorems for Wireless Relay Channels," in *Proceedings of the 41st Allerton Conference on Communication, Control, and Computer*, October 2003, Monitcello, Illinois.

[28]   Host-Madsen, A., "On the Capacity of Wireless Relaying," in *Proceedings of the Vehicular Technology Conference Spring*, 2002.

[29]   Laneman, J., and G. W. Wornell, "Energy-Efficient Antenna Sharing and Relaying for Wireless Networks," in *Proceedings of the IEEE conference on Wireless Communications and Networking*, September 2000, Chicago, Illinois, pp. 7–12.

[30] Esseling, N., H. Vandra, and B. Walke, "A Forwarding Concept for HiperLAN/2," in *Proceedings of the European Wireless 2000*, September 2000, Dresden, Germany, pp. 13–17.

[31] Mangold, S., "Analysis of IEEE802.11e and Application of Game Models for Support of Quality-of-Service in Coexisting Wireless Networks," *PhD Thesis*, Aachen University (RWTH), Aachen, 2003, http://www.comnets.rwth-aachen.de/.

[32] Schultz, D., et al., "Fixed and Planned Relay Based Radio Network Deployment Concepts," in *Proceedings of the 10th WWRF Meeting*, October 2003, New York.

[33] Song, M., and B. He, "Capacity Analysis for Flat and Clustered Wireless Sensor Networks," in *Proceedings of the International Conference on Wireless Algorithms, Systems and Applications (WASA)*, August 2007, Chicago, Illinois, pp. 249–253.

[34] FP6 IST Project CRUISE, Deliverable 240.1, "Contributions to the State of the Art in Wireless Sensor Transmission," December 2007, http://cordis.europa.eu/ist/ct/proclu/p/broadband.htm.

[35] Kini, A. V., N. Singhal, and S. Weber, "Broadcast Capacity of a WSN Employing Power Control and Limited Range Dissemination," in *Proceedings of the 41st Annual Conference on Information Sciences and Systems (CISS)*, March 2007, Baltimore, pp. 238–243.

[36] Rachlin, Y., R. Negi, and P. Khosla, "Sensing Capacity for Target Detection," in *IEEE Information Theory Workshop*, October 2002, pp. 147–152.

[37] Ekici, E., Y. Gu, and D. Bozdag, "Mobility-Based Communication in Wireless Sensor Networks," in *IEEE Communications Magazine*, Vol., 44, No., 7, July 2006, pp. 56–62.

[38] Grossglauser, M., and D. N. C. Tse, "Mobility Increases the Capacity of Ad Hoc Wireless Networks," in *IEEE Transactions on Networking*, Vol., 10, No. 4, August 2002, pp. 477–486.

[39] Mohr, W., R. Lueder, and K.-H. Moehrmann, "Data Rate Estimates, Range Calculations and Spectrum Demand for New Elements of Systems Beyond IMT-2000," in *Proceedings of the WPMC'02*, October 2002, Honolulu, Hawaii.

[40] Guey, J., et al., "Signal Design for Transmitter Diversity Wireless Communication Systems over Rayleigh Fading Channels," in *Proceedings of the IEEE Vehicular Technology Conference*, April–May 1996, Atlanta, Georgia, Vol. 1, pp. 136–140.

[41] Tarokh, V., N. Seshadri, and A. R. Calderbank, "Space-Time Codes for High Data Rate Wireless Communication: Performance Criterion and Code Construction," in *IEEE Transactions on Information Theory*, Vol. 44, No. 2, March 1998, pp. 744–765.

[42] Alamouti, S. M., "A Simple Transmit Diversity Technique for Wireless Communications," in *IEEE Journal on Selected Areas of Communications*, Vol. 16, No. 8, October 1998, pp. 1451–1458.

[43] Tarokh, V., H. Jafarkhani, and A. Calderbank, "Space-Time Block Codes from Orthogonal Designs," in *IEEE Transactions on Information Theory*, Vol. 45, No. 5, July 1999, pp. 1456–1467.

[44] Dohler, M., "*Virtual Antenna Arrays*," Ph.D. Thesis, King's College, London, U.K., 2003.

[45] Prasad, R., and A. Mihovska, editors, *New Horizons in Mobile Communications: Radio Interfaces*, Volume 1, Artech House, 2009.

[46] CELTIC Project WINNER+, at http://www.celtic-initiative.org/Projects/WINNER+/default.asp.

[47] Stefanov, A., and E. Erkip, "Cooperative Coding for Wireless Networks," in *Proc. Int. Workshop on Mobile and Wireless Comm. Networks*, Stockholm, Sweden, Sept. 2002, pp. 273–277.

[48] Hunter T. E., and A. Nosratinia, "Cooperative Diversity Through Coding," in *Proceedings of the IEEE ISIT*, Lausanne, Switzerland, June 2002, p. 220.

[49] Sendonaris, A., E. Erkip, and B. Aazhang, "User Cooperation Diversity—System Description and Implementation Aspects," in *IEEE Transactions on Communications*, Vol. 51, No. 11, November 2003, pp. 1927–1938.

[50] Laneman, J. N., G. W. Wornell, and D. N. C. Tse, "An Efficient Protocol for Realizing Cooperative Diversity in Wireless Networks," in *Proceedings of IEEE ISIT*, June 2001, Washington D.C., p. 294.

[51] Laneman, J. N., and G. W. Wornell, "Distributed Space-Time-Coded Protocols for Exploiting Cooperative Diversity in Wireless Networks," in *IEEE Transactions on Information Theory*, Vol. 49, No. 10, October 2003, pp. 2415–2425.

[52] Nabar, R. U., H. Bölcskei, and F. W. Kneubühler, "Fading Relay Channels: Performance Limits and Space-Time Signal Design," in *IEEE Journal on Selected Areas of Communications*, 2004.

[53] Azarian, K., H. E. Gamal, and P. Schniter, "On the Achievable Diversity-Multiplexing Tradeoff in Half-Duplex Cooperative Channels," in *IEEE Transactions on Information Theory*, Vol. 51, No. 12, December 2005, pp. 4152–4172.

[54] Yang, S., and J.-C. Belfiore, "Towards the Optimal Amplify-and-Forward Cooperative Diversity Scheme," in *IEEE Transactions on Information Theory*, Vol. 53, Issue 9, September 2007, pp. 3114–3126.

[55] Ahlswede, R., et al., "Network Information Flow," in *IEEE Transactions on Information Theory*, Vol. 46, July 2000, pp. 1204–1216.

[56] Cai, N., and R. W. Yeung, "Secure Network Coding," in *Proceedings of the IEEE ISIT*, Lausanne, Switzerland, June 2002, p. 323.

[57] Koetter, R., and M. Medard, "An Algebraic Approach to Network Coding," in *IEEE/ACM Transactions on Networking*, Vol. 11, No. 5, October 2003, pp. 782–795.

[58] Ming Xiao, et al., "Optimal Decoding and Performance Analysis of a Noisy Channel network with Network Coding," in *IEEE Transactions on Communications*, 2009.

[59] EU CELTIC Project WINNER+, Deliverable 1.3, "Innovative Concepts in Peer-to-Peer and Network Coding," January 2009, at http://www.celtic-initiative.org/Projects/WINNER+/default.asp.

[60] FP6 IST Project MEMBRANE, Deliverable 4.2.1, "Joint Routing, Scheduling and Power Control Algorithms," September 2006, http://cordis.europa.eu/ist/ct/proclu/p/broadband.htm.

[61] FP6 IST Project MEMBRANE, Deliverable 4.2.2, "Performance Studies of the New Routing, Scheduling, and Power Control Algorithms by Simulation and/or Mathematical Analysis," December 2007, http://cordis.europa.eu/ist/ct/proclu/p/broadband.htm.

[62] Bender, P., et al., "CDMA/HDR: a Bandwidth Efficient High-Speed Wireless Data Service for Nomadic Users," in *IEEE Communications Magazine*, Vol. 38, No. 7, 2000, pp. 70–77.

[63] Sharif, M., and B. Hassibi, "Scaling Laws of Sum Rate Using Time-Sharing, DPC, and Beamforming for MIMO Broadcast Channels," in *Proceedings of ISIT*, 2004, p. 175.

[64] Molisch, A. F., et al., "Capacity of MIMO Systems with Antenna Selection," in *IEEE Transactions on Wireless Communications*, Vol. 4, No. 4, 2005, pp. 1759–1772.

[65] Taesang, Y., and A. Goldsmith, "On the Optimality of Multiantenna Broadcast Scheduling using Zero-Forcing Beamforming," in *IEEE Journal on Selected Areas in Communications*, Vol. 24, No. 3, 2006, pp. 528–541.

[66] Zheng, L., and D. N. C. Tse, "Diversity and Multiplexing: a Fundamental Tradeoff in Multiple Antenna Channels," in *IEEE Transactions on Information Theory*, Vol. 49, No. 5, 2003, pp. 1073–1096.

[67] Ruifeng, Z., "Scheduling for Maximum Capacity in S/TDMA Systems," in *Proceedings of the IEEE Vehicular Technology Conference-Spring (VTC-Spring)*, 2002, pp. 1951–1954.

[68] Ramanathan, S., "Scheduling Algorithms for Multihop Radio Networks," in *IEEE/ACM Transactions on Networking*, Vol. 1, No. 2, 1993, pp. 166–177.

[69] Chen, L., et al., "Cross-layer Congestion Control, Routing, and Scheduling Design in Ad Hoc Wireless Networks," in *Proceedings of IEEE INFOCOM Barcelona*, Spain, 2006.

[70] Jain, K., "Impact of Interference on Multi-hop Wireless Network Performance," in *Wireless Networks*, Vol. 11, No. 4, 2005, pp. 471–487.

[71] Narlikar, G., G. Wilfong, and L., Zhang, "Designing Multihop Wireless Backhaul Networks with Delay Guarantees," in *Proceedings of IEEE INFOCOM*, Barcelona, Spain, 2006.

[72] "IEEE Std 802.11-1997 Information Technology—Telecommunications and Information Exchange Between Systems-Local and Metropolitan Area Networks-specific Requirements-Part 11: Wireless Lan Medium Access Control (MAC) and Physical Layer (PHY) Specifications," IEEE Std 802. 11-1997, p. i-445, 1997, www.ieee.org.

[73] "IEEE Std. 802.16-2001 IEEE Standard for Local and Metropolitan Area Networks Part 16: Air Interface for Fixed Broadband Wireless Access Systems," IEEE Std 802. 16-2001, pp. 0-322, 2002, at www.ieee.org.

[74] Mauve, M., A. Widmer, and H. Hartenstein, "A Survey on Position-based Routing in Mobile Ad Hoc Networks," in *IEEE Networks*, Vol. 15, No. 6, 2001, pp. 30–39.

[75] Murthy, C. S. R., and B. S. Manoj, *Ad Hoc Wireless Networks-Architectures and Protocols*, Pearson Education, 2004.

[76] Guerin, R., H. Ahmadi, and M. Naghshineh, "Equivalent Capacity and its Application to Bandwidth Allocation in High-Speed Networks," in *IEEE Journal on Selected Areas in Communications*, Vol. 9, No. 7, 1991, pp. 968–981.

[77] Schormans, J. A., K. W. Pitts, and L. Cuthbert, "Equivalent Capacity for On/Off Sources in ATM," in *IEEE Electronic Letters*, Vol. 30, No. 21, 2006, pp. 1740–1741.

[78] Awduche, D. O., et al., "Extensions to RSVP for LSP Tunnels," Internet Draft draft-ietf-mpls-rsvp-lsp-tunnel-04. txt, 2006.

[79] Guerin, R. A., A. Orda, and D. Williams, "QoS Routing Mechanisms and OSPF extensions," in *Proceedings of the IEEE Global Telecommunications Conference*, Vol. 3, 1997, pp. 1903–1908.

[80] Kodialam, M., and T. V. Lakshman, "Minimum Interference Routing with Applications to MPLS Traffic Engineering," in *Proceedings of IEEE INFOCOM*, Vol. 2, 2000, pp. 884–893.

[81] Perkins, C. E., E. M. Royer, and S. R. Das, "Quality of Service for Ad Hoc On-Demand Distance Vector Routing," IETF Internet Draft, draftietf-manet-odvqos-00. txt, 2000.

[82] Cruz, R. L., and A. V., Santhanam, "Optimal Routing, Link Scheduling and Power Control in Multihop Wireless Networks," in *Proceedings of IEEE INFOCOM*, Vol. 1, 2003, pp. 702–711.

[83] Li, Y., and A. Ephremides, "Joint Scheduling, Power Control, and Routing Algorithm for Ad-Hoc Wireless Networks," in *Proceedings of the 38th Annual Hawaii International*, 2005, p. 322b.

[84] Sundaresan, K., S. Lakshmanan, and R. Sivakumar, "On the Use of Smart Antennas in Multi-Hop Wireless Networks," in *Proceedings of the Third International Conference on Broadband Communications, Networks and Systems (Broadnets)*, 2006.

[85] A. K. Saha, and D. B. Johnson, "Routing Improvement Using Directional Antennas in Mobile Ad Hoc Networks," in *Proceedings of IEEE GLOBECOM*, Vol. 5, 2004, pp. 2902–2908.

[86] Yang, L., and M. Hong, "Analysis of Multipath Routing for Ad Hoc Networks using Directional Antennas," in *Proceedings of the IEEE Vehicular Technology Conference*, Vol. 4, 2004, pp. 2759–2763.

[87] Escudero-Garzás, J. J., C. Bousoño-Calzón, A. García-Armada, "An Energy-Efficient Adaptive Modulation Suitable for Wireless Sensor Networks with SER and Throughput Constraints", in *EURASIP Journal on Wireless Communications and Networking*, 2007.

[88] IEEE 802.15 Working Group for Wireless Personal Area Networks (WPANs), at www.ieee802.org/15/.

[89] Yuan, J., et al., "A Cross-Layer Optimization Framework for Multicast in Multihop Wireless Networks Wireless Internet," in *Proceedings of WICON*, July 2005, pp. 47–54.

[90]   De Sanctis, M., et al., "Energy Efficient Cross Layer Optimization for WPAN," *European Conference on Wireless Technology*, 2005.

[91]   Broustis, I., et al., "Multiband Media Access Control in Impulse-Based UWB Ad Hoc Networks," in *IEEE Transactions on Mobile Computing*, Vol. 6, No. 4, April 2007.

[92]   FP6 IST Project WINNER, Deliverable 3.4, "Definition and Assessment of Relay-Based Cellular Deployment Concepts for Future Radio Scenarios Considering 1st Protocol Characteristics," June 2006, at http://cordis.europa.eu/ist/ct/proclu/p/mob-wireless.htm.

[93]   Anghel, P. A., G. Leus, and M. Kaveh, "Multi-User Space-Time Coding in Cooperative Networks," in *Proceedings of the International Conference on Acoustics, Speech and Signal Processing (ICASSP)*, Hong Kong, China, April 2003.

[94]   Xue, Y., and T. Kaiser, "Cooperated TDMA Uplink Based on Hierarchically Modulated Alamouti Code," in *Proceedings of the International Zurich Seminar on Communications* (IZS), Zurich, Switzerland, February 2004, pp. 200–203.

[95]   Vitthaladevuni, P. K., and M. S. Alouini, "BER Computation of 4/M-QAM Hierarchical Constellations," in *IEEE Transactions on Broadcasting*, Vol. 47, No. 3, September 2001, pp. 228–239.

[96]   Hagenauer, J., "Rate-Compatible Punctured Convolutional Code (RCPC Codes) and their Applications," in *IEEE Transactions on Communications*, Vol. 36, No. 4, April 1988, pp. 389–400.

[97]   Hunter, T. E., and A. Nosratinia, "Performance Analysis of Coded Cooperation Diversity," in *Proceedings of the IEEE International Conference on Communications (ICC)*, Anchorage, Alaska, 2003.

[98]   Larsson, P., N. Johansson, and K.-E. Sunell, "Coded Bi-Directional Relaying," in *Proceedings of the Fifth Swedish Workshop on Wireless Ad hoc Networks*, Stockholm, Sweden, May 2005.

[99]   FP6 IST Project WINNER, Deliverable 3.2, "Description of identified new relay based radio network deployment concepts and first assessment by comparison against benchmarks of well known deployment concepts using enhanced radio interface technologies," February 2005, at http://cordis.europa.eu/ist/ct/proclu/p/mob-wireless.htm.

[100] Witrisal, K., et al., "Antenna Diversity for OFDM Using Cyclic Delays," in *Proceedings of SCVT*, October 2001, pp. 13–17.

[101] Bossert, M., et al., "On Cyclic Delay Diversity in an OFDM-Based Transmission Schemes," in *the Seventh International OFDM -Workshop (InOWo)*, September 2002, Hamburg, Germany.

[102] Davis, P. J., *Circulant Matrices*, Second Edition; New York: Chelsea, 1994.

[103] Verdu, S., "Optimum Multiuser Asymptotic Efficiency," in *IEEE Transactions on Communications*, Vol. 34, September 1986, pp. 890–897.

[104] FP6 IST Project WINNER, Deliverable 3.5, "Proposal of the Best Suited Deployment Concepts for the Identified Scenarios and related RAN Protocols," January 2006, http://cordis.europa.eu/ist/ct/proclu/p/mob-wireless.htm.

[105] FP6 IST Project WINNER II, Deliverable 3.5.1, "Relaying Concepts and Supporting Actions in the Context of CGs, " October 2006, at http://cordis.europa.eu/ist/ct/proclu/p/mob-wireless.htm.

[106] FP6 IST Project MEMBRANE, Deliverable 4.2.2, "Performance Studies of the New Routing, Scheduling and Power Control Algorithms by Simulation and/or Mathematical Analysis," December 2007, http://cordis.europa.eu/ist/ct/proclu/p/broadband.htm.

[107] Prasad, R., and A. Mihovska, Ed., *New Horizons in Mobile Communications Series: Networks, Services, and Applications*, Volume 3, Artech House, 2009.

[108] Hou, Y., and K. K. Leung, "A Framework for Opportunistic Allocation of Wireless Resources," in *Proceedings of the IEEE PacRim Conference*, Victoria BC, Canada, 2007.

[109] Lovasz, L., *Matching Theory*, North-Holland, 1986.

[110] Ramanathan, S., "Scheduling Algorithms for Multihop Radio Networks," in *IEEE/ACM Transactions on Networking*, Vol. 1, No. 2, 1993, pp. 166–177.

[111] Jain, K., "Impact of the Interference on Multi-hop Wireless Network Performance," *Wireless Networks*, Vol. 11, No. 4, 2005, pp. 471–487.

[112] Shannon, C. E., "A Mathematical Theory of Communication," in *the Bell System Technical Journal*, Vol. 27, July and October 1948, pp. 379–423 and 623–656.

[113] Boyd, S., and L. Vandenberghe, *Convex Optimization*, Cambridge University Press, 2004.

[114] Klein, T. E., K. K. Leung, and H. Zheng, "Improved TCP Performance in Wireless IP Networks Through Enhanced Opportunistic Scheduling Algorithms," in *Proceedings of IEEE Globecom*, Texas, 2004.

[115] Kelly, F., "Charging and Rate Control for Elastic Traffic," in *European Transactions on Telecommunications*, February 1997, pp. 33–37.

[116] Nguye, T. D., and Y. Han, "A Proportional Fairness Algorithm with QoS Provision in Downlink OFDMA Systems," in *IEEE Communication Letters*, Vol. 10, No. 11, November 2000.

[117] Bu, T., L. Li, and R. Ramjee, "Generalized Proportional Fair Scheduling in Third Generation Wireless Data Networks," in *Proceedings of INFOCOM*, Barcelona, Spain, April 2006, pp. 1–12.

[118] Girko, V. L., "A Refinement of the Central Limit Theorem for Random Determinants," in *Theory of Probability and Its Applications*, Vol. 42, No. 1, 1997, pp. 121–129.

[119] Telatar, I. E., "Capacity of Multi-Antenna Gaussian Channels," in *European Transactions on Telecommunications*, Vol. 10, No. 6, November–December 1999, pp. 585–595.

[120] Smith, P. J., and M. Shafi, "On a Gaussian Approximation to the Capacity of Wireless MIMO Systems," in *Proceedings of IEEE ICC*, New York, April 2002, pp. 406–410.

[121] Choi, J.-G., and S. Bahk, "Cell-Throughput Analysis of the Proportional Fair Scheduler in the Single-Cell Environment," in *Proceedings of the IEEE Transactions on Vehicular Technology*, Vol. 56, No. 2, March 2007, pp. 766–778.

[122] FP6 IST Project MAGNET Beyond, Deliverable 3.2.A.2, "Inter-PAN Communications," June 2008, at http://cordis.europa.eu/ist/ct/proclu/p/broadband.htm.

[123] IEEE Standard for Information Technology-Telecommunications and Information Exchange Between Systems-Local and Metropolitan Area Networks-Specific Requirements, IEEE Standard 802.15.3, 2003.

[124] http://www.ieee802.org/15/pub/2003/Jul03/03268r2P802-15_TG3a-Multi-band-CFP-Document.pdf.

[125] ftp://ftp.802wirelessworld.com/15/07/15-07-0693-03-003c-compa-phy-proposal.pdf.

[126] Tseng, Y. H., E. H. Wu, and G. H. Chen, "Maximum Traffic Scheduling and Capacity Analysis for IEEE 802.15. 3 High Data Rate MAC Protocol," in *Proceedings of VTC*, Vol. 3, 2003, pp. 1678–1682.

[127] Chen, X., et al., "An Energy DiffServ and Application-Aware MAC Scheduling for VBR Streaming Video in the IEEE 802.15. 3 High-Rate WPANs," in *Elsevier Computer Communications*, Vol. 29, 2006, pp. 3516–3526.

[128] Mangharam, R., et al., "Size Matters: Size-based Scheduling for MPEG-4 over Wireless Channels," in *Proceedings of the SPIE Conference on Multi-media Networking and Communications*, 2004, pp. 110–122.

[129] IEEE Draft Recommended Practice to Standard for Information Technology-Telecommunications and Information Exchange Between Systems-Local and Metropolitan Networks-Specific Requirements-Part 15.5: Mesh Enhancements for IEEE 802.15 WPANs, IEEE Draft 15-06-0237-02-0005, 2006.

Context Aware and Secure Personalized Communications

Wireless ad hoc communications are characterized by distributed users, which are moving between different activities and different contexts, and different networked interactions and relations. In addition, the novel approaches researched in relation to optimized and intelligent ad hoc communications, such as self-organization, autonomous decisions, and distributed protocols and algorithms, create new technical challenges and demands on issues relating to service and context management, security, trust, control, and privacy.

This chapter will describe the main contributions made by the European funded under the *Information Society Technologies* (IST) Framework Program Six (FP6) projects that were active within the *Mobile and Wireless* [1] and *Broadband for All* domains [2]. The FP6 IST project CRUISE identified related to *wireless sensor networks* (WSNs) open issues concerning energy efficiency, scalability, mobility, security, privacy, trust and middleware design [3]. The FP6 IST project E-SENSE [4] combined various enabling technologies to provide a seamless and nomadic user access to new classes of applications by capturing an unprecedented level of detail of context information for body, object, and environment sensor networks. The FP6 IST projects MAGNET and MAGNET Beyond [5] provided a complete service and context discovery framework including security, privacy, and trust mechanisms for *personal networks* (PNs) and *PN-federations* (PN-Fs). The FP6 IST project ORACLE [6] provided a framework for context information towards decision making in opportunistic networks; The FP6 IST project PULSERS provided the navigation capability for autonomous ultra-wideband (UWB)-based systems [7]. The FP6 IST project RESOLUTION provided an integrated system approach for wireless communication and local positioning with accuracy for mobile ad hoc networks [8]. Further details on many of the FP6 IST project achievements can be found in [9–11].

This chapter is organized as follows. Section 4.1 gives an introduction into the main challenges and requirements related to context-aware and secure ad hoc communications and services. Section 4.2 describes advancements in the state of the art in the area of context-aware service discovery and security for PNs and WSNs. Section 4.3 describes requirements and methodologies for data aggregation and fusion in the context of modern ad hoc networks. Section 4.4 describes similar novelty architectures designed and evaluated for WSNs. Section 4.5 concludes the chapter.

4.1 Introduction

Intelligent networking and connectivity refers to the ability of a network to adapt its services to the user's context and to support connections based on the application services attributes. These are considered the key technologies for fostering the mobility of users and the versatility of services. This is due to the fact, that when intelligence is contained in the networking protocols (e.g., information on security needs, user context and profiling, priority, and QoS issues) then the business applications can be built in a more efficient way and consequently to allow for broader functionality in the service layer [12].

The research towards cognitive packet networks goes in parallel with advances towards knowledge-enabled applications and services. Topics of special interest are issues regarding the availability of resources, the online adaptation in response to *quality of service* (QoS) needs, the perceived performance of the different network infrastructure elements, and so forth.

Ad hoc scenarios require the collaboration between network nodes, which can be *user terminals* (UTs), base stations (BSs), sensor nodes, and so forth, depending on the type of ad hoc network. The most important is the exchange of the sensing information, but also the exchange of policies, results of (local or individual) decisions or any other type of network or user related context information is required [13]. This in turn requires the support of efficient encoding and transport of sensing data as well as the support of generic and extensible (XML) data exchange.

Many protocols have been developed in the recent years trying to be established in the worldwide market in order to allow an extensive functionality that will include intelligence in the network functionality itself. One example is the establishment of the IPv6 and MIPv6 protocols that are including advanced functionality in terms of traffic characterization and security primitives and ongoing efforts from significant vendors, who are investigating in a *service-oriented approach* in the network service provisioning.

One approach is applying awareness to the fields of the protocols themselves and lowering thus the level of information assessment from the application level to the network level (with respect to the OSI layered approach). In such a way, the role of the "application semantics," that normally belongs to a layer that is usually above the application layer is taken over by the "network semantics layer," which is much lower in the architectural structure close to the network layer of the architecture. The purpose is to better meet the underlying needs of applications for real-time visibility, security, event-based messaging, optimized delivery, and other core integration and deployment services [12]. For example—an intelligent network provided with network semantics could enable the following:

- Communication between applications by routing application messages to the appropriate destination, in the format expected by that destination;
- Enforcement of consistent security policies for the application access and the information exchange;
- Visibility of the information flow, including the monitoring and filtering of messages for both business and infrastructure purposes;
- Enhancement of the application optimization by providing application level load balancing, offloading security, and XML operations as well as caching and compression services at the application message level.

Protocol capabilities, functions, and services required for next generation (e.g., opportunistic) ad hoc networks but usually not present in conventional wireless communication systems are the following [13]:

- Link layer and network layer protocols and relying on the same information elements (IEs) for link layer and network layer protocols;
- Support IE encoding for both binary encoded simple sensor data and XML encoded structured complex data;
- Assignment of "information flows" to dedicated communication channels according to data stream requirements;
- Support of both context change notifications and programmed context data requests.

The communication of sensing information or context, in general, in the course of distributed and collaborative sensing and decision-making must be suitable for both link layer and network layer transfers. Here, link layer protocols are fast and lightweight but suitable only for information dissemination in the direct vicinity, (e.g., by one-hop broadcast such as in ad-hoc scenarios). Network layer protocols can be designed for assured point-to-point and point to multipoint communications, such as in multihop communications for network wide context data exchange.

Both, fast transfers within the geographical or network-topological vicinity, as well as assured network-wide end-to-end communication of context information must be possible to minimize the network signalling load (i.e., to disseminate information only if and when relevant). To enable the forwarding of sensor information, for example, a common encoding of IEs or of complete *protocol data units* (PDUs) for link and network layer can greatly simplify the protocol architecture and will can help to avoid dedicated routing and gateway functions that must otherwise be colocated to each terminal [13]. The selected encoding format must be extensible because when it is not possible to foresee all the types of information required for the collaborative functions.

A full ad hoc/wireless multihop network, (e.g., resulting from direct PAN to PAN communication; see Chapter 3 of this book), is characterized by the dynamics of the network topologies caused by the device mobility and requires distributed and robust solutions. Instead of relying on a permanent proxy server, this could require the election of a master node to temporarily assume the role of a proxy server that would normally be implemented in a fixed network.

4.1.1   Personalization and Individualization of Devices and Services

Personalization of services allows the user for deciding, in which way and when to make use of services and how to manage these [14]. Service providers and operators, thanks to personalization, may obtain loyal customers by meeting their expectations in a more effective and efficient way.

To be able to achieve personalization, it is important to understand the concepts of *usability* and *user experience* [15]. The motivation behind the introduction of the 'user experience' concept is found in the lack of subjectivity related to using a product. The user experience includes emotional and aesthetic issues, which exceed the quality of use and put focus on the quality of experience. If a user's expectations are met, then there is a neutral user experience, if his/her expectations are not met, then there will be a negative experience and if the user's expectations are more than met there is a positive user experience. This understanding leads to a more relative measure of the user evaluation than many of the ones included in more traditional tests of usability [15].

It is more difficult to grasp the user experience than the usability of a given product. One reason is that well-established methods and measurements are developed within the area of usability [15]. Another reason is the difficulty in obtaining measures on highly subjective variables needed to evaluate the user experience. However both types of evaluations—usability and user experience—are difficult to make when dealing with mobile and networked systems as this implies changing contexts. Different contexts relates not only to location but also to the relations and interactions with other users. Other users may influence the usefulness and thus both the usability but especially the user experience related to the product.

In an ad hoc environment, the system is not only networked and mobile but also based on the covering of unlimited activities. These activities will be shifting and may include both known and unknown services and relations (to both technologies and people), which complicate the matter even further. In order to evaluate the usability and user experience for a scenario of changing rapidly different activities, different contexts, and for different networked interactions and relations, issues such as security, trust, control and navigating around the different possibilities, requests, and context become key.

Services can have a list of associated personalized parameters, (e.g., volume, QoS). During the service instantiation, the personalization returns the preferred values for the requested personalizable parameters to the instantiated service allowing it to personalize itself.

Services may need to personalize again when context changes occur. When this happens, the process of personalization will notify any updated preference outcomes to the appropriate services allowing them to repersonalize at runtime, dynamically changing to meet the user's needs [16]. Such context changes may also affect the architecture of a running service. Composite services, made from two or more services may need to be recomposed due to a context change. For example, if a composite service includes display service X, a change in context may mean the user now prefers to use display service Y. In this way personalization may trigger a service recomposition (to replace X with Y) to provide the user with dynamic services, which update to meet his/her current needs.

Personalization interfaces with the network layers to provide preferences for the network and the interface selection. Depending on the location and preferences of the user and the QoS requirements of the services, the network access will be dynamically selected based on this information [17].

Services are described by means of ontologies. Personalization ontologies are manually coded conceptual formalizations of user situations, activities, goals, and behavior patterns (e.g., "the user is at a business meeting," or "the user is going to take a person to the airport,") and preference outcomes and pro-active actions (e.g., "background color," or "start service of type T"), which can be used to describe the user preferences in a compact, readable, and meaningful way and to provide useful metadata when learning them automatically the from user action histories.

Services can be adapted in different ways to the user profile and the context information [18]. The following are summarizing some examples:

- The information presented to the user can be adapted based on profile and context information, (e.g., relevant information may be different depending on the user location).
- How the information is presented may differ according to the situation, preferences, and the device available. For example, navigation information could be displayed differently on a desktop computer than on a mobile phone, and also differently depending on the situation of the user, whether the user is standing or running, in which case the output could be reduced to easy-to-grasp arrows.
- Available services may be preconfigured with parameters used in the past or which are relevant in the current situation, (e.g., the wake-up call in a hotel could be preconfigured to the room of the user).
- Services may be executed automatically depending on the user profile and the situation, (e.g., calling a doctor in an emergency situation).

Thus, *context* is any information that can be used to characterize the situation of an entity. An entity is a person, a place, or an object that is considered relevant to the interaction between a user and an application, including the user and the application themselves [19].

The total set of user-related information, preferences, rules, and settings, which affects the way, in which a user experiences terminals, devices, and services, is defined as a *user profile* [18].

In the general case, it is a complex undertaking to decide, which part of the entire available user profile and context information is relevant and useful for performing a *service adaptation*. Organizing the information in an ontology supports reasoning and decision-making, but there is a lot more research to be done on how to combine this with an intelligent application logic and policies that provide a proper protection of the user privacy.

Figure 4.1 shows an ad hoc network scenario where a PN-user has full control over the resources that he or she wants to share with the PN-without exposing or revealing more personal information and content than needed.

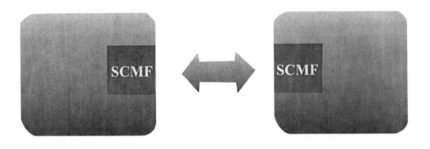

Figure 4.1   User control over the resources in a PN-F scenario [18].

When accessing foreign services (e.g., push or pull type), a PN-enabled user also has much better control of the personal information and can decide where the right balance lies between the protection of privacy and the revealing of personal information.

Today, a large number of Web sites offer users or subscribers a basic level of personalization. This is shown in Figure 4.2.

This can be initiated, when the user signs up for the first time, where typically a set of personal data such as name, address, e-mail address, phone number(s), may be requested, and the user chooses a user-ID and password to access the personalized services later on. Furthermore, the user is often given the option of ticking various preferences or areas of interest.

More sophisticated services will collect data about the usage history and based on this perform some "intelligent" processing in order to provide relevant information or offers to the user.

In an existing situation, the large number of third party profiles have to be handled independently by the user, and the personalization benefits are somewhat limited [18].

With a framework that makes possible that the service provider is informed when dealing with a "more sophisticated" user, a better personalization can be achieved. The template of the user profile might be publicly available, so a service provider would know, what kind of personal information is potentially available, and hence will be able to query the user for a certain part of this. However, the information may not have been filled in, or it may not be accessible because of the policies attached to the user profile. But if the information can be accessed, the service provider can use it to customize or personalize the service to this particular user. The FP6 Projects MAGNET and MAGNET Beyond [5] proposed a framework that can assist the user in filtering and navigating huge amounts of contents, services, and offerings. With such a framework, a better value for the user, and better revenue options for the service provider can be provided.

Figure 4.2 Enabling a basic level of personalization [18].

In a PN scenario, to fully take advantage of the network, a service provider would need to adapt services significantly, but some benefits in terms of personalization and service adaptation are then readily available.

In turn, to be able to personalize services (e.g., resources) of networks managed by a system administrator, the concept of a *service provider network* (SPN) can be used [20]. Then the sharing of resources and access to the network can be based on the principle of the PN-F [18].

Thus, in a next generation network, the user will play an increasingly central role not only as a complex consumer, but also as a new type of provider [21].

Another trend emerging with advances in the area of ad hoc networks is the *sharing of content*. Due to a proliferation of tools for content production and sharing, it has become quite easy for the average customer to become a content—in addition to being a service provider [22].

This proliferation of information, content, products, and services provided and owned by individual users, introduces new opportunities and challenges, namely, the following:

- Service integration instead of service provisioning;
- Privacy and identity management;
- Seamless interaction/access to the telecommunications infrastructure providing a high level of experience.

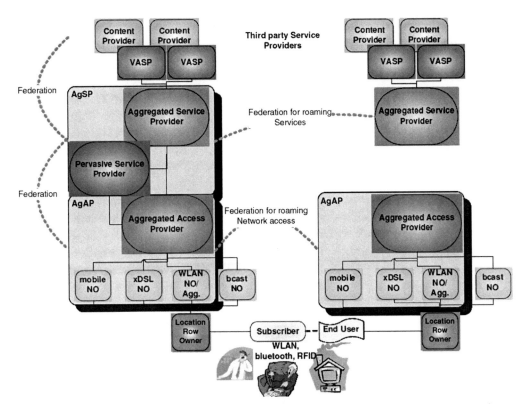

Figure 4.3 An example of a federated environment for the provision of seamless access and personalized services [21].

The next generation network can be viewed as a plethora of autonomous domains of various sizes that intersections that would need to cooperate based on dynamic service level agreements and associated trust relationships [21]. Such a trend imposes the requirements for interoperability of solutions and segments on horizontal and vertical levels. An example of a federated environment is shown in Figure 4.3.

Figure 4.3 shows that *third party* providers seek new opportunities in terms of acquiring roles in emerging complex value chains and systems, mainly targeting the exploitation of network capabilities opened-up by the network operators. *Third parties* can thus become providers of *enabling services* (platform/middleware providers) or *end-service* providers (content/application providers).

The wireless hot spots are rapidly becoming a dominant technology in urban environments. A major consequence is the high fragmentation of the low layers of the value chain, namely the venue (location owners) and the *hot-spot operator* layer. The hot spots a great opportunity for mobile network operators and *Internet service providers* (ISPs) to add additional services to their portfolios. The *aggregated service providers* serve as middlemen between the *content/application* providers, and the *enabling services* providers. These cover basic roles such as roaming between networks/services, variable end-to-end QoS provisioning, and service bundling [21].

4.1.2   Requirements for Ad Hoc Network and Service Architectures

Because ad hoc networks will be an inherent part of the network of the future, the supporting architectures must provide secure, personalized, and pervasive services across heterogeneous networks that would support a mobile user including the user needs for privacy. Devices will connect to the fastest and most efficient networks available at any given time. When a device moves out of range of a current connection, it will search for the next most-effective connection for a handover. Under a scenario such as this, ad hoc and cellular will be complementary technologies that simply serve different mobility demands [23].

The requirements that this imposes the following requirements:

- Ubiquitous and easy user access to services, networks, and content;
- Support of different network types, technologies, and services with view on migration aspects;
- Support of seamless mobility in order to make an optimal use of available resources and QoS provision. Techniques for seamless and controlled mobility should be optimized at all layers and levels to provide the best possible experience for the user.

Thus, in future network scenarios the need (originally from network considerations) to bring together mobility, *authentication, authorization, and accounting* (AAA), charging, auditing and QoS should be combined with the need to manage resources from the network side and the overall security concerns and should be seen relevant at network, service, and user levels [21]. The *virtual identities* (VID) concept was proposed by [17] to allow for universal user access while maintaining security and privacy requirements. The VID is linked to a (if needed, partially) reachable or pseudonymous identifier that allows the user network and service access, and anonymity when required. It is a concept that cuts across layers and allows access (and if needed reachability) independently of a particular device. From a privacy perspective, it is desirable for the user to disclose not more than a well-defined set (or "amount") of information when acting under a particular VID. This information, taken together as a collection of individual information items, conceptually constitutes the user's current VID. In other words, the VID is equivalent to all the information that one can gather through observation and interaction with the user [20]. The VID concept was found very suitable in the context of PNs and PN-Fs. In such a scenario, the VID includes the identifier(s), under which the user and his PN are known, the network addresses in use, the cryptographic material, user profile, context information that is disclosed as a result of incoming queries, the timing of events that are initiated by the PN, the user's behavior, the content of the messages that are sent from the PN, and so forth. The VID concepts was extended to the concept of a *federation* in [17], which allows user activities in a range of cases from extreme trust to very limited and managed trust between providers, and must not only work horizontally, (i.e., between network operators), but also vertically, (e.g., between network operators and content providers) [21].

A future network ad hoc architecture should properly address the concept of *pervasiveness,* to go beyond the traditional approaches that only focus on the *user interaction* aspects (e.g., proper design of applications and devices) to support ubiquitous, context-aware, and user-controlled access to services, as well as manage the access to context. The managed approach should support a controlled and chargeable access far beyond services, also addressing single devices.

Finally, integration at technology and service level is needed, which requires integration of disparate technologies [e.g., *digital video broadcasting* (DVB) with wireless and fixed access technologies]. Such separation of concerns allows for optimizations across the whole bandwidth of offerings, from the network via the service to the content [21].

From the WSN architecture viewpoint, there are several relevant research issues, mainly driven by the fact that the WSN is potentially very large and sinks are mobile terminals [24]:

1) A *multisink* scenario, with potentially many sinks distributed over wide and disconnected areas puts stringent requirements on connectivity and topology control;
2) Mobile sinks affect the MAC/routing/topology control because of the network dynamism;
3) Sinks are in unknown positions and this makes data aggregation, localization, and time synchronization mechanisms more complex.

Security requirements related to authentication and key management mechanisms for large WSN, secure mobility, secure data aggregation, secure service discovery, confidentiality, privacy, and anonymity also arise, due to the fact that a WSN has a large number of nodes, different resource constraints between levels of the hierarchy, and that sinks are mobile terminals.

Wireless hybrid networks (WHNs) are integrated WSns and cellular networks, and in turn, demand solutions on *traffic characterization*—a new type of traffic (bursty and periodic) is offered to the *radio access network* (RAN) of cellular systems, demanding proper characterization in order to drive the design of channel assignment (scheduling) techniques in the RAN; on the *access control*—depending on the MAC, the distribution of the active links at the different levels should be modeled to allow the proper control of the amount of information gathered at the sinks and reported at the infrastructure level, then given the requirements about the desired information, the amount of resources needed at the different levels of the architecture can be derived; internetworking at the network layer—the different levels may have different notions of addressing and different data forwarding paradigms (e.g., level 1 could be an IPv4 PAN that is bridged at layer 3, and level 2 could be running IETF protocols). The means of coupling between the different routing domains also needs to be examined. The support of *data-centric addressing* from level 4 up to level 1 is another issue.

[3] proposed a *hybrid hierarchical architecture* (HHA) for WSNs able to deal with the identified architectural requirements. It is shown in Figure 4.4.

The HHA addresses the integration of the WSN with other wireless communication technologies. It establishes some hierarchy levels; the lowest one embraces the sensors or actuators, objects with rather limited capacity, which normally requires the spontaneous formation of network topologies, including the need to employ multihop communications. In order to cope with the different types of necessities, as well as to foster the interconnection with other types of networks, some nodes belonging to the lowest level of the hierarchy take a more active role (i.e., similar

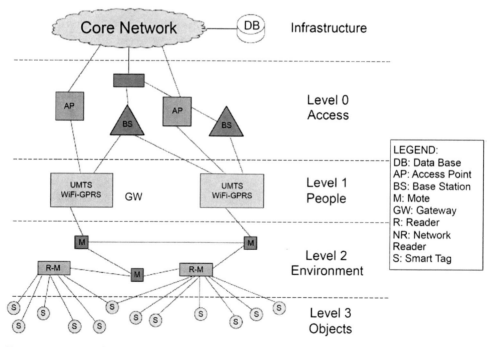

Figure 4.4   A HHA for WSNs [25].

to a clustering technique), and this technique is used within the different levels, to ensure the interconnection between them.

The HHA is a *particular* network architecture: it is strictly hierarchical involving four levels, the wireless nodes have some predetermined (and some unspecified) characteristics depending on the level they belong to. Other network architectures based on different paradigms and topologies, are also possible to use in a similar context. Some of the HHA features, however, make it useful as a realistic and innovative reference scenario. In other words, though specific and limited in the area of investigation, the HHA might represent a reference architecture for many WSN applications, and assist standardization work.

4.1.2.1   Data Aggregation and Fusion

In very dense ad hoc networks (e.g., WSNs), due to the high density of deployed devices and the type of monitoring or alerting applications supported, it is preferred to obtain an overall knowledge of the parameters of interest rather than individual ones. To this purpose *data aggregation* can be used. Data aggregation and/or fusion is the process of forwarding and fusing together data coming from different network nodes to provide a unified feedback to the sink(s) [25].

Data aggregation implies introducing some additional processing at the intermediate network nodes that are in charge of reducing the amount of data packets traveling throughout the network. This is made to counteract some redundancy effects, (i.e., conditions when the redundant and useless packets travel throughout the network, overloading the intermediate nodes and wasting energy and bandwidth resources). This is particularly true in the case of event monitoring applications where the sensor nodes are periodically required to send data to the sink to advertise the unusual behavior in the network. Accordingly, due to the high density of devices in sensor networks, it is likely that the nodes located in the closest proximity forward almost the same data to the sink, thus overloading and stressing uselessly the intermediate network nodes.

Another issue is related to the periodic monitoring of applications when the sensor nodes periodically report their data to the sink, thus causing temporal correlation in the received measurements when no significant variations in the monitored metrics occur. The impact of the useless transmissions is critical in terms of power consumption because transmission of a single bit over the radio causes a consumption of at least three orders of magnitude higher than the one required for executing a single instruction [28]. To counteract these spatial and temporal correlation events, aggregation can be an efficient approach.

When considering data aggregation, together with the reduction in energy consumption, positive effects on the limitation of the network congestion can also be observed [25]. Due to the so-called, *funneling effect*, when propagating data towards the sink, nodes few hops away from the sink are the most stressed under the perspective of both energy consumption and the amount of traffic to manage. Aggregation, would allow to reduce the amount of traffic that these funneled nodes should forward and, thus, reduces the possibility of congestion at these nodes and the possible network partitioning due to the misbehavior of the nodes [27]. Another problem, which data aggregation can help to counteract is related to the minimization of the effects of *error readings*. For example, in a sensor network it is common that possible compromised readings are propagated to the *base station* (BS), thus, causing unnecessary actions. That means that if a damaged sensor, say D, reveals that a carbon monoxide indicator overcomes the alert threshold and sends the alert message to the BS, this could cause the unnecessary propagation of backward control messages from the BS to other nodes in the proximity of node D in confirmation of the alert. With use of data aggregation a single malfunctioning device message is weighted and averaged with other messages sent by properly working devices and, thus, unnecessary actions are not invoked. The problem of minimization of the *error readings effect* could become even more problematic in the case of chemical monitoring, fire alert, and disaster relief networks; therefore, counteracting these unnecessary alerts is important and data aggregation gives a powerful means for this.

Data aggregation could lead, however, also to some drawbacks in the network behavior. On one side, aggregation, by implying the need for additional processing at network nodes, leads to an increase in the *complexity* of the operations performed by network nodes and in the *latency* for data delivery. This is because, due to the data fusion performed at the intermediate nodes, time should be spent in waiting for having the required number of packets in the nodes buffer before aggregating the data and sending them towards the sink. Complexity is, therefore, proportional to the delay in data delivery [26].

Although evaluating simple metrics such as the average and the sum of sensor readings is sufficient for the majority of the application scenarios, in the case of some others (e.g., chemical and biological applications) it is useful to obtain the information about the individual readings [26]. In order to estimate the distribution of the single values, it had been proposed there to use more sophisticated metrics such as *median, quantiles*, and *consensus*, which are also costly in terms of delay.

Aggregation, furthermore, implies the need for a strict *synchronization* among the network nodes [25]. When aggregating data packets, attention should be paid to the way the aggregation is performed. In fact, increasing the number of packets whose content can be aggregated together, is useful to reduce the energy *consumption* and *bandwidth resources overloading*; on the other hand, this could imply an increase in distortion and a possible loss of information, because compressing too many packets leads to losses in terms of data integrity and information distortion [29].

Accordingly an appropriate trade-off should be searched in the way aggregation is achieved.

Figure 4.5 shows the process of data aggregation for an example based on the HHA shown in Figure 4.4. Figure 4.5 is an example of a simulation scenario that is HHA-compliant, for the monitoring of objects in a small office. At level 0 (see Figure 4.4), the *radio access ports* (i.e., fixed stations covering the area forming a RAN (e.g., based on a GPRS, UMTS, or a WiFi standard) provide access to mobile terminals (i.e., the *mobile gateways* shown in level 1), carried usually by people. These mobile devices can also connect through a different air interface (e.g., ZigBee, or Bluetooth) to a lower level of wireless nodes (level 2), with limited energy and processing capabilities, which can access the fixed network only through the

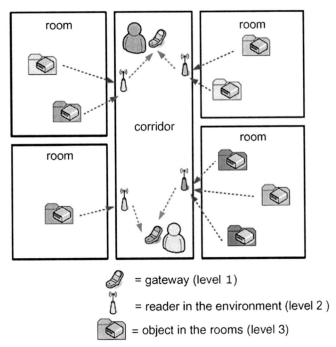

= gateway (level 1)

= reader in the environment (level 2)

= object in the rooms (level 3)

Figure 4.5   Data aggregation in a WSN scenario [25].

gateways. The wireless nodes are distributed in the environment and collect information from it (e.g., sensors providing localization data); moreover, they interact through different air interfaces with tiny devices at level 3 (e.g., smart tags, or very-low-cost sensors), which are part of some movable objects (e.g., printers, books, tickets, etc). The hierarchy is thus composed of four levels.

Under a network topology perspective, this scenario defines a forest of possibly disjoint trees, with heterogeneous radio interfaces at the different levels. Note that if the environmental level is connected through a tree-based topology, then the number of levels in the hierarchy further increases as level two is subdivided into sublevels.

The mobile terminals, using WLAN in indoor environments or cellular networks outdoor, might be used to gather the data sensed by the sensors and transport them towards the data storage systems and application servers. The user terminals in these networks might therefore act as mobile gateways collecting the data from the WSN and forwarding it.

This scenario can be also considered as an extension of the traditional WSN scenario where a sink collects information from the sensor nodes distributed in the environment, through wireless links. In this case, there are multiple sinks, and their locations are not known. Moreover, the sinks are gateways forwarding the information collected to higher levels through heterogeneous wireless interfaces characterized by different parameters like transmit power, capacities, and so forth.

In the example in Figure 4.5, nodes at level 3 are mobile and can be modeled as IEEE 802.15.4 *reduced function devices* (RFDs) with a device ID or *radio frequency identification* (RFID) tags. The nodes at level 2 are fixed and can be modeled as IEEE 802.15.4 *full function devices* (FFDs) or RFID readers. The nodes at level 1 are mobile and can be modeled as IEEE 802.15.4 FFDs with UMTS/GPRS (or WiFi) interfaces.

Assume that it is important to determine the position of the objects in the room scenario. There are a lot of objects (level 3) moving in the room. The fixed readers (at level 2 of the HHA) in the environment can be used to monitor the position of the objects. Finally, these readers will send the data sensed in the environment to some mobile gateways (level 1 of the HHA), to be eventually delivered to a remote user [25].

Interactions Between Data Aggregation and Topology Control

In WSN reducing energy consumption is a main concern because of the need to increase the network lifetime. Accordingly, topology control techniques have been implemented to allow for reducing the overall energy expense. These approaches basically require that nodes alternate between *active* and *sleep* states, while also guaranteeing the reliability in the transmission and the network connectivity.

Only few works had reported the relationship between the operating states of the nodes and the data aggregation performance. In [30] it had been investigated how the aggregation behavior is impacted by the *sleep/active* dynamics and consequent variation in the channel contention. To this purpose an analytical model had been developed. The results showed that in order to increase the fidelity in the aggregation process, both in conditions of medium to high reporting rate, use of *shorter* and *fatter* aggregation trees would allow for reducing the delay and the energy consumption, both, thus guaranteeing the prompt delivery and increased network lifetime. Similar considerations had been carried out also in [31] where the focus was on the evaluation of the impact of the network density on the performance of the network aggregation exploiting greedy and opportunistic forwarding. The conclusion was that *greedy forwarding* exhibits high energy performance in the case of very high density networks with a reduction in the energy consumption in the orders of 45% with respect to opportunistic routing approaches, such as *directed diffusion* [32].

Networking Paradigms for Data Aggregation

The driving idea behind data aggregation is to combine the data coming from the different nodes *en route*, eliminating the redundancy, minimizing the number of transmissions and, ultimately, saving energy. Accordingly, network routing and data aggregation are strongly interconnected. The typical task is to find routes from multiple sources to a single destination that allows for efficient in-network filtering of redundant data. Although data aggregation results in fewer transmissions, it potentially results in greater delays. Therefore, in addition to the transmission cost, the fusion cost can (and must) significantly affect the routing decisions. The following describes some networking paradigms, which can positively be used for performing data aggregation [25].

Gossip-Based Paradigm   *Gossip-based* algorithms for data aggregation are able to support distributed aggregation of sensed data in large-scale sensor fields [33–36].

In addition, *gossip-based* algorithms offer simplicity and robustness in noisy and uncertain environments [25].

The *gossip-based* paradigm can be considered as an evolution of flooding. In fact, for reliable data delivery, flooding is usually the chosen approach. Flooding is based on the forwarding of data packets to all network nodes, until all devices in the network have been delivered the data. This implies that multiple copies of the same packet can be delivered to a single node, thus wasting unnecessarily network bandwidth and energy resources.

The gossip-based paradigm has been proposed to deal with the problem of avoiding the unnecessary flooding of routing messages, which could lead to the problems discussed above. These are particularly crucial in the case of sensor networks where batteries cannot be recharged. *Gossiping* differs from *flooding* because each node, upon receiving a route request issued by a source or another adjacent node, decides to forward or not the request according to a probability *p*. Gossiping can be both applied to query delivery or flooding of events aimed at establishing a gradient, as required, for example, by the *directed diffusion approach* [32]. Gossiping typically exhibits a bimodal behavior in the sense that either it dies out quickly and hardly any node receives a message or a high fraction of nodes receive the message. The behavior strongly depends on both the topology of the network, the potential mobility of the nodes, and the value of the parameter *p*.

Gossiping can save up to 35% of the resources spent for propagating the overhead messages.

Epidemic routing is another variant of gossiping as proposed in [37]. There, the delivery of data packets from a sender to a receiver in a low connectivity network is dealt with. In particular, when the sender and the receiver are not in range and a path between them does not exist, epidemic routing is used to distribute the messages to certain hosts, denoted as carriers, in charge of propagating data within the connected portions of the network. In this way, the chances that the messages are delivered to the destination increase significantly at the expense of the delivery delay.

Another similar approach is proposed with the *ant* algorithms [38]. In this case, the mobile agents, denoted as *ants*, dynamically learn good quality routes. At each hop, these agents choose the next hop using a probabilistic methodology driven by the direction of the good routes. This allows for restricting the research mainly to the highest quality areas of the network. The convenience of this approach with respect to the previous ones consists in the better performance achievable in the case of failure due to the routing search performed around good areas. The key drawback is represented by the scarce reliability to very dense sensor networks.

Gossip-based algorithms exploiting the flooding principle, however, can waste significant amounts of energy by essentially passing the redundant information throughout the network.

This inefficiency had been addressed in [39] by proposing a geographical-based gossiping protocol when applied to the *data-averaging* problem in data aggregation. *Data averaging* is basically a distributed and fault tolerant data aggregation algorithm, by which all nodes can compute the average of all *n* sensor measurements. Gossip algorithms solve the averaging problem by having each node randomly pick one of their neighbors and by exchanging their current values. Each pair of nodes computes the pair-wise average, which then becomes the new value for both nodes. By iterating this pair-wise averaging process, all nodes converge to the global average in a completely distributed manner within the predefined bounded sensor field. Further investigation into gossip algorithms show that with this approach nodes are basically *blind* and require repeatedly the computation of pair-wise averages with their one-hop neighbors, with information only diffusing slowly throughout the network. [39] extends the gossip-based data aggregation to include localization information, which can make the data dissemination process more efficient in terms of radio communication. The idea is not to exchange information with one-hop neighbors, but to use geographical routing to gossip with random nodes that are at a further distance in the network. Then the extra cost of multihop routing is compensated by the rapid diffusion and aggregation of information.

With reference to the HHA in Figure 4.4, the application of *gossip-based* routing can be appropriate for levels 2 and 3. Gateway nodes can also take part in data aggregation with the local nodes, although mobility is not a functional requirement. Gossip-based paradigms for data aggregation can be partially applied also to the HHA level 1.

Rumor-Based Paradigm The *rumor-based* paradigm proposed in [40] focuses on the identification of, events that are localized phenomena occurring in a fixed region of space. Rumor routing was proposed, like gossiping, to limit the flooding in query propagation and event notifications going back to the querying nodes. The concept is based on the building of paths leading to events within the network by the random distribution of event data throughout the network. To design these paths some long-lived agents are considered.

Agents are packets responsible for the rumoring about the events happening across the network. These agents are used to create paths towards the events met along the network, where paths are intended as states in nodes functioning. Path states can be also aggregated; in fact when an agent creates a path leading to event x and crosses a path leading to another event y, the path states can be unified. The paths can be updated in the case shorter or better ones are identified. As event data interacts with the intermediate nodes, it synchronizes the event tables and reinforces the data paths to regions of interest in the network. In this way, when a query is generated it can be sent on a random walk until it finds the event path instead of flooding it throughout the network. As soon as the query discovers the event path, it can be routed directly to the event.

Rumor-based protocols were proposed as data-centric data dissemination and aggregation techniques that offer significant power cost reductions, and improve the network lifetime in large-scale WSNs.

In networks that span multiple hops, data dissemination algorithms must be highly localized as the large distance transmissions are expensive and reduce the networks overall lifetime. Applications deployed within such large networks must be able to gather data efficiently from each part of the network without significantly impacting the limited bandwidth and node energy. Furthermore, data dissemination across large distributed networks is prone to retransmission and, hence, increases the network load due to noisy channels and inherent failure rates on the wireless channels. Although the flooding of data events throughout the entire network is usually regarded as nonoptimal, there are cases, in which flooding can produce acceptable performance. For applications, where there are few events and many data queries, it may be more optimal to flood the events and set up gradients towards the event region of the network.

As an example, in the HHA architecture, the application of rumor-based protocols could be useful in large distributed networks to facilitate the efficient data aggregation by periodic transfer of event data throughout the specific regions of the sensor field at levels 2 and 3 [25].

Geographical-Based Paradigm   Geographic paradigms [41] had been proposed as methodologies for forwarding the data from a sender to a destination, progressively reducing the distance between the source and the destination. A local optimal choice is performed where each relay node chooses as next hop the best neighbor in its proximity, (i.e., the node in its neighborhood that is geographically closest to the destination). The greedy choices are performed based only on the local information about the neighbors of the nodes. In the case, when holes are met in the network and, thus, greedy forwarding is impossible to apply, *perimeter forwarding* can be invoked. A data aggregation protocol based on the *Greedy Other Adaptive Face Routing* (GOAFR) geographical protocol had been proposed in [42]. The protocol that is called *Distributed Data Aggregation Protocol* (DDAP), builds upon the GOAFR strategy to provide efficient data aggregation using node localization information. The objective of the algorithm is to find an efficient data aggregation method that reduces the number of messages needed to report an event to the sink node. In principle, the minimum number of transmissions required is equal to the number of edges in the minimum Steiner tree in the network, which contains all the alerted sensors and the sink node.

Finding an optimal aggregation tree in the network requires global information on the nodes and available communication links among them [25]. One simple approximate solution is to use the *shortest path tree* (SPT) for data aggregation and transmission. In this data aggregation scheme, each source sends its data to the sink along the shortest path between the two, and overlapping paths are combined to form the aggregation tree. DDAP is a self-organizing algorithm, that uses randomization to distribute the data aggregator roles among the nodes in the network. Nodes can elect themselves to be local *aggregator nodes* (ANs) at any given time with a given probability. These ANs then broadcast their status only to their neighbors. The task for all the nodes is to take notes, which of their neighbors act as AN. The optimal number of local AN in the system needs to be determined a priori. This will mostly depend on the network topology.

The GOAFR geographical routing protocol had been modified to incorporate DDAP [43]. The extension, the *Geographical Routing with Aggregator Nodes* (GRAN), works as follows. When looking for the next hop, instead of trying to find the neighbor that is closest to the sink, the task is to find the neighboring AN that is closest to the sink. If there is no neighboring AN at all, the routing continues as in the original GOAFR algorithm. If there is one (or more) AN(s) within a radio range but farther away from the sink than the actual node, the routing switches back to the original GOAFR as well. If there is (at least) one AN that is closer to the sink than the actual node, the message is sent to that AN even if there are neighboring nodes closer to the sink. This routing can result in a longer path than the optimal path would be, and the efficiency of opportunistic data aggregation is increased significantly by preferring the ANs as relay nodes.

Semantic-Level Data Aggregation
When performing data aggregation, it is not sufficient to deal with networking aspects, (i.e., to define the metrics to build the most appropriate aggregation paths); instead, in order to completely design the data fusion process, the semantics needed to correctly disseminate queries and collect and fuse sensor data should be specified.

The chronologically first and most simple framework the semantics behind performing an aggregation process had been considered was [44]. There aggregation is regarded as the provision of a core service through the system software. By use of a high-level programming abstraction users can request data from the network without taking into consideration any programming problems.

The *tiny aggregation* (TAG) service was developed to assist this for sensor networks. TAG provides an interface for the data collection and aggregation driven by the selection and aggregation facilities in database query languages. Moreover, TAG allows for distributing evenly the aggregation process among the network nodes, thus also reducing the energy consumption. Queries are expressed through an SQL-like language over a single table. Queries are in the form:

```
SELECT {agg(expr),attrs) FROM sensors
WHERE {selPreds}
GROUP BY {attrs}
HAVING (havingPreds)
EPOCH DURATION i
```

Similarly to the SQL language, the SELECT clause is used to specify the expression and the attributes by which sensor readings are selected. The WHERE clause is used to filter the specific sensor readings. The GROUP BY clause specifies a partition of the sensor readings driven by some interest attributes. The HAVING clause filters some groups, which do not satisfy specific predicates. The EPOCH DURATION clause specifies when the updates should be delivered.

Only few differences between the TAG and SQL queries exist. The output of the TAG queries is a stream of values and not a single value because sometimes, in monitoring applications, it is preferred to have more data than a single value, which better characterizes the network behavior. The aggregation is performed here by using three functions: a merging function f, an initializer I, and an evaluator e.

F has the structure $<z>=f(<x>,<y>)$ where $<x>$ and $<y>$ are the partial state records that register intermediate computation states at certain sensors required to evaluate the aggregation; $<z>$ is a partial state obtained by applying f to $<x>$ and $<y>$. The initializer is needed to say how to instantiate a state record for a sensor value; the evaluator instead takes a partial state record and estimates the value of the aggregate.

TAG queries are also characterized through *attributes*, where each node is provided with a catalog of attributes. Attributes represent sensor-monitored values, such as humidity or temperature, or sensor parameters such as the residual energy. When a TAG sensor receives a query, the named fields are changed into local catalog identifiers. If a node does not know the attribute, it labels it in its query as NULL.

The TAG aggregation is performed in two steps: in the first one, called *distribution phase*, the aggregation queries are inserted into the network. In the second phase, called the *collection phase*, the aggregated values are routed into the network. The query semantic is based on the consideration of epochs of a given length, during which a unique aggregate of the devices different readings should be produced. A key aspect here is represented by the selection of the duration of the epoch and the time a sensor node should wait for its children answers and then sending this data up to the sink. In fact, on the one hand, in order to support a good efficiency level a node should wait enough to collect many answers. On the other hand, the delivery delay to the sink should be minimized, even in the case that there are possible high delays in reporting. Another aspect to be considered is related to the *network partition* or *grouping*, (i.e., the partition of nodes) into groups to make the aggregation a more limited process and avoid emitting more than an aggregate per epoch per group at each node.

The main advantages of TAG are:

- Lower communication overhead with respect to a centralized aggregation approach;
- Long idle times are implemented through the epoch mechanism, which reduces the energy consumption;
- Reduction in the number of messages transmitted by a node during each epoch, independently of its depth in the tree. This allows for saving energy resources at nodes closer to the sink.

As an evolution of the semantic approach proposed in [44], a *semantic sensor data fusion* was proposed in [45]. In this work the synthesis between the use of the MicroToPSS middleware constrained by some knowledge on the application expressed through an appropriate application ontology was discussed. The idea was that a certain application could be interested only in having knowledge of the overall status of the environment and not in the specific parameters monitored by sensors, distributed throughout the environment. In fact, only recently some research focused on the methodology for processing and providing of low-level sensor data to high-level applications appeared. This implies that low-level data should be appropriately filtered, aggregated and correlated from heterogeneous and various sensor network sources. The final target would be to allow the applications to exploit the sensor data while disregarding the low-level language and interfaces. Accordingly, applications will express their interest in semantic events that are transformed into sensor network level events. To provide this semantic communication a publish/subscribe [46] messaging approach was used. The employed middleware, denoted as MicroToPSS, allows the management of data flows to support the specifically distributed automation applications.

The most interesting aspects of this middleware are:

- SQL-like language;
- Task abstraction similar to the object-oriented abstraction;
- Distributed run-time data flow optimizer working transparently from the application.

Another system for semantic data fusion was proposed in [47] based on the use of a *content-based publish/subscribe* (CPS) technique [46] to couple the application scenarios to semantic data fusion of low-level sensors data. The proposed system had been inspired from the MicroToPSS middleware and defined semantic level events independently of the sensor interfaces and applications.

The CPS is a messaging model where notifications are delivered to clients based on their interests.

Different roles are distinguished in the CPS: subscribers, publishers, and brokers. Subscribers emit subscriptions, that are filtering constraints on *tuples*, which describe the kind of data they are interested in. These subscriptions are stored by *brokers*. Publishers issue publications, that are sets of attribute-based tuples, to brokers who take care of forwarding the publications to subscribers based on the received notifications.

Security Aspects of Data Aggregation

Security is a major concern in ad hoc networks. For example—due to the deployment of nodes (e.g., sensors) in open unattended environments, as well as the vulnerability of cheap network devices, whose cryptographic keys can be kept by adversaries, false actions can be invoked in the network driven by unreliable data received by the sink(s) [25].

Possible compromised aggregated data can imply that the sink has wrong information about events happening in a portion of the entire network. This problem can be additionally aggravated by the position of the aggregator node in the network. Untrustworthy, high-level nodes produce more serious problems than low-level nodes because the aggregated result calculated by a high-level node derives from the collection of the measurements performed by a higher number of nodes.

Assuming that an aggregator node collects the information sent by multiple sensor devices, processes, and sends it again towards the sink(s), it is important that the final information collected by the sink is reliable. In existing proposals, aggregator nodes are usually assumed to be *honest* (i.e., to send throughout the network reliable and secure data). However, in a realistic scenario it should be considered that aggregators are very prone to physical attacks, which could compromise the data reported by them, thus leading to false alarms. Accordingly some techniques aimed at verifying the content of the data received at the sink should be devised.

The following three types of security threats can be discussed when considering data aggregation [25]:

- *T1*: an aggregator node simply drops the information it is expected to relay to the sink;

- *T2*: an aggregator node falsifies its readings so as to modify the aggregate value it is expected to relay to the sink;
- *T3*: an aggregator node falsifies or modifies the aggregate generated by messages received by its children.

While T1 can be identified with the condition when a node failure is met, thus causing loss of information generated by a node, the T2 can be solved using fault tolerance approaches.

As an example, [48] proposes an estimation algorithm to be performed at the fusion center for localizing events by combining the binary measurements received by the sensor nodes. The main idea behind the *subtract negative add on positive* (SNAP) is to fuse the observations of all sensors (positive or negative) to efficiently construct a likelihood matrix by summing contributions of ±1. In other words, sensors with positive observations add their contributions, while sensors with negative observations subtract their contributions. The power of this algorithm lies in the simplicity. Also, by giving an equal importance to the sensors that did and the ones that did not sense the event, we do not allow any particular sensor to change the estimation results and this is the basic reason for the fault tolerant behavior. Moreover, the algorithm is energy efficient since for the construction of the likelihood matrix only single bits need to be transmitted to the sink. The SNAP algorithm works in the following major phases:

- *Grid formation*: The entire area is divided into a grid. The number of cells is a trade-off between the estimation accuracy and the complexity. Each sensor node is associated with a cell (i, j) based on its position.
- *Likelihood matrix construction*: For each cell of the grid the likelihood of a source occurring in the particular cell, based on the binary data received from each sensor node is calculated. To this purpose the *region of coverage* (ROC) is defined as a neighborhood of cells around (i, j). This is shown in Figure 4.6. There are various ways that one can define the neighborhood. The optimal shape of the ROC depends on the signal propagation model. Given a uniform propagation model, one could assume that the ROC is the set of all cells whose centers are within a distance R from the center of cell (i, j). However, for computational efficiency, a square approximation can be used as shown in Figure 4.7. Each alarmed sensor adds a positive one (+1) contribution to

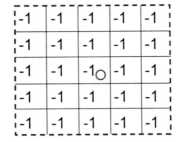

Alarmed Sensor *n*   Non-alarmed Sensor *n*

Figure 4.6   Region of coverage (ROC) [25].

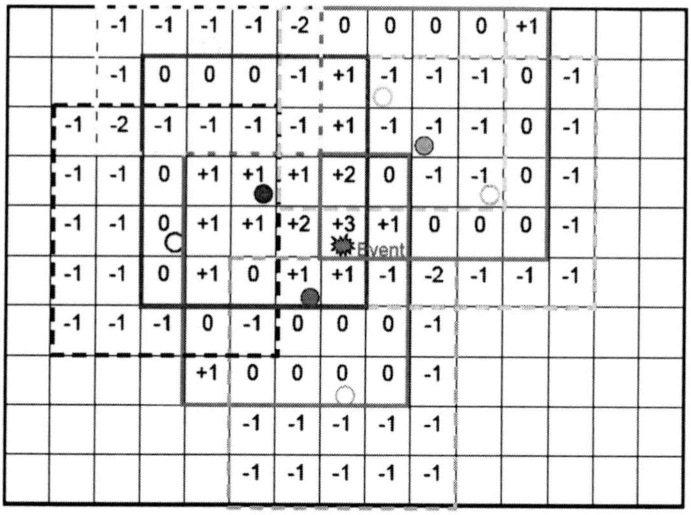

Figure 4.7 *L* resulting from SNAP with 8 sensor nodes, 3 of which are alarmed and are shown in solid color. The event is correctly localized in the grid cell with the maximum value +3 [25].

the elements of *L* that correspond to the cells inside its ROC. On the other hand, every non-alarmed sensor adds a negative one (–1) contribution to all elements of *L* that correspond to its ROC as shown in Figure 4.6. Thus, the elements of the likelihood matrix are obtained by spatially superimposing and adding the ROC of all the sensor nodes in the field. Figure 4.7 shows an example of the resulting likelihood matrix after adding and subtracting the contributions of 8 sensors using SNAP.

- *Maximization*: The maximum of the constructed likelihood matrix L points to the estimated event location. If more than one elements of the likelihood matrix have the same maximum value, the estimated event position is the *centroid* of the corresponding cell centers.

The fault tolerance of SNAP, can be tested by an experiment assuming a variation in the number of faulty sensor nodes [25]. The sensor nodes exhibiting erroneous behavior are randomly chosen and their original belief simply reversed.

An example is shown in Figure 4.8.

Figure 4.9 shows the results for the fault tolerance analysis of the SNAP, a centralized *maximum likelihood* (ML) [49], and a *centroid estimator* (CE).

In the absence of faults the performance of ML and SNAP is very similar. As the number of faults increases, however, ML is very sensitive to the sensor faults and loses accuracy continuously. For sensor faults as few as 10, it starts losing accuracy and for sensor faults greater than 20, its performance is worse than the CE. SNAP displays a fault tolerant behavior and loses very little in accuracy even when 25% or 50 out of the 200 sensor nodes exhibit erroneous behavior.

CE is especially sensitive to the presence of false positives because of noise or malfunction due to overheating. The algorithm essentially treats all alarmed sensor nodes with equal weight, therefore, it is especially sensitive to false positives that occur far away from the true event location. These can result in large errors when

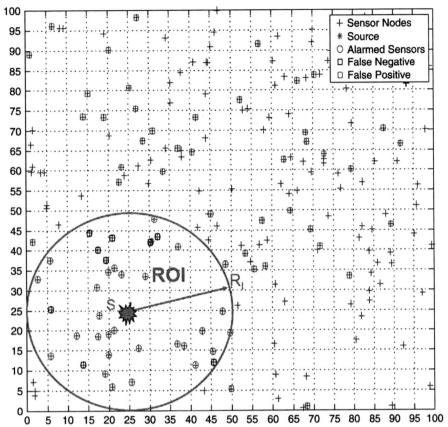

Figure 4.8 A field with 200 randomly placed sensor nodes and a source placed at position (25,25). Alarmed sensors are indicated on the plot with circles inside the disc around the source (ROI). 50 of the sensor nodes exhibit faulty behavior and are indicated [25].

calculating the centroid of the alarmed node positions. ML is extremely sensitive to false negatives. Even a single faulty node inside the neighborhood of the source can completely throw off the estimation results. This is a direct result of the construction of the likelihood matrix of ML.

Among the three security threats discussed above, T3 is the most critical. No stable solutions were available at the time of writing of this book. Some preliminary approaches to the solution of T3 can be found in [50–52].

In [51] a secure hop-by-hop data aggregation protocol was proposed. Differently from [50], which was one of the first works where malicious nodes were considered, here security relies on collaboration among network nodes. More specifically, by exploiting the divide and conquer paradigm, network nodes are probabilistically partitioned into multiple logical trees or groups. Accordingly, since a reduced number of nodes will be grouped under a logical tree routed at a high level aggregator node, the threats related to a compromised high level node have a reduced impact. An aggregate packet is generated by each group and once a user receives aggregates by all groups it uses a *bivariate multiple-outlier* detection algorithm to figure out if suspicious packets were received. When a suspicious group is identified it takes part in an attestation process to prove

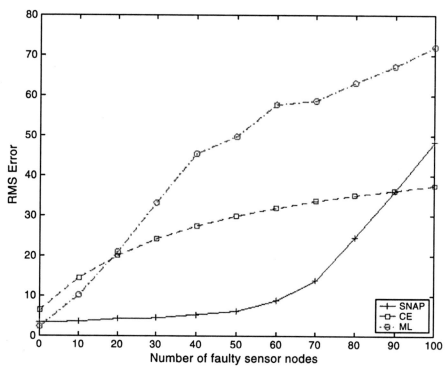

Figure 4.9 Fault tolerance analysis for a 100x100 field with 200 randomly deployed sensor nodes. (RMS Error reported is the average for 500 Monte-Carlo simulations). SNAP displays a fault tolerant behavior and loses very little in accuracy even when 25% of the 200 sensor nodes are faulty [25].

its honesty. If a groups fails in attesting its honesty, its packets are discarded by the user who, finally, takes into consideration only packets received by verified groups.

A similar problem is dealt with by [52]. There, the focus is on use of *multipath networking* paradigms for performing the aggregation, which are more reliable in case of network failures with respect to tree-based approaches but still present some problems of reliability in case of duplicate-sensitive aggregate metrics such as count and sum. To this purpose the *synopsis diffusion* routing approach had been proposed. However, *synopsis diffusion* does not provide any support for security and thus in [52] an evolution of the synopsis was presented. In this, a subset of the network nodes provides an authentication code as well as data packets responding to a possible query. These authentication codes are propagated up to the sink and used to estimate the accuracy of the final result so as to filter and discard possible compromised packets.

4.1.2.2 Middleware

The design of efficient management software architectures plays a very important role in terms of real functionalities. For example, in a WSN, in order to decouple the development of the applications from the hardware features of the mote platform, it is necessary to design an efficient *middleware layer* that can perform various tasks, such as query requests, command dissemination, wireless mote re-programming, and so forth is key [24]. Namely, existing real-world applications of ad hoc networks have been designed for a specific application or purpose. In many instances, however, it may be possible to exploit the ad hoc infrastructure also for other applications rather than have the system lie idle. A suitable middleware layer can allow for the rapid deployment of new applications or for mechanisms for self-composing applications; then such a network could serve a number of purposes and facilitate an evolution of services.

A pervasive and seamless services access architecture cannot exist without a huge amount of information exchanged among all entities playing complementary roles [57]. The main components of such kind of environment are based on the ability of service, network resources, and terminals discovery mechanisms, and notifications of several significant events related to network and terminals behavior. The middleware is the enabler for this interaction by providing the intelligence needed for the services in a system to act rationally. A system is considered *rational* if it does the "right thing," given the available information [58].

An overall view of a pervasive service platform is shown in Figure 4.10. It provides a set of *application program interfaces* (APIs) towards third party service providers, which are mostly based on open but controlled web standards, and relate to the two areas of functionality described above. Service and Identity Management APIs provide functionality for the VID management and personalization, service discovery and composition, dynamic session management and deployment. The *user experience management* APIs provide access to individual functionalities such as learning user preferences, negotiation of privacy when accessing services, and refined context provisioning. Such a layered approach could be an attractive transition path to third party service providers who have already invested in Web standards [21]. One essential point is that these components of the pervasive

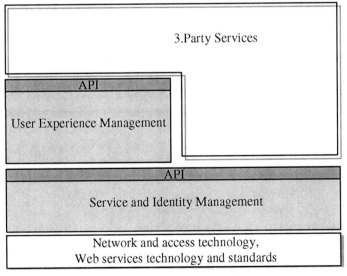

Figure 4.10   General view of a pervasive service platform [21].

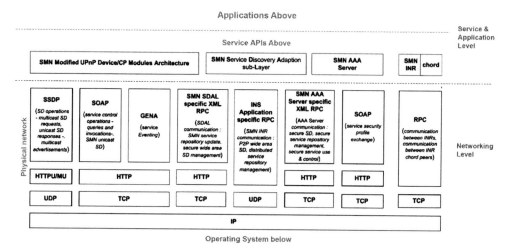

Figure 4.11 Protocol stack of middleware components [59].

platform are distributed across entities, but may be also provided by specialized operators.

Figure 4.11 shows a the protocol stack of a *service session management module* that is incorporated for the monitoring and controlling service sessions between clients and servers in a PN-F scenario.

PN-F participants publish, register, update, and discover the information on their shared services within the PN-F through a *PN-F agent service overlay*, by relying on an intentional name format providing the needed service descriptions. The PN services are discovered by the *service management platform* (SMP), therefore, it is possible to divert the signaling and, when required, the normal service flows through the SMP [59]. In this case, the SMP overhears signaling messages between the clients and servers to achieve monitoring, or sends control messages (to either clients, servers, or both) to harness service sessions. The service session control in the simplest case may include termination of a service session because of the policy enforced by other components of the system (e.g., a policy engine), or the termination and reinitiation of the service sessions in case of mobility. A PN user will interact with one of its cluster *service management nodes* (SMNs) to perform a service discovery request. The SMP service discovery mechanisms are normally designed to provide *name-based* queries on any combination of service attributes. In the case when one of the service discovery attributes is PN-F scoped, the PN cluster SMN receiving the query will forward this query to the appropriate PN-F agent of its PN, in a transparent way from the user standpoint. Otherwise the SD query will be propagated within the PN SMN service overlay.

Semantic Data Spaces
Semantic data spaces are used as interaction medium in pervasive computing environments to deal with the highly dynamic and unreliable character. Taking up the idea of the *service-oriented architecture* (SOA), the service provider and requester are loosely coupled with regard to time, space, and request representation. Figure 4.12 shows an exemplary interaction sequence.

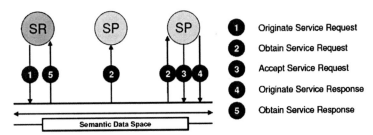

Figure 4.12 Exemplary interaction of service providers (SP) and requester (SR).

The *service requester* (SR) originates a *service request* and publishes it to the *semantic data space*. Suitable *service providers* (SP) are notified, whereas one of them accepts the request and fulfils it. The originated *service response* is then obtained by the *service requester*. The SPs are not addressed by their name, but by the functionality that they are supposed to provide. In opposite to normal publish-subscribe mechanisms, it is also possible to subscribe for and get a *service response* that was published to the *semantic data space* before. Furthermore, SRs can use the data space environment to directly retrieve the needed information. Regarding this, it is also possible to monitor communication through the *semantic data space* [53].

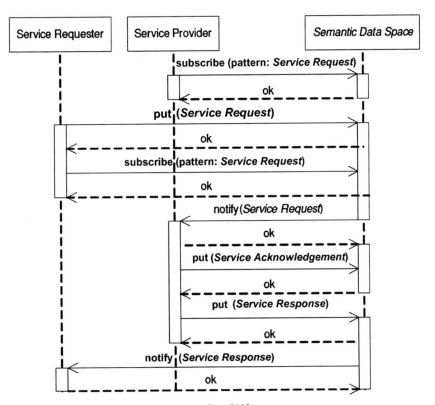

Figure 4.13 Simplified illustration of the data flow [53].

A REST-conform peer-to-peer framework had been developed in [54], serving as a base for the realization of the *semantic data spaces* approach. Initially, the *semantic data space* had been implemented in a centralized manner on a single peer, but it is also possible to distribute it over multiple nodes to increase reliability. The semantic data space, consists of multiple disjoint data spaces, whereas each data space is used for one particular interaction subject. A data space handle is used for the identification.

Exchanged documents through the data space are represented in RDF/OWL [55] to facilitate a semantic description as well as the retrieval of data from the semantic data space with the RDF query language SPARQL [56]. A complete example of an interaction sequence in a realization proposed in [53] is given in Figure 4.13. There, the semantic data space appears as a separated component to simplify matters.

Ontologies and the Semantic Web

The *semantic Web* is an evolution of the current *World-Wide Web* (WWW) [61]. In the WWW vision for the semantic Web, the semantic metadata that describes the contextually relevant information is used to play a pivotal role in the adaptation or the transformation of the delivery of the service capabilities to end users [60]. In the WWW semantic service scenarios, the semantic metadata including all domain-specific information about the *content in a specific context* or a setting in a *domain-specific representation* of the concepts for the specific domain using an ontology language, or a similar conceptualization representation language, is used as an implicit input to the operations of the system.

To enable the semantic Web, the following three key tasks must be solved:

1. The development of suitable *ontology languages* (OL) that allow for the representation of the semantic metadata;
2. The development of suitable *ontology query languages* (OQL) to query over the data models created in the ontology languages;
3. Implementation of *efficient reasoners* that support inference and a logic based on entailment.

An *ontology* is a data model. It is a set of machine interpretable representations used to model an area of knowledge or some part of the world, including software. An ontology represents a set of concepts within a domain and the relationships between those concepts. It is used to reason about the objects within that domain.

Ontology consists of several concepts. Three key ontology concepts are classes, instances, and relations. Classes are the types of instances. This is similar to *object oriented programming* where one can define one's own class and instantiate an object, the type of which is the class of the object.

Classes can also be interpreted as abstract groups or collections of instances. A subclass relation is a common tool for building class hierarchies. These classifications are sometimes called taxonomies.

Ontology without instances can be said to form a structured type system, a schema.

Instances (individuals) are the basic components of the ontology. An instance belongs to a certain class. The type of the instance is the name of the class. Because of the subclass relationship between classes, an instance can belong to several classes: if it is an instance of a child class, it is also an instance of the parent class. There can be classes that have no instances.

The relations define how the concepts can relate to each other. The sub-class relationship mentioned above is one example. The relations have a domain and a range. A relation maps one concept of the domain to a concept in the range.

Relations can map an instance to another instance, an instance to a class or a class to another class. In the relationship, the domain is a *class* and the scope is a class as well.

Existing ontologies can be used as *building* blocks for new ontologies.

It is encouraged to use existing ontologies as much as possible. The aim is to avoid having several ontologies representing the same part of the world.

Ontologies can be presented in several formats. What is the most suitable format for a specific task or a reasoning process is system specific. However, the ontology interchange format between systems and between a system and a user of the system is valuable to be commonly agreed. This allows systems to communicate and new systems to build on top of the multiple existing ones.

The knowledge presented in the ontology can have much longer lifetime than the system it was originally designed for. This imposes certain restrictions to the serialized format of the knowledge. Self-descriptiveness is the most notable.

An *ontology language* is a formal language dedicated for presenting ontologies. The *ontology Web language* (OWL) is one example of an ontology language. It is based on the *Resource Description Framework* (RDF) and W3C URL technologies [62, 63].

Ontology languages play a very important role in the semantic Web, because these are formal languages that allow for making the information available to both humans and computers by constructing ontologies [60].

[17] developed a *context management ontology* (COMANTO), that can enable the various stakeholders (i.e., end users, context providers, service providers, resource providers, and manufacturers) to share a common vocabulary, and to collaborate and synchronize their knowledge. COMANTO can establish efficient and lightweight runtime ontology mapping mechanisms to achieve a resource-aware automatic deployment and reconfiguration, based on common semantic principles. With the fast evolution in the hardware and software industry, it is important that solutions provided today regarding the context formalization are adaptable and extensible. In the COMANTO framework, a generic upper-level context ontology was defined, which describes general context types and interrelationships that are not domain-, application- or condition-specific, and may address the requirements of pervasive computing environments.

The root class of the COMANTO ontology is the *semantic context entity* (SCE). Several subclasses that extend this major super class were identified. The core classes of the designed ontology that are subclasses of the SCE class are shown in Figure 4.14.

The context model includes all the classes that model the context information to be retrieved, exchanged, maintained, and managed.

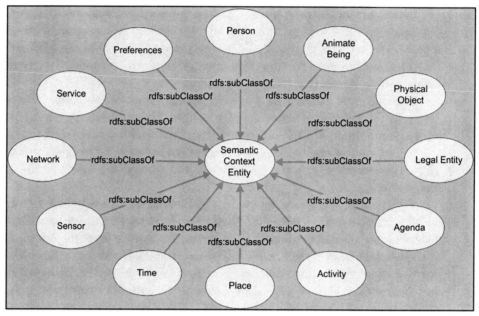

Figure 4.14   The COMANTO ontology core semantic context entities [21].

4.2   Context-Aware Service Discovery

Context-aware service discovery is based on a concept where a service score is associated to all discovered services, which reflects the relevance to the user in his/her given context (considering the user's preferences) [59]. Context, together with the user profile information is the basis for the personalization of services and a better user experience (see Section 4.1.1 of this book).

4.2.1   Context-Awareness for PNs

A PN scenario was briefly described in Section 4.1.2.2 of this book and in [64] in the context of service platforms. In a PN, the context management has to work on the personal nodes that are currently available. Nodes may go down, and connections may become unavailable. For the context management system this means the following [65]:

- It cannot rely on a dedicated context infrastructure;
- It has to dynamically adapt to changes in the availability of the nodes.

A detailed representation of an ad hoc PN-F scenario with the information exchange and other interactions is shown in Figure 4.15.

The scenario is initiated by the user who defines the PN-F profile. This action is done through the *graphical user interface* (GUI) provided by the *PN manager*. Besides setting the PN-F Profile parameters, the user has to decide whether it would be an ad hoc, infrastructure, or combined kind of publication [66]. For the ad hoc

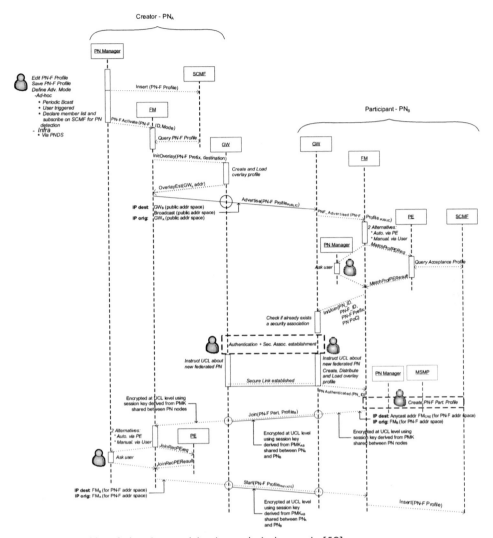

Figure 4.15 Ad hoc federation participation technical scenario [59].

case, the policy for sending the advertisements must also be defined. Three possibilities were considered in [59]: periodically, at once triggered by the user action, or to specific PNs. For the latter case, the *federation manager* (FM) should register on the SCMF for the PN detection event.

The FM gets the PN-F profile from the SCMF and initiates the advertisement. If the advertisement is to be a unicast packet, the FM has to ask the corresponding GW about the address, to which the advertisement should be sent. This address belongs to a public address space (e.g., provided by a DHCP server on a hot spot). In any of the cases, the FM instructs the GW to set-up the network overlay for the specific PN-F in order to be prepared for its future use.

When the advertisement reaches the participant GW, the latter informs its FM about the event. Upon this indication, the FM may contact the *policy engine* (PE), in order to automatically decide whether the PN-F is interesting or not, or it may contact the user by showing an alert on the PN Manager GUI. For the first case, the PE in turn checks a predefined acceptance profile (can be part of the user profile) that contains the interests of the user and the kind of PN-Fs she is interested in.

In any of the cases, upon an affirmative response, the FM indicates to the corresponding GW the willingness to form a PN-F with the peer PN. The security block on the GW, first checks if there is a security association already established with this PN, and if this is not the case, starts the authentication and security association establishment. This procedure is shown in Figure 4.17.

The two PNs in Figure 4.16 are connected via the *PN-formation protocol* (PFP). The protocol is used on the MAC layer if the gateways are on the same link, and an encapsulated PFP can be used for any other case. The result of this process is a *primary master key* (PMK) shared between both PNs, which forms the basis for securing any communication between two nodes belonging to these PNs. If a valid PMK is already available, this process is not needed. The PMK is passed to a *universal convergence layer* (UCL), which can use it to establish the secure link between both GWs. The second step after securing the connection between ad-hoc PNs is to exchange certificates. These are *self-signed* certificates, and the same public keys are derived in *certified PFP* (CPFP).

The CPFP protocol provides means to have new nodes join the PN, to revoke nodes, or to trigger the establishment of bilateral long-term keying material between nodes that form part of the PN [68].

If the UCL of the GW can establish a secure link with the peer PN one, then it reports this to the FM, which requests the user to create the specific PN-F *partici-*

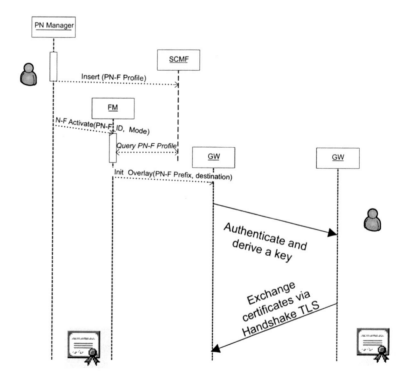

Figure 4.16 Ad hoc certificate exchange during a PN formation [59].

pation profile. This profile organizes the participation of the PN in that particular PN-F. For this process, the FM interacts with the SMP (as controller of PN services) and the SCMF (as a provider of the context information and a final holder of the PN-F participation profile). The PN-F participation profile editing step should be kept very simple for the user.

When the PN-F participation profile is ready, the FM sends it to the creator's FM. As shown in Figure 4.15, a *Join* message is encapsulated in an IP datagram destined to an anycast address representing the creator's FM, and originated on the own FM. Both of these will belong to the proper PN-F address space [59].

The FM and GW do not have to be on the same node, therefore, the route of this packet may include the links between the personal nodes within the same cluster. This part of the route is encrypted at the UCL level, using intra-cluster security mechanisms. When the packet reaches the GW, the routing block on it (see Figure 4.16) forwards it to the creator's GW, and from there it is in turn forwarded to the creator's FM.

After an analysis of the participant's PN-F participation profile (whether automatically via the PE or manually by alerting the user), and if the result is positive, the creator will issue a *Start* message to the participant, concluding the participation phase.

From the security point of view, the *Join* message contains a federation token [67, 68]. The token is used for accessing all the resources shared in the federation. The messages are all signed with *cryptographically generated IDs* (CGI) to provide a proof of uniqueness.

Service information can be handled by the SMN in the SMP but among all the service-related data the ones related to the service availability/accessibility, and maybe to the service session status, are viewed as *context information*. This means that these have to be forwarded to a *security context management framework* (SCMF) [65] by the SMP, (i.e., by the SMN to the local context agent). In the extreme case, the SCMF has to work on a single node, providing access to context information available from local context sources as well as to user profile information available from the local storage.

Therefore, a *context agent* (CA) is running on every node. It provides applications and services with access to all context and user profile information that is currently available. The SCMF in a PN is composed of the interacting CAs on all the nodes.

According to the SCMF architecture, this implies that the SMN acts as a data source forwarding the service-related context data to the *data source abstraction layer* (DSAL) of the context agent [59]. A high-level architecture of a context agent is shown in Figure 4.17. Requests for context information originate at the local *context aware components* (CACo) formulated in the *context access language* (CALA) and are sent via the *context management interface* (CMI) $[I_a]$. They then go through the *context-aware security manager* (CASM) $[I_b]$ that may enforce privacy policies to the *context access manager* (CAM) $[I_c]$. The CAM has an index information about the context information that is available locally, either from the *data sources* through the DSAL or the *processing and storage* module (P&S). If the context information for answering the request is available locally, the relevant parts of the request are forwarded to the DSAL $[I_f]$ or the P&S $[I_e]$.

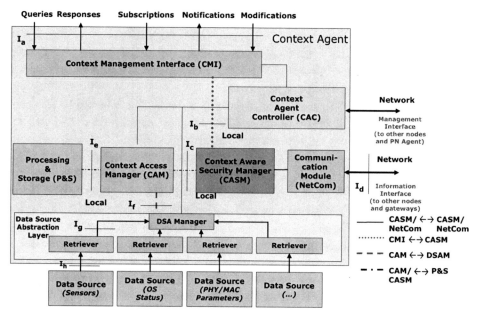

Figure 4.17 High-level architecture of a context agent [65].

Depending on the role of the node in the SCMF and the scope of the request, the request is forwarded to other nodes through the CASM that may again enforce privacy policies, and the *communication module* (NetCom) $[I_d]$. Finally, the CAM integrates and filters the returned results and returns them to the requesting CACo, again through the CASM and the CMI.

The *context agent controller* (CAC) is responsible for the configuring and dynamically reconfiguring of the CA that might become necessary due to changes in the availability and roles of other CAs.

The purpose of the *DSAL* is to provide a uniform interface to context information from all data sources to the CAM. For each data source, there is a retriever that implements the proprietary data source interface $[I_h]$. The retriever also converts the data from its proprietary representation into the common SCMF representation that follows an ontology-based context model [66] and, on request, provides it to the DSA Manager (DSAM) $[I_g]$. The DSAM manages the different retrievers and provides a single interface for synchronous and asynchronous requests to the CAM $[I_f]$, thereby, hiding the details and the distribution aspects of the DSAL.

The *CAM* is responsible for the processing of the CALA requests. To efficiently do this, it keeps the index structures about what context information is available from the DSAL, the P&S module, and, depending on its role, possibly also what context information is available from other nodes in its cluster. In general, the processing of CALA requests consists of the following steps, not all of which may be applicable for all types of requests:

- Decomposition of requests into subrequests based on the index information;
- Forwarding the subrequests for gathering the context information from local and distant sources depending on the scope of the request;
- Integrating the results;

- Filtering the results according to restrictions or/and subscription conditions in the original request.

The P&S module is responsible for the following:

- Storing of context and user profile information. Information stored there has either been put there by applications, (e.g., user profile information, or the information is the result of some processing or replication). The storage can also be used for storing histories or tendencies.
- Processing of context information, (e.g., to derive higher-level context information like the current situation of the user or certain part of the meta data such as confidence).

The *CMI* handles the interaction between any CACo and the SCMF for accessing or modifying context information. These interactions are performed using CALA over the I_a communication interface. This interface supports both synchronous and asynchronous access to information. Hence, the purpose of this module is to convert from the XML-based CALA representation used by CACos to the CA-internal representation and also to manage the call-back information in the case of asynchronous requests. [5] developed a prototype implementation for the SCMF and, there, the interface I_a is implemented using XML-RPC [69].

The purpose of *CASM* is to enforce the user's privacy policies with respect to accessing the context and the user profile information. This is especially relevant in the case of interactions in a PN-F or with external components. The location of the CASM is selected so that all information going in and out needs to pass through this module. This ensures that all the information and the request messages are authenticated, authorized (and potentially accounted), and that information going out of the CAM is ensured to fulfill the privacy requirements of the owner of the device [65].

There are the following two types of CAs depending on the type of device:

- The *basic context node* (BCN) is a CA with rather limited functionality;
- The *enhanced context node* (ECN) is a full functionality CA.

The BCN nodes are *autonomous* nodes, able to have access to the context information that they are able to gather. Nevertheless, due to their low processing capabilities, the BCNs are not able to derive complex pieces of context information and rely on other nodes for this kind of requests.

The following two roles are typically fulfilled by stronger nodes in the cluster and the PN, respectively:

- *Context management node* (CMN): The full version of the CA, but with extra responsibilities of keeping the index structures of what context is available inside a cluster, and to act as a broker to other CAs in that cluster. Only one CMN per cluster is allowed.
- *Context management gateway* (CMG): A dedicated node-enabling context being shared across different PN's in the case of PN-federations.

The CMN—similarly to the SMN—serves also as a *broker* node within each cluster. Two modes of operation are supported, the locator-mode and the context-access mode. In the locator mode, a CMN only returns the locators pointing to the nodes that have the actual information [65]. This information is available from its index structures. The requesting node then directly contacts the nodes that have the information and retrieves it from there. In the context-access mode, the CMN will gather the actual context information and return it directly.

The CMG is responsible for the forwarding of requests to and for receiving requests from its peers in the other PNs participating in the PN-F.

An example of how the SCMF could be organized in a network, with the above-mentioned entities is shown in Figure 4.18.

The organization of the SCMF is hierarchical and specifically f\design to comply with the PN network architecture as proposed by [5]. The steps in organizing the SCMF and which are controlled by the CA are shown in Figure 4.19.

Initially, when a user switches on the device, the CA would start by reading its node information (e.g., node identifier and IP address) in the PN. It would then build its index structures based on the information that is available at the node level. This refers to *step 1)* in Figure 4.19.

When the local indexes are built, the CA looks for the cluster CMN, referring to *step 2)*. The selection and the maintenance procedures are implemented for assuring that there is always one CMN on a cluster upon any network circumstances (CMN node switched off, cluster splitting, cluster merging, etc.). In summary, the procedure is divided in two main tasks/phases, 1) the actual selection or discovery of the most capable node; and 2) the maintenance procedure [65]. While the first phase is based on a *challenge-response* procedure, triggered by the nodes upon their

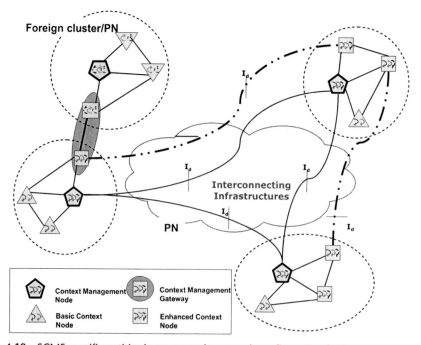

Figure 4.18 SCMF specific entities in an example network configuration [65].

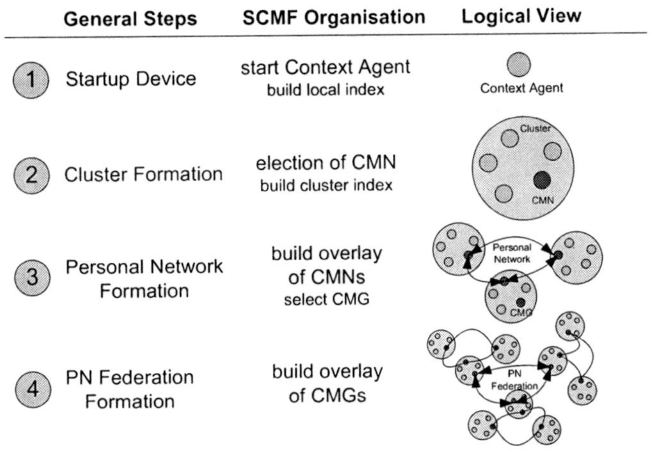

General Steps	SCMF Organisation	Logical View

① Startup Device   start Context Agent / build local index   Context Agent

② Cluster Formation   election of CMN / build cluster index   Cluster / CMN

③ Personal Network Formation   build overlay of CMNs / select CMG   Personal Network / CMG

④ PN Federation Formation   build overlay of CMGs   PN Federation

Figure 4.19 Main steps for the SCMF operation [65].

start-up or when they detect the absence of the CMN; the second is carried out by periodic messages sent by the CMN. These advertisements inform the rest of the nodes about the CMN identity and the current capabilities, enabling the role hand-overs between the nodes in the case when the CMN capacities have gone below a predefined threshold. In such a situation, the new CMN will request the former one to synchronize its internal database whenever possible (i.e., soft-handover like situation). Besides that, a CMN can withdraw its role. This situation will involve a transitory state where the remaining nodes will challenge to take the CMN role, but in all cases the selection algorithm will end up in the definition of a unique CMN within the cluster. Whenever a CMN has to be selected, the nodes will multicast/broadcast their capabilities to the rest of potential competitors. To normalize and reduce the complexity of the selection procedure, a cost function has been defined in order to assess the suitability of a node being the CMN. In a generic manner, each of the parameters are assigned a weight depending on the characteristics desired for the cluster head to be selected resulting in a value named *context node weight* (CNW) [70].

When the requests go beyond the scope of the local cluster (i.e., the scope is PN) the CMNs will interact with their peers in other clusters to find all the nodes in the Personal Network that have information which is potentially relevant for the given request. Prior to this, the CMN's from each cluster needs to organize them selves, and need to become aware of each other. This refers to *Step 3*). In this process the main features that have to be assured are the network *self-organization* and *self-healing*. A dedicated PN Agent framework can be developed [71] and made responsible for handling the PN cluster registration/deregistration. This procedure consists of registering (deregistering) the name of the cluster gateway (which is formed by the gateway and the PN identifiers) and its IP address to the PN agent. The register-ing/deregistering functionality is useful for the SCMF in the process of locating the CMNs from other clusters. Following the CMN election procedure in the cluster, the elected CMN will register itself in the PN agent framework. In turn, the PN agent will notify this registration to the already registered CMNs. In this way, the rest of the clusters will be aware not only about the existence of a new cluster, but also about the identity of the elected CMN within the cluster. It is also possible to follow a reactive approach, on which CMNs queries the PN agent to get the information about the other CMNs only when a PN level query is to be done.

4.2.1.1 Election of a CMN

In a self-configuration step, the CAs on all the PNs in a cluster select a CMN [65]. Additionally, at the PN level, all the CMNs from the different clusters will form a P2P network in order to allow for intercluster context information provisioning. The node, which has been elected, must be qualified to assume this role because it imposes additional responsibilities that require extended capabilities. As the CMN keeps an index of the overall cluster context information with locators or pointers to the desired information, it is also mandatory to ensure the existence of one unique CMN in each cluster to avoid the possible race conditions in the operation of the SCMF.

Within the SCMF, any cluster node (i.e., BCN or ECN) can participate in the challenge to become a CMN. The ECNs will typically be more qualified to take this role as these nodes are provided with full set of functionalities.

The following additional parameters can be considered for the CMN selection [65]:

- Computational power (the type of processor, speed);
- Remaining energy (battery level);
- Available communication interfaces (number of interfaces, bandwidth, coverage, connectivity, number of direct neighbors);
- Load of the node.

Whenever a CMN has to be selected, the nodes will multicast/broadcast their capabilities to the rest of the potential competitors. To normalize and reduce the complexity of the selection procedure, a *cost function* should be defined in order to assess the suitability of a node becoming the CMN. On a generic manner, each of the parameters are assigned a weight depending on the characteristics desired for the CMN, resulting in a value named *context node weight* (CNW):

$$CNW = \sum_i \alpha_i \cdot \Psi_i$$

$$(4.1)$$

where α_i is the parameter weight and Ψ_i is the parameter value. Considering the node connectivity $C(n)$ and the resources $R(n)$ as the two main parameters, then:

$$C(n) = \sum_{i=1}^{N_1(n)} \Phi_i$$

$$(4.2)$$

where $N_1(n)$ is the number of neighbors at one hop distance and Φ_i is a combination of link quality and binary rate of the i-th neighboring link.

For $R(n)$, (2.3) holds:

$$R(n) = P_n \cdot \left[E_n + \sum_{i=1}^{N} I_n^i \right]$$

$$(4.3)$$

where E_n is the remaining battery, P_n is the node processor, and I_n is the weight of the i-th interface of the node.

The node with the highest weight is the one that will be selected as CMN.

Figure 4.20 shows the selected values for these parameters.

Additionally, the saturation limits for both $C(n)$ and $R(n)$ can be defined, because if a high weight value is reached, the capabilities of the nodes do not differ significantly. These saturation limits are the ones that fix the values of parameters λ, β, and so forth.

The load of the node should also be added to the parameters, because the same node should not take all responsibilities within the cluster (e.g., service discovery and provisioning management, context provisioning), if other suitable nodes are available. This is useful to balance the distribution of responsibilities avoiding that slightly less powerful nodes do not take any important role. Therefore the final value of the CNW, depending on the type of CA, can be calculated in the following way:

$$CNW(n) = \text{type of } (CA) \cdot \frac{C(n) + R(n)}{\text{load}(n)} \qquad (4.4)$$

As there always has to be a CMN within a cluster, the selection procedure is launched upon the cluster creation. In [72], the implementation of a *cluster head selection algorithm* used to select the most appropriate node within a WPAN had been presented. This selection algorithm is used for the selection of the node to take the CMN role. Summarizing, the procedure is divided in two main tasks/phases, (1) the actual selection or the discovery of the most capable node and (2) the maintenance procedure.

While the first phase is based on a challenge-response procedure, triggered by the nodes upon their start-up or when they detect the absence of the CMN, the second is carried out by periodic messages sent by the CMN. These advertisements inform the rest of the nodes in the cluster about the CMN and its current capabilities, enabling the role handovers between the nodes in case the CMN capacities have gone below a predefined threshold. In such a situation, a new CMN will request the former one to synchronize its internal database whenever possible (i.e., soft-handover like situation). Besides that, a CMN can withdraw its role. This situation will involve a transitory state where the remaining nodes will challenge to take the CMN role, but in all cases the selection algorithm will end up in the definition of a unique CMN within the cluster [65].

Resource	Stages			
Battery Level $E = \lambda \cdot v$	From 100% (1) to 20% (0.2)			
System Architecture $P = \beta \cdot v$	Workstation Desktop (4)	Mobile device (3)		Embedded device (1)
Air interfaces $I = \gamma \cdot v$	802.15.3 (6)	802.11abg (5)	Bluetooth (2)	802.15.4 (1)
Channel quality $\Phi = \delta \cdot v$	SNR > 30dB (5)	30 dB > SNR > 15 dB (3)		SNR < 15dB (1)

Figure 4.20   Node resource evaluation table [65].

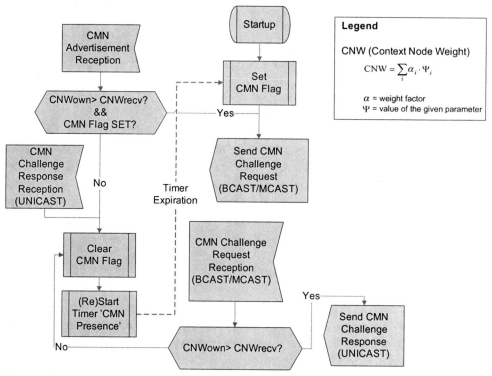

Figure 4.21 CMN challenge and maintenance protocol flow [65].

Figure 4.21 shows the flow diagram for the discovery and maintenance of the CMN.

Whenever a CA is started, it will challenge to become a CMN. It will multicast/broadcast its capabilities to the rest of the potential competitors in its cluster. This implies the use of a cluster-wide multicast address, which will be interpreted by the routing protocol in order to make the message reach the whole cluster.

Upon sending the CMN challenge request, the node considers itself as the CMN of the cluster. This state is kept until any other node answers the initial request. This CMN *Challenge Response* is only sent by the CAs with a higher CNW, indicating the existence of another node within the cluster more suitable to be the CMN. The reception of a response leads to the node withdrawing its CMN state. This could result in transitory stages, where more than one node would consider itself as the CMN. However, this situation will last less than the period of CMN advertisement, which can be considered as a stabilization time, and the rest of the Nodes will never detect more than one CMN.

Upon the reception of a CMN *Challenge Request*, the CNW included in the request will be processed, comparing it with its own one. As the cost function is the same for all the nodes, they apply the same standards. If the received CNW does not surpass its own, the CA will reply by sending a CMN *Challenge Response*. This response denies the request as it is indicating that the responding CA has better attributes to be a CMN than the ones of the requester. Therefore, the reception of a CMN Challenge Response is a sign of the presence of, at least, one CA in a better position and the internal CMN state of the requester has to be quit, becoming an ordinary CA. On the other hand, if the received CNW surpass its own, the node will not send any response and will immediately become a CA, leaving the CMN state if it was the former one. This way of implicitly acknowledging the role of CMN reduces the signaling within the network.

These procedure ensures that only one CMN will exist within the cluster.

The node that holds the CMN role informs of its presence by periodically sending a CMN *Advertisement* message. The reception of this message triggers the activation of a timer, reinitialized each time when another advertisement is received. The expiration of this timer means that the current CMN is no longer available, either because it leaves the cluster area or shuts down, or because its capacities have gone below a predefined threshold and it can no longer keep the CMN role. This situation will involve a transitory state where the remaining CAs will challenge to take the CMN role. It is important to mention that the reconfiguration time is shorter than the advertisement period; this means that at least the Cluster gets a new CMN before two advertisement period times. In order to avoid situations, in which no CMN is present, which might happen when all CAs are below their threshold level, this value could be configured so that each node could settle its own limit and, even more, if none is above the threshold the most capable one will become a CMN.

Figure 4.22 (a) and (b) shows the results for the election process for a scenario implemented and simulated in NS-2 with the purpose of determining the behavior within the cluster depending on the timer [65].

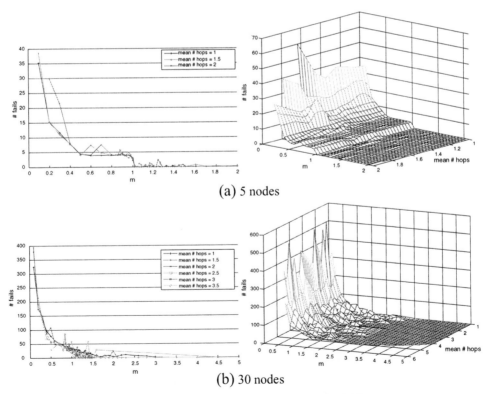

(a) 5 nodes

(b) 30 nodes

Figure 4.22   (a, b) Cluster CMN selection process evaluation results [65].

The number of failures decreases as m increases, where m is the optimal value very close to 1. For the case of a typical small personal cluster (see Figure 4.22(a) 5 nodes) the first value of m for zero faults is approximately between 1.02–1.05, which means that the stabilization timer should be at least 1.02 times the negotiation timer.

The results obtained for a bigger cluster of 10 nodes (not shown) reports quite similar values for parameter m ranging from 1.02 for 1 hop to 1.15 for 2.5 hops. In this case, the number of hops affects the value of m as the packets cover more distance. For the case of 5 nodes this behavior could not be shown as the random positioning does not offer the possibility to obtain layouts with high number of hops.

For the case of 30 nodes [see Figure 4.22 (b)] the optimal value of m is bigger than in the case of lower number of nodes, ranging from 1.1 to 1.35. For the 1-hop scenario, the value of m is really high with a value above 1.5. The reason for this unexpected behavior is the simulation environment itself. As the number of nodes is big and they all send broadcast packages simultaneously, lot of collisions, retransmissions, and even errors take place leading to a bigger latency, which at the end implies a higher number of failures. When the nodes are spread over a wider area, the number of neighbors is smaller, thus less packages are transmitted. The above also explains why the slope of the curves is smaller.

The values obtained are the minimum and optimal values considering the simulation environment. For real implementation it is advisable to increase the value by some small percentage in order to avoid problems such as misbehaviors or all environment variables that may have some impact on the values.

4.2.1.2 Replication of Index Data

Replication is important for achieving a good balance between the efficient access to information and the system-internal overhead [65]. There are two types of data elements that can be replicated within the SCMF, namely (a) index information (i.e., information of the location of certain context information) and (b) the context information itself.

To ensure efficient access to context information, the index information for all cluster nodes is replicated in the CMN, so any context information in the cluster can be accessed with two one-hop requests on the overlay network in the locator-mode case and one two-hop request in the context-access mode case.

Different options should be investigated regarding the replication of index information on the PN level; some of the options are the following:

- *No replication*: The index information is not replicated between the CMNs in different clusters, so all requests with a PN scope have to be forwarded to all CMNs in the PN.
- *Complete replication*: CMNs exchange the complete index information between themselves, so they have a complete map of the context information within the PN.
- *Differentiated replication*: Depending on the type of the context information, it may not make sense to replicate outside the cluster, while others may. Only

the context elements, which indicate a global relevance, will have their index replicated to other CMN's, thus with this approach, context is differentiated between a *local* and a *global* accessibility flag, however, it will require additional meta data related to the context element itself.

PN-Fs typically behave in an *on-demand* manner. Also, replicating the index information about all the context information that is potentially accessible to the requester still has different privacy implications than providing context information on a specific request, where policies can be enforced and changed on a case-by-case basis.

4.2.1.3   Access Strategies to Context Information

The complete replication of the context information within or between clusters would lead to an immense overhead, especially when most of the information may not currently be needed on a regular basis.

If the CA has more than one client trying to access the same context information, a purely reactive strategy may not be the best option at all time. Thus, a CA may desire to switch to other access strategies, (e.g., a subscription, when the information demand becomes high, thus implementing a replication strategy). Basic interaction models to access remote and dynamic context information are the following:

- *Proactive access*:
  - *Periodic*: With some time interval specified in the subscription condition, the remote node sends an update of the value to the subscribing CA (and application);
  - *Event-driven*: As a part of the subscription condition, the remote CA sends a notification, to the subscribing node, if this condition is met.
- *Reactive access*:
  - For every request by a *context aware client*, the CA sends a request to the remote node.

Network Usage of SCMF Context Access Strategies
The performance of the introduced earlier SCMF was evaluated in [5] in terms of the network usage that the context queries incur. The purpose was to compare the implemented access strategies with other possible alternatives.

The analyses were performed in terms of the number of hops that are necessary to successfully complete a SCMF transaction in the form of a query-response action [20, 65]. This metric shows the amount of nodes in the network that are involved on a single query-response operation and provides a good insight on the complexity of the routes that are needed.

Although, the mean number of hops represents a good showcase for the network load generated with a single context query, it is also interesting to see what the actual traffic volume of a query-response operation is. The analyses done to measure this parameter were aimed at understanding the impact of the relation between the packet sizes of the different access modes.

A PN was defined, with a variable number of nodes that were equally distributed over a random number of clusters, such that on the average there were a fifth of the nodes in each cluster.

The performed analyses studied the behavior of a query-response operation. The initiator of the query was randomly selected among all the nodes in the PN. For the destination, two strategies were modeled. In the first one, the model simulates a complete randomly selected target. In real-world terms, this would mean that the context is typically fetched from all the nodes equally in the PN. Thus, a *uniform* distribution was used for selecting the query destination.

The second strategy takes the destination from the nodes that are near the initiator. In real-world terms, this would mean that the context is typically fetched from nodes in the neighborhood. This translates into a *normal* distribution with mean at the initiator. In most of the cases, this would lead to querying for a piece of context that is served by a node in the same cluster.

PN clusters are in essence multihop ad hoc networks, therefore, the distances must be defined in terms of hops that exist within the cluster. Again, two different models were used. In the first one, the number of hops between two nodes in a cluster was defined using a *uniform* distribution with variable mean value. This represents the fact that a PN cluster configuration is loosely connected and with a varying spread of nodes. The second strategy uses a normal distribution where the mean is set to one hop while the variability is controlled with the variance. In real-world terms, this represents tightly coupled clusters, which can be spread, but typically, nodes in them are interconnected through short routes.

Finally, in order to investigate the difference between intra-cluster and inter-cluster communications, a third model is needed. Remote clusters are interconnected through tunnels across the Internet, therefore, it is not realistic to assume the same model as the one for communications within a cluster. In this sense, the complexity of the Internet access could be modeled adding a number of hops each time an access through the Internet is necessary. This leads to some complexity that might affect the system behavior.

The number of hops used by a SCMF query using the context-access mode can be defined as follows:

$$\#hops = \begin{cases} \#hops_{CMN} + \#hops_{dest} & \text{if } init\_cluster = dest\_cluster \\ \#hops_{CMN} + Internet\_Access + \#hops_{dest} & \text{if } init\_cluster \neq dest\_cluster \end{cases} \quad (4.5)$$

Depending on whether the initiator and destination are in the same cluster or not, the total number of hops is different. In the context-access mode, the initiator contacts the CMN for this to directly answer the query. Thus, when both nodes are in the same cluster, the total number of hops corresponds to the number of hops between the initiator and the CMN plus the number of hops between the destination and the CMN. It is important to note that this model considers the necessary hops to populate the CMN information base. Although the CMN directly responds to the initiator, it needs the destination to inform the actual piece of context queried. When the destination is in a different cluster, it is necessary to add one Internet access to the hop count. These three variables are modeled for the simulations using different approaches. $\#hops_{CMN}$ and $\#hops_{dest}$ are *uniform* and *normal random* variables, while the *Internet_Access* is varied to show its impact.

The number of hops when the SCMF *locator* mode [65] is used can be defined as follows:

$$\#hops = \begin{cases} \#hops_{CMN} + \#hops_{init} + \#hops_{dest} & \text{if } init\_cluster = dest\_cluster \\ \#hops_{CMN} + 2 \cdot Internet\_Access + \#hops_{init} + 2 \cdot \#hops_{dest} & \text{if } init\_cluster \neq dest\_cluster \end{cases}$$

$$(4.6)$$

In this case, the initiator only contacts the CMN in order to retrieve a pointer towards the destination. After this, the initiator interacts directly with the destination to get the context information. Therefore, this interaction must be added between the initiator and the destination to the calculations. This represents the $\#hops_{init}$ term when both nodes are in the same cluster and the complete inter-cluster communication path (i.e., $\#hops_{init}$ + Internet_Access + $\#hops_{dest}$) when they are in different clusters.

At this point, it is important to mention the difference between the locator mode and the context-access mode packets. While the first ones are typically reduced in size because they only contain the indexes of the piece of context queried, the latter contains the full context information description. Hence, although in terms of total number of hops, the size of the packet does not affect when looking at the actual traffic volume, it must be differentiated, which of the interactions involves which kind of packet.

Equations (4.7) and (4.8) define the traffic volume calculations, where $Size_{CA}$ is the size of the packets corresponding to the context-access mode and $Size_{LO}$ is the size of the packets corresponding to the locator mode:

$$traffic\ volume = \begin{cases} Size_{CA} \cdot \#hops_{CMN} + Size_{CA} \cdot \#hops_{dest} & \text{if } init\_cluster = dest\_cluster \\ Size_{CA} \cdot \#hops_{CMN} + Size_{CA} \cdot Internet\_Access + Size_{CA} \cdot \#hops_{dest} & \text{if } init\_cluster \neq dest\_cluster \end{cases}$$

$$(4.7)$$

$$traffic\ volume = \begin{cases} Size_{LO} \cdot \#hops_{CMN} + Size_{CA} \cdot \#hops_{init} + Size_{LO} \cdot \#hops_{dest} & \text{if } init\_cluster = dest\_cluster \\ Size_{LO} \cdot \#hops_{CMN} + (Size_{LO} + Size_{CA}) \cdot Internet\_Access + \\ \quad + Size_{CA} \cdot \#hops_{init} + (Size_{LO} + Size_{CA}) \cdot \#hops_{dest} & \text{if } init\_cluster \neq dest\_cluster \end{cases}$$

$$(4.8)$$

Equation (4.7) shows that all the interactions are done using the context-access packets. In (4.8), the first query to the local CMN, the interaction between the CMNs through the Internet, and the interaction between the destination and the remote CMN are done using *locator mode* packets, while the direct interaction between the initiator and the destination uses *context-access mode* packets. It is important to note that all the analyses involve the access to the actual context information. Thus, in locator mode the phase, in which the context information is actually retrieved is considered and not limited to the retrieval of the index to the information storing point.

The behavior of the SCMF context access strategies is evaluated by a centralized approach [65]. Typically, centralized approaches have an agent that is located on the Internet but in the case described here, the agent is randomly placed in one of the PN's clusters. A fully distributed approach, although interesting, is not possible for the adopted scenario, because it would not be possible for the initiator to know where to retrieve the piece of context information it is interested in.

The total number of hops and the traffic volume for a centralized context access strategy using context-access mode is defined as follows:

$$
\#hops = \begin{cases}
\#hops_{Agent} + \#hops_{dest} & \text{if } init\_cluster = agent\_cluster = dest\_cluster \\
\#hops_{init} + \text{Internet\_Access} + \#hops_{dest} & \text{if } init\_cluster = dest\_cluster \neq agent\_cluster \\
\#hops_{Agent} + \text{Internet\_Access} + \#hops_{dest} & \text{if } init\_cluster = agent\_cluster \neq dest\_cluster \\
\#hops_{init} + \text{Internet\_Access} + \#hops_{dest} & \text{if } init\_cluster \neq agent\_cluster = dest\_cluster \\
\#hops_{init} + 2 \cdot \text{Internet\_Access} + \#hops_{dest} & \text{if } init\_cluster \neq agent\_cluster \neq dest\_cluster
\end{cases}
$$

$$(4.9)$$

$$
\text{traffic volume} = \begin{cases}
Size_{CA} \cdot \#hops_{Agent} + Size_{CA} \cdot \#hops_{dest} & \text{if } init\_cluster = agent\_cluster \\
 & \qquad\qquad = dest\_cluster \\
Size_{CA} \cdot \#hops_{init} + Size_{CA} \cdot \text{Internet\_Access} & \text{if } init\_cluster = dest\_cluster \\
\quad + Size_{CA} \cdot \#hops_{dest} & \qquad\qquad \neq agent\_cluster \\
Size_{CA} \cdot \#hops_{Agent} + Size_{CA} \cdot \text{Internet\_Access} & \text{if } init\_cluster = agent\_cluster \\
\quad + Size_{CA} \cdot \#hops_{dest} & \qquad\qquad \neq dest\_cluster \\
Size_{CA} \cdot \#hops_{init} + Size_{CA} \cdot \text{Internet\_Access} & \text{if } init\_cluster \neq agent\_cluster \\
\quad + Size_{CA} \cdot \#hops_{dest} & \qquad\qquad = dest\_cluster \\
Size_{CA} \cdot \#hops_{init} + 2 \cdot Size_{CA} \cdot \text{Internet\_Access} & \text{if } init\_cluster \neq agent\_cluster \\
\quad + Size_{CA} \cdot \#hops_{dest} & \qquad\qquad \neq dest\_cluster
\end{cases}
$$

$$(4.10)$$

All the different possibilities are computed and depending on the location of the different nodes, the count is different. The access to the agent is given by $\#hops_{Agent}$ and the one for populating the agent from the destination is $\#hops_{dest}$. When only the initiator and the destination are in the same cluster, the access towards the agent comprises a number of hops within the cluster ($\#hops_{init}$) that refers to the distance to the cluster gateway plus one *Internet_Access* and the actual access to the information ($\#hops_{dest}$). A similar situation is given when initiator and agent are in the same cluster, but not the destination. The initiator reaches the agent with $\#hops_{Agent}$ hops but the destination needs $\#hops_{dest}$ plus one Internet_Access to populate the agents database. The same number of hops results when the destination and the agent are on one cluster and the initiator, in another one. Finally, if the three nodes are in different clusters, the number of access to the Internet is increased since both the initiator and the destination have to access the Internet to query and populate the agent, respectively.

Following a similar analysis, when a hybrid mode similar to the SCMF locator mode is used with a centralized approach the number of hops and traffic volume are the ones derived as shown in (4.9) and (4.10):

$$\#hops = \begin{cases} \#hops_{Agent} + \#hops_{init} + \#hops_{dest} & \text{if } init\_cluster = agent\_cluster = dest\_cluster \\ 2 \cdot \#hops_{init} + 2 \cdot Internet\_Access + \#hops_{dest} & \text{if } init\_cluster = dest\_cluster \neq agent\_cluster \\ \#hops_{Agent} + 2 \cdot Internet\_Access + \\ \qquad + \#hops_{init} + 2 \cdot \#hops_{dest} & \text{if } init\_cluster = agent\_cluster \neq dest\_cluster \\ 2 \cdot \#hops_{init} + 2 \cdot Internet\_Access + 2 \cdot \#hops_{dest} & \text{if } init\_cluster \neq agent\_cluster = dest\_cluster \\ 2 \cdot \#hops_{init} + 3 \cdot Internet\_Access + 2 \cdot \#hops_{dest} & \text{if } init\_cluster \neq agent\_cluster \neq dest\_cluster \end{cases}$$

$$(4.11)$$

$$traffic\ volume = \begin{cases} Size_{LO} \cdot \#hops_{Agent} + Size_{CA} \cdot \#hops_{init} + Size_{LO} \cdot \#hops_{dest} & \text{if } init\_cluster = agent\_cluster \\ & = dest\_cluster \\ (Size_{LO} + Size_{CA}) \cdot \#hops_{init} + (Size_{LO} + Size_{CA}) \cdot Internet\_Access & \text{if } init\_cluster = dest\_cluster \\ \qquad + Size_{LO} \cdot \#hops_{dest} & \neq agent\_cluster \\ Size_{LO} \cdot \#hops_{Agent} + (Size_{LO} + Size_{CA}) \cdot Internet\_Access + & \text{if } init\_cluster = agent\_cluster \\ \qquad + Size_{CA} \cdot \#hops_{init} + (Size_{LO} + Size_{CA}) \cdot \#hops_{dest} & \neq dest\_cluster \\ (Size_{LO} + Size_{CA}) \cdot \#hops_{init} + (Size_{LO} + Size_{CA}) \cdot Internet\_Access + & \text{if } init\_cluster \neq agent\_cluster \\ \qquad + (Size_{LO} + Size_{CA}) \cdot \#hops_{dest} & = dest\_cluster \\ (Size_{LO} + Size_{CA}) \cdot \#hops_{init} + (2 \cdot Size_{LO} + Size_{CA}) \cdot Internet\_Access + & \text{if } init\_cluster \neq agent\_cluster \\ \qquad + (Size_{LO} + Size_{CA}) \cdot \#hops_{dest} & \neq dest\_cluster \end{cases}$$

$$(4.12)$$

In this case, the agent is first queried for the actual node that serves the piece of context information. After the pointer is fetched from the agent, a direct communication is established between the initiator and the destination.

By evaluating the response of the SCMF to a varying number of hops within a cluster, the behavior of the system when facing different cluster configurations ranging from tightly coupled and homogeneous clusters to spread heterogeneous ones, can be assessed.

Figure 4.23 and Figure 4.24 show the results when the destination was selected in the same Cluster as the initiator and the mean number of hops per cluster ranged

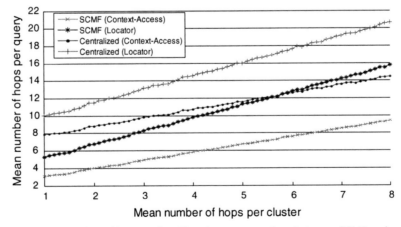

Figure 4.23   Mean number of hops and traffic volume comparison between SCMF and centralized approaches (destination: Normal distribution - # hops per cluster: Uniform distribution) [65].

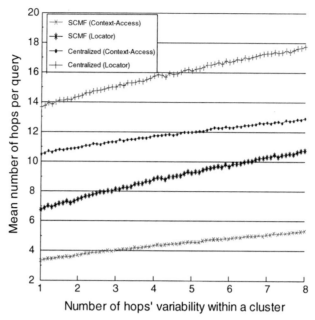

Figure 4.24 Mean number of hops and traffic volume comparison between SCMF and centralized approaches (destination: Normal distribution—# hops per cluster: Normal distribution) [65].

from 1 to 8 using a uniform random variable and when a normal random variable was used with the variance ranging from 1 to 8 and the mean always 1, respectively.

For all the cases, the PN was composed of 100 nodes, the complexity of an Internet access was fixed to be 5 times higher than intra-cluster communications and the context-access mode packets were 50% bigger than the locator mode packets. For the centralized approach, the agent role was randomly assigned to any of the PN nodes.

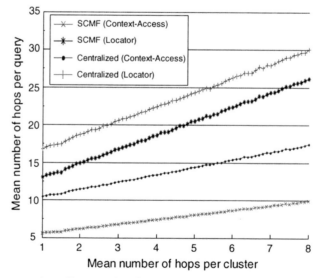

Figure 4.25 Mean number of hops and traffic volume comparison between SCMF and centralized approaches (destination: Uniform distribution—# hops per cluster: Uniform distribution) [65].

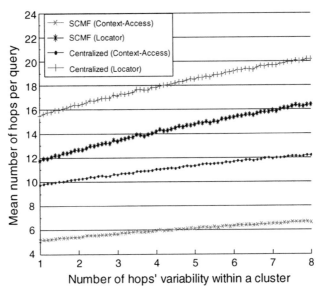

Figure 4.26   Mean number of hops and traffic volume comparison between SCMF and centralized approaches (destination: Uniform distribution—# hops per cluster: Normal distribution) [65].

Figure 4.25 and Figure 4.26 show the case, in which the destination was randomly selected from all over the PN and the mean number of hops per cluster ranged from 1 to 8 using a uniform random variable or when using a normal random variable with the variance ranged from 1 to 8 and the mean always 1.

The mean number of hops per query grows linearly with the cluster complexity proving the scalability of the SCMF approach. In absolute terms, the SCMF approaches get advantage of not having to access the Internet in most of the cases when the destination is located nearby, while in the centralized approach, the agent is typically in a remote cluster. On the contrary, when using the uniform distribution for selecting the destination, the situation is even better for the SCMF approaches since the query does not follow any pattern and for the centralized approach not only the initiator but also the destination needs to access the Internet for querying and populating the agent. With the SCMF approach, there is always a CMN in the cluster and although the anarchy of the queries affects similarly to both approaches (it increases the mean number of hops per query) when looking at the traffic volume the impact is much higher in the centralized approaches.

When only the number of hops is considered for measuring the network usage per query, an important parameter is diminished. The size of the packets used for the different modes rules the actual traffic volume caused by a context query. Although the mean number of hops represents a good showcase for the network load generated with a single context query, it might be a little bit biased against the locator mode because it does not take into account the reduced size of the packets used in this mode. While in the context-access mode, the CMN directly responds to the query with the context information (i.e., meaning larger packets) in the locator mode, the context information is directly exchanged between the peers, while the interaction with the CMN is based on the indexes towards the actual holder of the context information (i.e., meaning shorter packets). Therefore, it is interesting to

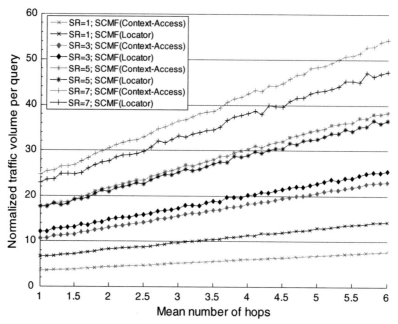

Figure 4.27 Effect of SR on traffic volume (destination: Normal distribution—# hops per cluster: Uniform distribution [65].

see what the actual traffic volume of a query-response operation is depending on the packet size and what the impact of this parameter on the overall performance is.

The *size ratio* (SR) is the relation between a packet used in the context-access mode and one used in the locator mode. By varying the SR, it can be shown which

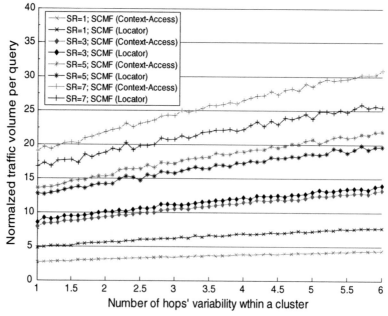

Figure 4.28 Effect of SR on traffic volume (destination: Normal distribution—# hops per cluster: Normal distribution) [65].

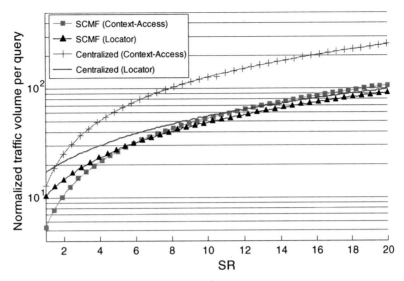

Figure 4.29 Effect of SR on the traffic volume [65].

of the four strategies tolerates the increasing size of the context-access mode packets best. The effect of the SR on the strategies is shown in Figure 4.27 and Figure 4.28.

When the SR is increased for the same network configuration, the network utilization also increases. The locator mode results in better system behavior when the SR is so high that it penalizes the excessive use of context-access mode packets. It is interesting to see that this happens irrespectively of the cluster dispersion. For SRs higher than 5, all locator mode strategies outperform their context-access counterparts independently of how spread the cluster is.

Although the duplicity of modes (i.e., context-access and locator) might result bizarre, the growth in terms of size of the packets containing the whole context description instead of simply the pointers towards the actual context agent makes the locator mode perform better under certain circumstances. This benefits the support

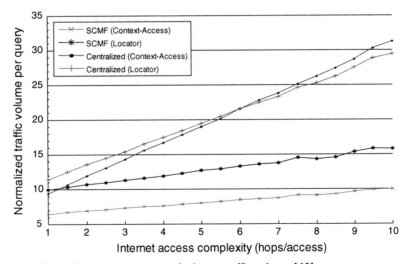

Figure 4.30 Effect of Internet access complexity on traffic volume [65].

of other more subjective features where the locator mode has certain advantages compared to the context-access mode (e.g., complexity within the CMN, storage room within the CMN, etc.).

When it comes to the comparison between strategies, it can also be seen that the traffic volume is always smaller in the SCMF cases. Figure 4.29 shows that the centralized approach shows quick degradation on its performance when using context-access mode and for an increased SR.

The effect on a varying Internet access on the performance of the SCMF is shown in Figure 4.30. The SCMF context access strategies limit the access to the interconnecting infrastructures so that they are more robust against increases in the Internet Access complexity and thereby minimizing the total traffic volume per query.

4.2.1.4 Impact and Usage of Mismatch Probability for Context Aware Service Discovery

Several approaches had been investigated on how to deal with dynamic context information accessed remotely, if knowledge of mismatch probability is available [73, 74]. It was shown that *service score errors*, (i.e., deviation from a true score and the actual calculated score can be reduced effectively by eliminating or even using estimated context values), when the mismatch probability is higher than a given threshold. The *service score error* can be defined as [65]:

$$S_\Delta = \left| S_{SSF}(\bar{x}_s, \bar{x}_u, \bar{r}) - \tilde{S}_{SSF}(\tilde{x}_s, \tilde{x}_u, \bar{r}) \right| \tag{4.13}$$

The score errors can be evaluated as a function of the network delay using different function types. A function using numerical-based context information can be given in a linear form as follows:

$$f_{CSF}(x_s, x_u, r) = a_1(\| x_s x_u \| - r) + a_0 \tag{4.14}$$

Equations (4.15) to (4.17) give the definitions for a sigmoidal, bell-shaped, and a general bell-shaped functions, respectively:

$$f_{SCF}(x_s, x_u, r) = \frac{\alpha}{1 + \beta \exp(-\gamma(\| x_s x_u \| - r))} \tag{4.15}$$

$$f_{CSF}(x_s, x_u, r) = \exp\left(-\frac{(\| x_s x_u \| - r)^2}{2\sigma^2} \right) \tag{4.16}$$

$$f_{CSF}(x_s, x_u, r) = \frac{\alpha}{1 + \left| (\| x_s x_u \| - r)/\beta \right|^{2\gamma}} \tag{4.17}$$

The errors due to the usage of outdated information are mapped differently with the above four different score functions, and for the bell-shaped and general bell-shaped function, there exist parameter settings where the score functions lead to access errors to produce the correct score value, hereby concealing the fact that outdated information has been used [75]. There, it was shown that the *score errors* simply increase as a function of the delay. However, as the functions are limited (the linear function has been hard limited to produce output in the range of 0 and 1, while the others has a natural limit due to their function nature), the knowledge of

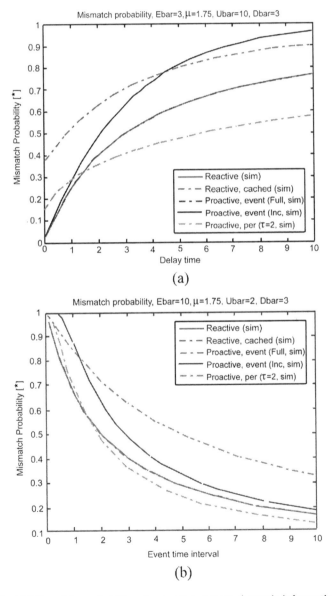

(a)

(b)

Figure 4.31   (a, b) Mismatch probability of accessing remote dynamic information in a two-node scenario with varying delay and event rates (assuming all involved processes being Poisson) [65].

mismatch probability can be used actively to adjust the weight of the information element. This leads to a much less overall service score error, and finally improves the *reliability* of the context-aware service discovery system to the end user. However, there is still a long way for such a system to be efficiently implemented and researched as the above examples put a large range of assumptions on what information is available [75].

Figure 4.31 shows the mismatch probability with an increased (a) delay rate (assuming exponentially distributed delays with rate ν) and (b) the event arrival rate (assuming exponential distributed event inter arrival times with rate λ).

A high variance in delay and event arrival times, generally leads to a lower mismatch probability, than for events that arrive with deterministic time intervals.

The selection of an access strategy is complicated by the fact that there are many parameters to select between.

For a reactive strategy a cache period must be determined. [76] shows how to choose a cache period optimally under the given constraints on the performance metrics.

For a proactive, periodic update strategy an update time period must be selected. In [75] it is shown how the update time period can be optimally chosen under given constraints on the performance metrics.

For a proactive, event-driven approaches it must be decided whether to update only the difference between events that change the context value or send the full information every time (this differentiation impacts the mismatch probability and potentially also the network overhead).

The evaluation of the right choice of strategy can be done either statically, (i.e., if the statistics of the delay, event process, and expected information demand is known a priori). However, this is typically not the case, so the context agent would need to estimate these metrics at runtime, and adapt thereafter. The results shown in [75] are a vital step towards such an adaptation, and, in general, the understanding of the access to a remotely located dynamic information, if successfully deployed, lead to a context management system, which can provide *quality of context* requirement, which in combination with the CALA language could include quality requirements that ensure reliable context aware systems. The perspectives given here are important to a successful deployment of context aware services and applications.

4.2.1.5 Self-Organizing Context Management

The PN and the PN-F are networking concepts for the user [65]. This means that typically, no system administrator manages the network. Furthermore, the PN and its clusters are dynamically established and may split or merge any time. As a consequence, the connectivity and networking layer must be designed in such a way that they can react in an autonomic and self-organized way. In the same way, the SCMF should organize itself and adapt to the network structure and the availability of nodes. Table 4.1 contains the major characteristics of the self-organized context management system:

Table 4.1 Comparison of Features for Different Context-Management Systems [65]

Features	SCMF	Other Examples	General Context Systems
Context Framework Management	Node election determines index server in a cluster. Context management node (CMN) from multiple cluster connect with each other.	Older context systems do not have the notion of a broker or index servers. They use statically build context processing structures. Some newer system provide a broker and some have index structures	Centralized context broker provides lookup means and keep index tables. Usually no self-organization, but centralized management.
Scalability Support (e.g., Self-Organization, Index Server)	Replication strategy allows balancing communication overhead to reaction time. Scoping of requests prevents querying the whole system, if not necessary.	Usually context systems do no take replication into account. Also, scoping of requests is not foreseen in many systems.	Main system still needs to check for all available data sources. Limited scalability support when using a SoA-based approach.
Communication Means	Synchronous queries, asynchronous subscribe/ notify, insert, modify, delete.	Most context systems concentrate on one request type.	Most context systems offer at least query mode, though many offer also subscribe/notify.
High-Level Query Language	SCMF interprets a high-level query language and determines automatically which context services of other nodes to access (service composition). High abstraction of declarative qQuery language.with support for optimization (scope, filter).	Context toolkit: provides widgets that a developer can plug together. No high-level query language and no (automatic) service composition. Comparably lower abstraction level.	Many context systems operate on a comparably low level of abstraction. Context sources are exposed and accessed directly. No support for declarative high level queries.

4.2.2 Distributed Service Discovery for WSNs

A strong requirement for WSNs is to capture the heterogeneity and complexity of possible scenarios. Therefore, the middleware layer should be designed to abstract the device-specific and distributed network aspects of the system and allow the application to view the entire system as a single platform [77].

A distributed middleware implementation for WSNs was proposed by [4], [77] as a context-aware service discovery approach to optimize the efficiency of the discovery. The approach exploits ontological reasoning for semantic-based service matching, integrating cost metrics in service descriptions, and optimizing the discovery process for low latency in order to exploit short-lived network access. The implementation of the approach is shown in Figure 4.32.

The middleware in Figure 4.32 comprises the following three main aspects:

- A set of distributed services;
- A set of distributed data processing functions (including brokering);
- A data centric resource management functions.

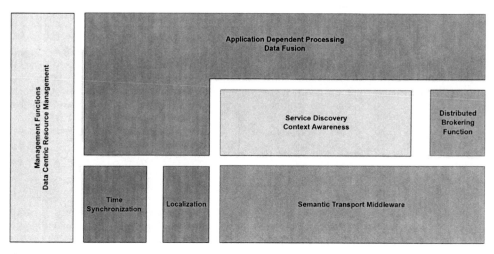

Figure 4.32 Distributed processing middleware architecture [68].

4.2.2.1   Middleware Elements

Distributed Services
The distributed services are related to the middleware's ability to unify the functionality of each individual device into a single platform abstraction. The distributed nature of the system resources, however, introduces a number of areas of complexity. Three main areas relate to *time*, *space*, and *capability* [77].

A common perception of *time* is essential if individual devices are to synchronize their functionality. Whether synchronisation is required for the operation of the devices, or whether it is required for time stamping of sensed data, an abstraction of the multitude of devices into a single platform view requires all devices to synchronize to the system time. The characteristics of the synchronization, as well as the meaning of the system time are application dependent, and must be part of the service request parameter.

Although most distributed systems attempt to hide the notion of the geographical location of resources, sensed data is often only meaningful if the location of the sensor can be determined. If the middleware is to provide the application with a single platform view of the system, it must also be able to interpret location parameters in the context capture requests from the application, and to establish and attach a notion of location to the context information from the system.

The source of the context information cannot always be fully known a priori.

The nature and capability of the devices may be variable, due to the evolution of the deployment of resources and to the mobility of users through environments with a varying richness (capability and density) of context sensing. The service discovery mechanisms should identify and locate the capability of the underlying infrastructure and exercise the relevant functionality in the system to deliver the required service to the application.

Timing and *synchronization*, localization, and service discovery are three fundamental distributed services required in sensor networks, and their complexity is dependent on the deployment characteristics, linked to the scenario and the application.

Timing and Synchronization   Establishing and maintaining a common view of time is an underlying issue for any distributed system. There is a trade-off between the effort required to establish and maintain synchronization, and the stability of the clock in each device. The clock stability is directly linked to its cost, therefore, sensor networks must be able to operate with very low clock stability [77].

The amount of resources available to establish and maintain synchronization is, however, also low in sensor networks, which makes the traditional approaches to synchronization inappropriate for the sensor network context. Given the wide variety of scenarios and application domains for sensor networks, a variety of mechanisms are required, which are tailored to suit particular requirements. For example, a *body area network* (BAN) typically has a star-topology centered on the gateway device, which in turn has wide-area network connectivity.

The gateway can achieve global synchronization either through the wide-area network (e.g., the *network time protocol*—NTP [78]), or through some external means as it has higher energy and capability than the sensor node (e.g., Global Positioning System—GPS [79]). The situation is very different for a large-scale environmental sensor field, typically deployed as a large diameter peer-to-peer mesh network. In such a scenario, synchronization cannot be achieved by traditional means and requires a WSN-specific solution.

Time synchronization is an important piece of infrastructure in any distributed system, whether to distinguish the order of events in the system or to record the time events that have happened.

In WSNs, a host of factors make efficient and flexible time synchronization particularly important, while making it more difficult to achieve than in a traditional distributed system [77].

If very precise clocks (e.g., long-term stability atomic clocks), are used on sensor devices, all sensor devices will then be able to share the same notion of time for long periods and there will be no need to maintain time synchronization, once it has been established. Unfortunately, such clocks are large and very expensive, and thus, not suitable for WSNs.

In WSNs, time synchronization is required at a number of levels. At the MAC level, the time information is needed to schedule the *sleep/wake* cycles (i.e., duty cycling) of the communication interface of the sensor device to reduce the energy spent on idle listening.

At the application level, the so-called *time stamps* are generated to keep a track on the time, at which certain events occurred. As sensor devices are normally deployed in a batch, rather than used in a stand-alone fashion, these need to be kept time-synchronized so that the sensed information share the same timescale for a meaningful analysis. In addition, the sensed information relating to the same events can be identified, thereby, allowing for aggregation, fusion or enrichment.

The time information can also be used to assist with localization. For example, one of the localization techniques uses the signal propagation time for ranging and then computes the position in (x, y, z) coordinates.

Some of the constraints relate to hardware and application. Sensor devices range from passive RFID tags to powerful *personal digital assistant* (PDA)-type sensor devices. Sensor devices generally have a very small form factor, low energy, low processing capability and a limited memory. Together with the drive to keep the cost of these sensor devices low for mass deployment, only low quality clocks can be afforded.

Different application spaces of WSNs pose different constraints and freedom on time synchronization, and should be incorporated in a time-synchronization solution.

Hardware clocks are a fundamental issue [77]. Most types of clocks are built around some form of an oscillator, an arrangement that goes through an endless sequence of periodic state changes, and designed to provide a continuous and stable reference frequency. The periods of this oscillator are then counted and converted into the desired time representation.

The optimum oscillator as a clock source for a particular application is dependent on the cost, accuracy, power consumption, and the environmental parameters. In general, there are the following three types of oscillators:

1. *RC oscillator*—this is based on electronic components (e.g., resistors and capacitors);
2. *Resonator-based oscillator*—(e.g., crystal and ceramic resonators);
3. *Atomic oscillator*—this is based on the transition of chemical elements such as rubidium, caesium.

In practice, the oscillators in each of the devices in a distributed system will run at slightly different frequencies, causing the clock values to gradually diverge from each other. This divergence is called the *clock drift* and can lead to an inconsistent notion of time. An example of a clock drift is shown in Figure 4.33.

The clock drift is not constant, due to the aging of the clock and the environmental influences. The time difference between two clocks at a given instant is known as the *clock offset*.

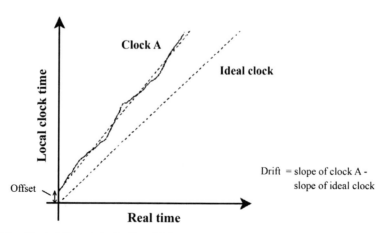

Figure 4.33   Clock drift and clock offset [77].

The granularity of the *clock accuracy* very much depends on the *clock frequency*. The higher the clock frequency, the higher the clock accuracy it can potentially achieve. For example, a 4-MHz clock oscillates once every 0.25 *μs* (i.e., one clock tick). That means that the clock can only be accurate up to one clock tick error of 0.25 *μs*. However, at a high frequency, the clock may suffer from stability problems, making the clock less reliable as an accurate clock source.

There are generally three major characteristics of the communication that can affect the time synchronization in a distributed system.

Firstly, communication can be *unicast, multicast,* or *broadcast* (see Chapter 3 of this book).

Unicast communication limits the time synchronization to take place between only two end points at a time. The implication of this is that $O(n)$ messages are needed to keep *n* devices time synchronised. Multicast or broadcast communication can, on the other hand, allow for sending one message to many devices at a time. The time synchronization with this form of communication, therefore, will use less energy for communication and would keep the network load lower.

Secondly, a communication link can be *uni-* or *bidirectional*. A uni-directional link limits the communication to one direction only (either transmit or receive), whereas the bidirectional link allows for communication in both directions (both transmit and receive). This means that, for example, a time synchronization mechanism that relies on the ability to measure the *round-trip delay* will require a bi-directional link.

Finally, the *delay characteristics* of a communication link are of great importance to time synchronization. Let consider a simple time synchronization scenario with two devices. A source sends to its neighbor its current view of the time, and the neighbor either monitors it or corrects its clock accordingly. In the process of sending the timestamp message to the source's neighbor, a variable amount of delays are incurred before the timestamp message arrives at the receiver. This is shown in Figure 4.34.

These delays can be detrimental to the clock accuracy. The variable delays can be attributed to four major sources, as summarized in Table 4.2.

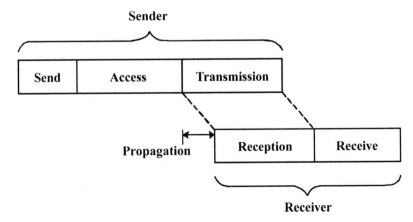

Figure 4.34   Decomposition of message delivery delay (over wireless link) [77].

Table 4.2   Sources of Delay in a Message Transmission [77]

Time	Distribution
Send and receive	Nondeterministic, depends on the processor load.
Access	Nondeterministic, depends on the channel contention.
Transmission and reception	Deterministic, depends on message length and speed of the radio.
Propagation	Deterministic, depends on the distance between sender and receiver.

In a multihop network, especially a large one, it may not be necessary that time synchronization is achieved to the same order of accuracy across the network. It is, therefore, beneficial that the scope of synchronization required, whether pair-wise, local, or global, is identified to allow for the optimum time synchronization.

The following two methods can be used to achieve time synchronization can in a network:

1) *Master-slave*: A single master reference clock is used or elected and all the slaves synchronize their clock to the master clock, or the neighboring clock one-hop closer to the master. Hence, there is a global notion of time (defined by the master).
2) *Peer-to-peer*: Any device can synchronize with every other device in the network. This allows for devices to be locally synchronized with each other, (i.e., there is no synchronization to a global notion of time).

Some time synchronization approached designed specifically for WSNs can be found in [80–82]. The majority of the research reported in the literature had focused on improving the aspects of time synchronization such as accuracy, stability, event ordering, with very relative performance measurements. [4] developed a novel time synchronization solution for WSNs that is not described here due to patent-related confidentiality.

Although efforts to minimize the energy cost of time synchronization by techniques like post-facto synchronization can satisfy the application time-stamping requirements, the synchronization of the MAC is often unavoidable from the point of view of duty cycled sensor nodes [77]. It is necessary because in order for the nodes to be able to communicate, in an energy-efficient and timely fashion, their communications modules need to be scheduled to wake up at roughly the same time.

[4] proposed a rate adaptive, multilevel duty cycling approach including scope control to time synchronization in WSNs. The approach involves multiple pre-defined duty cycles, thereby allowing for different duty cycles to be used at different times for any given node. The rationale behind doing this was to allow for more frequent communications, using a higher duty cycle, when needed. The nodes only return to a *sleep* state if they consider their task accomplished in the short term. This enables a great degree of flexibility in the utilization of resources, so that nodes can be highly available when required, and sleep for long periods of time when idle. Through *scope control* is used to deal with the requirements for more intensive

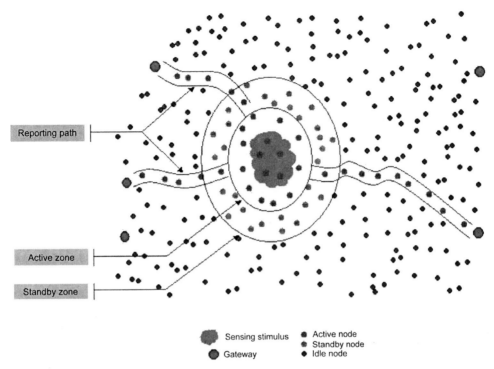

Figure 4.35   Sensor network with scope control [77].

monitoring of some of the sensor nodes. Figure 4.35 shows an example where a few sensor nodes sense something happening around the grey area.

In a large sensor field, there is no real need for every node to participate even if a true alarm is sensed, apart from those close to the stimulus where tasks such as distributed local processing for example needs to be done. Even when there are a couple of gateways, only nodes along the path from the source to the gateways need to get involved to relay the reports. The zone and path scope control are used to the sensor nodes with *multilevel duty cycling*. The advantage of this is that sensor nodes do not need to stick to just one rate of duty cycle throughout the network. The sensor nodes close to the stimulus, (i.e., within the active zone), and those on the reporting path between the active zone and the gateways (referred to as active nodes in Figure 4.35) can operate at the highest duty cycle available, providing high availability to participate in the distributed operations. The sensor nodes that are remote from any distributed tasks need not participate and can continue to operate at the lowest duty cycle. As for the nodes, which are located in the standby zone in Figure 4.35, they are put on *standby* mode, (i.e., operating at medium duty cycle), so that they can be called upon for action with less delay when needed. Scoping is performed in a distributed fashion within the network, and is application driven.

Localization   Knowing where devices are is essential for sensor networks in order to add the location metadata to the sensing information. Localization methods range from manually established *look up* tables (established during the deployment or through a post deployment survey) to fully automated and dynamic localization of nodes (deploy and let the devices work out where they are). Localization information may take a variety of forms, which will be appropriate to different scenarios: (x, y, z). Information may be required in some cases, while semantic localization (e.g., *first-floor, main bedroom ceiling*) may be required for others. Similar to timing and synchronization, localization in sensor networks cannot be achieved through a "one size fits all" solution. For example—the BAN is strategically deployed around the body (predetermined semantic locations dictate the deployment), and remains in a fixed relative position.

The position of the BAN in a global frame of reference is determined by the position of one of its devices. As the gateway has more resources than the sensor devices, and has a wide area network connectivity, it may be localized either through GPS or *ultra-wideband* (UWB) indoors, or by means of *cellular network triangulation*. A large-scale environmental sensor field on the other hand has very different localization constraints. The large number of devices precludes the strategic positioning, and the restricted resources on the devices cannot support external localization, such as GPS. Sensor network specific solutions must therefore be tailored to the requirements of the scenario.

Localization approaches require a combination of the following three components:

- A mode, defining which physical stimulus is being measured and processed to derive the positions;
- A metric, defining which measurement is applied to the particular mode;
- An algorithm, which combines the measurements into positions.

Localization algorithms can be *centralized, distributed,* or *localized.* The *centralized* algorithms are algorithms where all the computations are performed at a single central point. The central node has to collect all the data from the network, compute the positions of the other nodes and eventually distribute the position information back to them.

The disadvantages are a huge communication costs and computational power required. Nevertheless, the precision of the position estimates computed by centralized localization algorithms is always superior to other algorithms. This is due to the fact that the central node has a global view and can apply refinement phases to improve the computed positions.

The distributed localization algorithms make use of several computing and communication capabilities, with the usual advantages of not relying on a single point of failure, not requiring a specialized central node and balancing the load by making use of a geographically distributed resource. A localized algorithm is not only distributed, but makes use only of local data, each node having to communicate just to nodes situated in a limited region.

Localization algorithms can also be classified as *range-free* algorithms (algorithms that do not make use of specialized hardware to perform distance measurements) and *range-based* algorithms (algorithms making use of measured distances between the nodes to compute the positions). Intuitively, the *distance-based* algorithms should perform better that their distance-free ones. This fact is not always true, due to the fact that any measurement contains errors as well. If the precision of measurements is bellow a certain threshold, then not using measurements at all provides better results.

Another way to look at localization algorithms is whether they provide *local*, *relative*, or *absolute* coordinate systems. An absolute coordinate system has global coherence and is desirable in most of the situations. These algorithms are in general expensive in terms of communication cost. Relative positioning establishes positions in a coordinate system specific to the network. The obtained locations still provide network wide coherence. The local positioning returns the relative coordinates to a limited set of nodes (usually the neighbors) and helps the communicating parties to position themselves relatively to each other.

The localization algorithm running on top of a Zigbee protocol stack [78] should make use of the characteristics of the networking protocol in order to achieve an efficient design [77]. The integration between the two building blocks will have to consider also the following communication aspects:

- *Network topology*—the Zigbee standard specifies clearly the possible network topologies that will be in use. The dataflow between the devices will be directly influenced by the topology used, so the localization algorithm should consider the data flow path when exchanging information between the nodes. At the same time, physically close devices may not be directly linked to each other with a communication path, fact that is one of the most important constraints to be addressed in this integration.
- *Message delivery delays*—mesh networks are usually networks that accept large delays in the message deliveries. Also, the order of the arriving messages between two points cannot be usually guaranteed and needs to be taken into account.
- *Low data rate communication*—the Zigbee standard aims at low data rate networks, so transmitting large quantities of information (as in the case of centralized algorithms) will not be an option. To achieve maximum efficiency, the localization algorithm should use all the data that is overheard from the network and from the networking layer.

The ability of *self-localization* is fundamental to unattended and ad hoc deployable hardware dealing with position-specific problems. Thus, it is very beneficial for the sensor nodes to know their own position in case their sensor network is intended to perform *event localization* (e.g., determining the position of an accident).

Manual location determination of sensor nodes is very expensive or sometimes completely nonapplicable, (e.g., in dynamic scenarios where the positions of single nodes change quickly). Hence, the ability of automatic self-localization is desirable. Existing systems addressing this problem (e.g., GPS) are unfortunately unsuitable for many application areas because of the high cost and energy consumption, complexity, and so forth; besides not being available for indoor applications.

[77] proposed to use *ambient audio noise* in order to infer the position of the devices. The audio noise from the environment is a very rich source of information that can be exploited. For example—distinctive events (such as the slamming of a door, the sound of a horn, the ringtone of a mobile phone, etc.) can be separated

from the context and used by the nodes hearing them to estimate the distance and synchronization information. Both of these can be used in order to develop a new localization algorithm or enhance the existing ones.

In order to capture the ambient noise, the sensors need to be equipped with an additional hardware: usually audio-recording hardware in the form of a microphone and if needed a sound processing DSP. This additional hardware (and software) will have to capture the ambient noise, separately from the background noise, or it has to distinguish the events, analyze them, and provide the sensor nodes with their characteristics (signatures, features obtained via a feature extraction algorithm, etc.).

The sensor nodes will exchange information about the heard events and will identify the common ones. Based on the stored information (such as local time when the event was heard, intensity of the signal, etc.) the nodes will be able to infer distances and time information by running a localized algorithm.

Thus, such a mechanism would contain the following three stages:

- *Detection of distinctive events*—if an event is detected, the event will be transformed into a feature vector marked with an accurate time information and signal strength characteristics.
- *Detection of common events*—this mechanism implies a collaborative algorithm, in which the nodes agree on sets of the commonly heard events. At this step, the distances and the synchronization details can also be computed in a localized manner.

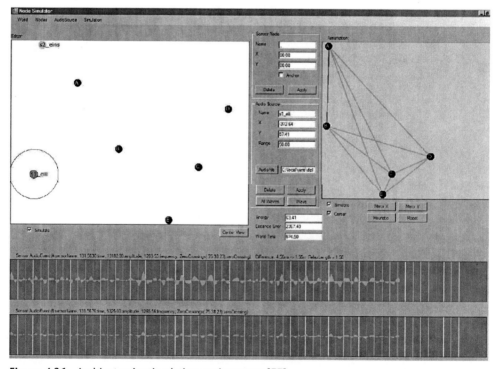

Figure 4.36 Ambient noise simulation environment [77].

• *Determination of range*—based on the available data from the previous step, a distance-based algorithm can compute the range between the nodes. Issues, such as the rate of the position computation, the amount of data traffic, latency, etc. are to be determined and decided here.

The results can be visualized in an environment called the *world model* [77]. Converting the relative positioning to the absolute locations is quite straightforward. Figure 4.36 shows a simulator used for ambient noise localization, in which the left side is the deployment environment and the right side the *world model*.

Service Discovery   In order to achieve high-level tasks in a distributed environment, there is a need to identify what services can be provided by each part of the overall system. For scalability reasons, a centrally held mapping may not be appropriate, and given the dynamics of the network, such a mapping may not be static. The devices, therefore, must be able to dynamically discover and request services. This implies a common way of describing and advertising the capability of each node, the mechanisms for formulating and disseminating the service requests, and the means to invoke and parametrize the services. In certain scenarios, there may be a one to one map between services and devices (e.g., a BAN would be designed for no redundancy in the sensing services), while others will require a selection of the service provider (e.g., data storage services may be provided by many nodes). In the latter case, the issues of the selection arise, both in terms of the advertising of capability (e.g., in a large-scale environmental sensor field, a node must not be overwhelmed by the number of candidates for providing a required service) and in terms of service invocation.

This requires a model for the overall cost associated with each service provider in order to assist with the selection.

When the configuration of the network is not previously known (e.g., WSN), the devices would need a standard way of informing each other of the services they provide. Such a *universal* description must be able not only to describe what kind of a service is provided, but also how it can be controlled, and how the data delivered looks like.

Additionally, it must be structured in a way that allows an efficient processing of a service query, which is the search for a specific service that some application wishes to use.

In order to spread the description over the network, a *service discovery protocol* can be used [77]. It is responsible to inform the network of a service provider, such that a service client may connect to it, if it wants to use the service. Matching the service clients to the providers has a certain similarity to the process of *routing*, as routing can be seen as a special type of service discovery—one wants to find the recipient of a packet. Due to this similarity, some approaches to service discovery pair service discovery with a *route request*.

The service discovery protocol includes not only the mechanism of matching a service client with a provider, but also how services are invoked. As services may be executed only once, or keep doing something, the client may also subscribe to a service, and thereafter will continuously receive some data.

The dissemination protocol must also consider what service information to deliver to which device. This implies some context-awareness of the service discovery system. On one hand, it does not make sense to inform a small sensor node that there is a GUI available if it does not have any means of using it. On the other hand, body sensors collect very private information, and what kinds of services some body-worn devices offer may be information that a user does not want others to know.

Sensor devices in the networks are designed to be as small as possible in order to be embedded just about anywhere. Advances in technology point at a trend of miniaturization, while keeping the processing, memory, and energy resources constantly at the minimum needed for their task. This must be taken into account when the services are advertised, in order to balance the load according to the available resources. The nodes that have more resources should also answer to more service requests.

One of the crucial points is how many messages are needed to exchange the service information. This follows from the fact that radio communication is very expensive in terms of power consumption compared to processing on the local processor. The number of messages should therefore be kept as low as possible and the information contained in every message transmitted should be optimized.

Small memories on the nodes prohibit the use of complex descriptions of the services. This is not generally true; for example, mobile phones have now several hundreds of megabytes of memory available to store such information. The discrepancy to the small devices may pose a problem for a service discovery.

In heterogeneous networks, not all devices are able to communicate with others. The different networks may have different kinds of constraints or may even require different approaches to service discovery. Some *translation mechanisms* must be found.

Finding a *universal* description without a large overhead is difficult. Some networks may need to implement only a subset of the description to be able to handle the processing. Mobility, in turn, dynamically changes the network topology and the number of available devices and services. Keeping the service information up to date is challenging.

Restricting the service discovery to a certain scope without having a central manager assigning those scopes leads to questions as how the devices can know where they belong to, and what kinds of groups other devices are member of.

The *service description* and the *dissemination protocol* can be defined with some degree of independency.

The fundamental issue about the service description is how flexible it should be. One of the main approaches is to have some *intelligence* in the system that knows what it is dealing with by partitioning the services in classes, explaining their attributes, and how they relate to each other. This forms an *ontology* (see Section 4.1.2.2). The advantage is that the actual names of the services do not matter anymore; whether a name is given in German, French, or English is irrelevant, the attributes and the relations remain the same. The ontologies are usually built by strings identifying elements and, thus, tend to have relatively large memory requirements. Additionally, the processing overhead also tends to be significant.

To reduce the amount of data that needs to be stored in a single node, an ontol-ogy could be implemented in a *local domain*, or section of the network, and trans-lating it to some globally valid ontology when interacting between domains.

Another approach is to uniquely identify a service by assigning it a globally valid number. This has the advantage of being very efficient for known services. Usually, an identified service can be described in more detail by giving values for certain attributes also defined by unique identifiers. The problem arising with this kind of description is that there must be some institution that assigns such globally unique identifiers to ensure that they are unique.

Another approach is to give a unique identifier only to the main services and use the description language to describe more details.

One of the major service discovery design decisions is whether or not to use *service directories*. Service directories are central repositories of services where service providers and service clients connect to, to be matched with each other. Service di-rectories can be placed on devices with more resources, (e.g., mobile phones, PDAs, or similar). It is important that such devices are able to be constantly reachable and have the capability of serving enough clients when the need arises.

In the case when no service directories are used, the service clients and the pro-viders must find each other by themselves. This approach is similar to the routing needs. If a client searches a service, it sends a message that needs to be delivered to a service provider, that is the same situation as if it has a message that has an address of a specific device—just with the difference that a service might be provided by a number of devices, and any one may be the searched one. It, thus, makes sense to include service information in routing tables, and that service providers issue service advertisements, which are kept in caches of its neighbors. These will then route incoming service requests to their neighbor that is closer to the one that provides the service. There are many different ways to improve the behavior of such service discovery algorithms [83].

In a *semantic* network, the messages are not routed by their addresses but by the data they contain. In an addressed network, a node sends a message and speci-fies the receiver by adding an address, which specifies to which node it should be sent. In a semantic network on the other side, it specifies what kind of data it is sending, and submits it to the network middleware. A receiver tells the middleware what kind of data it wants to receive.

A semantic network thus already handles many of the service discovery tasks. However, service discovery is still needed to control the specific devices, if the need arises. An application may, for example, want to set different parameters of indi-vidual sensors when measuring.

The *service directories* could exploit the semantic addressing by requesting the middleware to receive the service advertisements and the service requests. It can keep an overview of the services provided by the network and allow for reducing the number of advertisements needed for the whole network to be informed about services provided.

Some of the most prominent existing solutions for service discovery that are already available are in [78, 84–88].

The main issue with existing service discovery protocols is that they are de-signed for networks that contain devices with much larger processing resources than the sensor devices. Additionally, the WSN is a much more dynamic environment that leads to a number of requirements that are not very well addressed by the above-mentioned protocols.

Distributed Data Processing

The distributed data processing functions [89–90] provide an abstraction of the individual devices data processing capability, and allows for the application to view the system as a single processing entity. The following components are required to achieve this:

- *Algorithm specification*—In order to exploit the processing power available within the network, the middleware layer must be able to break down an *application-defined* data processing algorithm into *device-specific* processing tasks (depending on the devices processing capability and the data it has at his disposal). To achieve this, an *algorithm specification language* must be defined to allow the middleware layer to accurately break down the application algorithm into low-level processing tasks in order to produce the correct overall output.
- *Distributed and adaptive data processing schemes*—Exporting raw data out of the system for processing at the edge of the network has inherent scalability limitations, both on the communication bandwidth and the central storage and processing power requirements. The distributed processing schemes are used for the pre-processing of the context data as it flows through the network.
- *Collaborative dependable information processing*—The context will not be derived from simple combinations of sensing reports, but from a much more complex signal processing across groups of sensors, in different places and using different sensing modes. Collaborative algorithms can be used to request, retrieve, process, and report information to provide high added value output from the system.
- *Context awareness*—Translating sensed information into context is a non-trivial task. The final step in providing context involves *interpretation*. To achieve this in some meaningful way, there is a need to define a modular approach to context recognition, defining a context alphabet with a suitable mapping to the sensing data, in order to build the context description and awareness from these atomic primitives.

Data Centric Resource Management

The data centric resource management functions map the application-specific requirements to individual devices. Given the inherent heterogeneity of the device capability and the general scarcity of energy, data rate, processing power, the resources must be used in the most efficient manner to provide the services required by the application. At the device level, *reprogrammability* and *reconfigurability* are essential. At the system level, efficient selection and configuration of resources are needed to ensure the optimal system-wide survivability. This is a tightly coupled, application-driven, resources-constrained environment, in which a data-centric approach can maximize the service efficiency [77].

4.2.2.2   Context-Aware Service Directory Clusters

The service discovery explores and describes the functionality that is provided by a WSN [77]. The functionality can either be offered by single nodes or by a collaborative effort of multiple nodes in the network. The functionality information can be collected on a single, or multiple service directories in the network. Such service directories maintain a database of the functionality that is available in the part of the sensor network surrounding it.

Each of the network parts that connect to the same service directory forms a *service cluster*.

The service directories of every service cluster may contact each other to provide the additional service information to their clients.

By using service directories, the nodes that join a network can quickly access the service information by contacting the service directory of the service cluster. The new node can then receive information about all the nodes in the neighborhood without communicating with each and every one. This results in a low number of messages to be exchanged to search for services as well as low latency, because only two nodes are involved in a service search.

Such a fast access is particularly interesting for BANs, where the sensor nodes are attached to a person's body. When the person is moving about, the BAN keeps connecting and disconnecting from the sensor networks in the environment.

The body area sensor nodes can access the services of the environment network by only connecting to the service directory.

Service Descriptions

Services provided by the sensor nodes are seen as black boxes of functionality. The service directory only saves the information about what black boxes are available, and what type of inputs they read, and the outputs they generate. The services and the data types on their input and output ports are described by using *globally unique identifiers* (GUID). An application can use the GUIDs to request for a service at a service directory. The services can be described in more detail by *attributes*, which are specified either by the GUIDs or by service specific identifiers. Besides the identifier, attributes also contain values, which quantify the attribute.

The attributes contain information about the different parameters of the service, can describe the requirements for the service, or also advertise it for certain contexts.

For a more complex configuration, input and output specifications may group a set of GUIDs, and use the descriptions *ONE-OF* to indicate only one of the group can be selected, or *ANY-OF*, to indicate that zero or multiple of these can be selected as output, providing a different set of outputs. The number of ports can depend on some parameters of the service, which can be referred to with the description of output ports. This is shown in Figure 4.37.

Besides the service information, the service directory also collects information about resources available on the surrounding sensor nodes. Resources can be processing capabilities, transmission channels, memory sizes available, etc. The resources are described by the GUIDs and the attributes, as well. Some attributes of the services can be matched to resource attributes to determine whether the

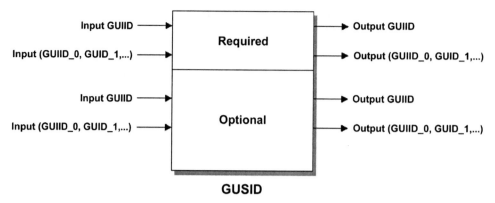

Figure 4.37   Service description scheme. A service is specified by its GUIDs [77].

resource can execute the service or not. This allows the service requestor to search for services and the resources, which can actually execute the services.

The services can also describe the dependency on other services. A service group specifying a set of services can be indicated to be required at input ports. If no service of a service group is available in the network, the first service cannot be executed. By recursively resolving these dependencies, a service graph is formed, which contains all required services, and how they need to be interconnected. This forms a *distributed* application that needs to be executed by the whole sensor network in order to perform a service.

Semantic Service Discovery
The service response can contain a single service, a service graph, or a list of services that can be used, depending on the requirements of the request. Service directories periodically advertise themselves to their neighborhood. If nodes receive such advertisements, and they already know another service directory, the advertisement is forwarded to the already known service directories. Service directories with knowledge of other directories may forward service requests, or parts of them to these directories.

The service discovery process can be improved by using *reasoning* on *ontologies* to the service directory. Introducing the semantic descriptions of the services allows for the process to derive a knowledge of the comparable services. The ontologies describe the classes of services and their properties through a semantic network as a form of knowledge representation. To derive the service descriptions from the service requests, a *best matching* algorithm is employed to classify the matching services. As a last step, a ranking phase selects the services matching the service request the best. Figure 4.38 shows the whole semantic service discovery process.

The semantic match strategies use the ontologies to address the concepts and associated semantics, allowing to recognize the related services despite of possible syntactic and modeling differences. The process consists of performing *inferences* on a consumption hierarchy in order to identify a subset of services that can perform the requested actions. For this purpose, *domain-specific* ontologies are needed where concepts and relations among the local services are well defined. This makes it possible to perform reasoning on a submitted query, the services descriptions and

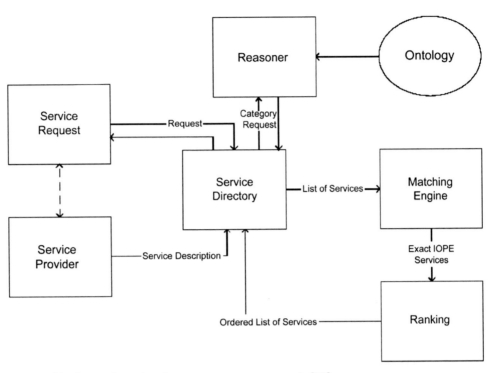

Figure 4.38   Semantic service discovery process components [77].

the context data. To carry out a reasoning process, the functional properties, the *input and output parameters, the precondition, and the effects* (IOPE), are mapped to concepts on an ontology, which are rated reaching a taxonomy of functional services. This is accomplished by a *reasoner engine*, which implements the calculus of the description logic.

The service matching is the core part of the whole semantic service discovery process. It takes inputs from the specified parameters into the service query and uses them with service descriptions and ontologies to create a list of services that fulfill the received request. The three phases of the matching process are shown in Figure 4.39.

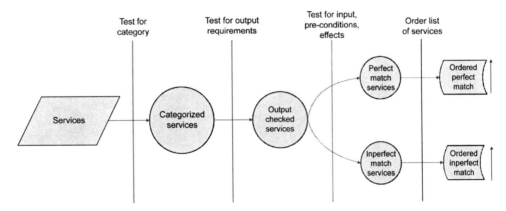

Figure 4.39   Phases of the matching process [77].

The service matching process can select between five different degrees of matching. The perfect matching lists include the services classified as an exact match or a plug-in match. On the contrary the imperfect list contains subsume and intersection matches. The last category is called a disjoint match and is assigned to services that do not match the requirements.

Setting Up Service Clusters

The network is clustered to keep the services offered by nodes moving together in a service directory, such that the information stored in the service directory is accurate as long as possible. The different states of the protocol are shown in Figure 4.40.

The algorithm starts when a node is started up from the *Boot* state into the *SearchSD* state or has lost connection to its service directory. In this case, it tries to find a new service directory by broadcasting a *service directory search message* (SDSM). If a neighbor receives such a message, it checks whether it knows a service directory, and replies to the SDSM with a *service directory location message* (SDLM), which also contains a timestamp of when the service directory has last been successfully accessed. If the neighbor does not know a service directory, it forwards the broadcast, adding its address to the recipients of the SDLM.

Upon receiving an SDLM, the node searching for the service directory changes into the *Register* state and registers with the service directory specified in the SDLM. A service directory may choose whether or not to accept a new node.

This allows a service directory for keeping only the information of particular nodes. If the service directory acknowledges the registration, the searching node changes into the *Ready* state and is ready to query for the service information. If no answer is received, the node falls back into the *SearchSD* state and the search is repeated.

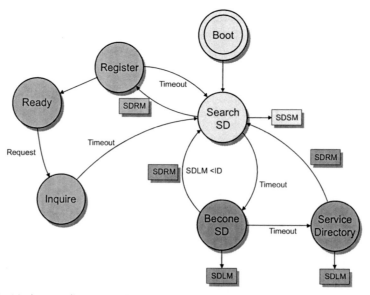

Figure 4.40 Node state diagram, with initialization states, service clients, and service directory (SD) states [77].

Every node keeps track of different service directories it sees and always tries to join the service directory, which it knows for the longest time. This is done by keeping a list of service directories in the vicinity and keeping track of the announcements they periodically emit. The service clustering message formats are shown in Figure 4.41.

If no SDLM replies are received on an SDSM, the searching node waits for a period *SearchTime* and then tries to become a service directory itself by changing into the *BecomeSD* state and advertising its role to the network using SDLM messages that it emits periodically after a time specified by *AnnouncementPeriod*. If it does not receive a notice of an already existing service discovery for a time *SDInitTime*, it changes into the *ServiceDirectory* state, where it stays until it chooses to resign. If is resigning from being a service directory, it issues a *service directory resignation message* (SDRM), which can optionally contain a replacement candidate for the service providers to register to.

Distributing the service directories and discovery mechanisms allows for scalability [90]. Additional context-aware clustering can then be employed to group services together into meaningful clusters.

In every one of those clusters, one of the nodes is selected to act as a service directory and collects the services provided by its cluster peers. The service directories periodically broadcast the presence messages, which are also forwarded to the service directories of the adjacent clusters, forming the distributed service directory.

The WSN service discovery protocols designed by [4] can handle mobility of clusters of sensor motes. To this end the motes moving together are gathered into the same cluster and kept together. This has the advantage that the distributed processing within these clusters can be done with higher reliability, as it can be assumed that the sensor motes will not move out of range as easily as when arbitrary motes are chosen. The result is that fewer adaptations of the distribution of the processing have to be made. An additional advantage of single points of access within clusters is that service discovery requests can be directed at a single point instead of being broadcasted, allowing a low latency to the service reply, which is crucial for networks with high mobility [90].

The idea of the context-aware clustering is to cluster together sensor nodes, which share the same context information (e.g., located on the same person, can sense the same object, or similar). Such clusters would provide an ideal platform

SDSM: Service Discovery Search Message

Type:SDSM	Source Address	Path Length (L)	Path[L]

SDLM: Service Discovery Location Message

Type: SDLM	Directory Address	Sequence ID	Hops

SDRM: Service Discovery Resignation Message

Type: SDRM	Directory Address	Replacement Addr	Last Used	Hops

Figure 4.41 Service clustering message formats [77].

for the inference of more complex context information, such as what this specific person is doing.

The main difficulty in using context information for building the service clusters is that the context information can only be computed with a certain *confidence*, meaning that the decisions on two nodes sharing the same context can only be taken with a certain *success probability*. The approach in solving this issue is to use a history of *past context decisions* to *lower the fluctuation rate* of the cluster membership [90].

Context Dependant Clustering Evaluation Algorithm

Each node v periodically computes the confidence of sharing the same context with its neighbors. If the confidence with a neighbor u exceeds a certain threshold, then v considers that it shares the same context with u for the given time step. The final decision for the sharing of the same context with u is founded on the confidence values from a number of previous time steps, called the *time history*. This history stores the last decisions of the evaluation of the shared context and is only stored for nodes having the role of a clusterhead to reduce the amount of memory needed [90].

The algorithm constructs a set of one-hop clusters, based on the context information shared by the nodes. A node v can be: (1) unassigned, where v is not part of any cluster, (2) root, where v is clusterhead, or (3) assigned, where v is assigned to a cluster where the root node is one of its neighbors. Any arbitrary node v in the network changes or chooses its root at every time step in the following cases: (1) v is unassigned, (2) v does not share a common context with its root, (3) the root of v is no longer a root and (4) v is root and there is another neighbor root, sharing the same context with v, that has a higher priority number. In one of these cases, v chooses as root node the neighbor root u, with which it shares a common context and which has the highest priority number. If such a neighbor does not exist, v competes for clusterhead or becomes unassigned. The decision is based on the current status of the neighbors and tries to minimize the effect of the following erroneous situation: due to the context fluctuations, an assigned node v may loose its root node and cannot join another cluster because none of its neighbors is root. Therefore, v may become root, form a new cluster and attract other nodes in that cluster. To avoid this undesirable outcome, a node declares itself root only if all its neighbors, with which it shares a common context are unassigned. If there exists at least one neighbor u with which v shares a common context and u has a valid root node, then v becomes unassigned.

The evaluation of the clustering algorithm was based on a simulation of nodes attached to walking human bodies [90]. A *context recognition* algorithm decides whether two sensor nodes are attached to the same body. p denotes the probability of the correct detection of the common context and q is the probability of the correct detection of the different contexts.

A network of 20 nodes was simulated, grouped in two clusters and the context recognition probabilities p and q were varied from 0.8 to 1. Figure 4.42 shows the performance of the clustering algorithm depending on the two probabilities that correspond to the accuracy of the context inference algorithm. The *stability* measures the amount of simulation time the nodes are being assigned to the correct clusters.

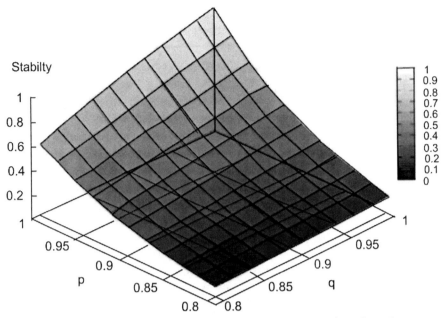

Figure 4.42 Cluster stability depending on the probability of p correct detections of common context and the probability q of correct detection of different context [90].

In order to achieve stability of more than 0.9, the two probabilities have to exceed 0.98. As an example, the probabilities computed from the results reported in [91], who propose a context detection algorithm to deduce if two devices are worn by the same person, are $p \approx q \approx 0.96$, which gives a cluster stability of 0.64. It is interesting to see how these probabilities change if the time history would be involved in the decision process [90]. The probabilities p_h and q_h of the correct detection of the common context for a minimum time history h_{min} out of a total of H time steps is given by the CDF of the binomial distribution as follows:

$$p_h(h_{min}, H) = \sum_{k=h_{min}}^{H} \binom{H}{k} p^k (1-p)^{H-k} \tag{4.18}$$

$$q_h(h_{min}, H) = \sum_{k=h_{min}}^{H} \binom{H}{k} q^k (1-q)^{H-k} \tag{4.19}$$

If $h = 3$, $H = 5$, then $p = q = 0.99$ to reach a cluster stability of 0.96. A more in-depth analysis of the clustering algorithm is available in [92].

The behavior of the service discovery protocol was analyzed for the personal and community application spaces situations of different events involving people carrying local WSNs on their body and contacting networks on other people's body or the environment [90].

In all three situations the simulation starts by placing two groups of 10 sensor motes in the simulated area such that the two groups can only communicate within each other. This gives the protocol the opportunity to create the service clusters. In a subsequent step, the groups are moved according to the situations shown in Figure 4.43.

Figure 4.43(a) gives the indications on how stable the two clusters remain, while crossing other clusters and for how long the two service directories located on the cluster heads know of the presence of each other. In the ideal case, none of the motes would change their cluster membership and the service directories instantly notify each other and remove the contact information upon disconnection. Figure 4.43(b) shows how long it takes for the two clusters to merge and form a single cluster. This join operation will only work up to a certain cluster size. On a too high number of motes, multiple clusters will be formed, such that the hop-distance between the motes and the cluster service directories is minimized. Figure 4.43(c) shows how fast the exchanged mote joins the new group and how long it takes for the service directories to add/remove the information stored on that node.

Each of the situations was simulated 100 times with random placement of the nodes within the groups. The clustering phase of the simulation was analyzed as well as the maintenance of the service information during the three situations. The important values in the analyzes were the *number of messages* generated during the initialization and the maintenance of the service clusters, and the *accuracy* of the service information stored in the service directories. The *accuracy* is measured by whether the services listed are actually reachable in the network at the specific instance of time and whether all the available services are available in the service directory. Another point that was analyzed is the *number of messages* per location, leading to an indication of contention problems that might arise on the motes close to the service directories. The simulations also served to determine the values for different parameters used in the protocol, such as the expiration time of services in the service directory.

Figures 4.44 and 4.45 show the number of messages transmitted during the activation and the maintenance phase of a field of nine motes that are not all in the transmission range of each other.

The activation time is the time, during which the different sensor motes are randomly turned on and try to join existing service clusters. There might be a higher number of service directories during this time, as some parts of the

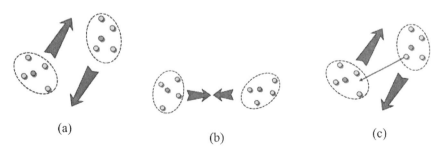

(a)   (b)   (c)

Figure 4.43 Situations simulated for a service discovery protocol evaluation [90]: (a) two service clusters pass by each other; (b)two service clusters move over each other; and (c) two clusters move across each other.

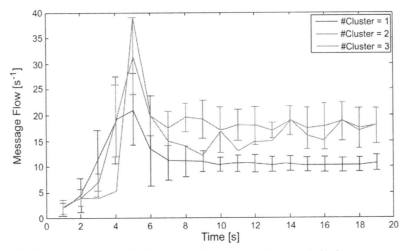

Figure 4.44   Messages generated with an activation phase of 5 seconds [90].

wireless network may be out of reach of other parts. In such a case, multiple service clusters are formed, leading to a higher message overhead. When the clusters connect due to newly active nodes, the service directories negotiate over, which one will cease to act as a service directory and join the service cluster of the other. When all motes have joined a service cluster, the protocol enters the maintenance phase, where the service directories do not give up their role as easily as during activation phase thus stabilizing the existing clusters. The number of stable clusters formed depends on the length of the activation phase. In a 5 second activation period, 9% of the simulation runs showed more than a single cluster, while this number increased to 13% for a 10 second activation period. In the maintenance phase for a single cluster, the number of the overall messages

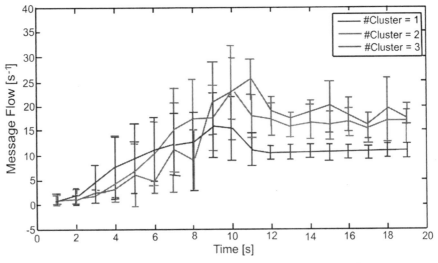

Figure 4.45   Messages generated with an activation phase of 10 seconds [90].

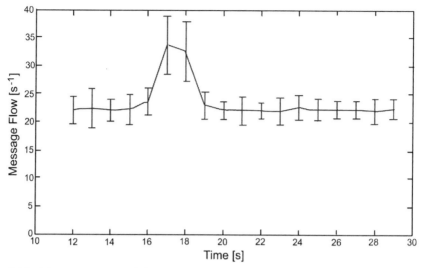

Figure 4.46 Messages generated during a cluster crossing as in situation (a) [90].

generated is about 1 per second per mote and increases to about 2 in the case of multiple stable clusters.

The message flow during the maintenance phase of the simulations of situations 1 is shown in Figure 4.46.

There is a temporary increase of messages during about 2 seconds, in which the crossing occurs. During this time, the communication paths between the service directories are built and a mutual registration occurs. The number of messages per node increases during this period from 1.10 to an average maximum of 1.65. Afterwards, the number stabilizes again at the level of separate clusters.

When two service clusters start communicating with each other, the service directories establish a communication path with each other, over which they can

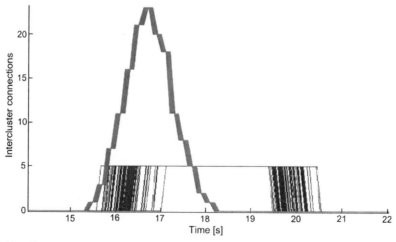

Figure 4.47 Cluster interconnections established during crossing of two clusters and the time of mutual registration of the service directories.

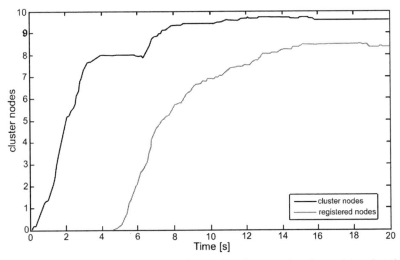

Figure 4.48 Comparison of actual nodes in the service cluster and nodes registered at the service directory [90].

communicate. As service clusters pass by each other, several communication paths between the two service directories are possible, and the effective path must be constantly updated to ensure a reliable communication path.

Figure 4.47 shows the number of interconnections between the nodes of the two clusters.

During the time that the clusters move across each other, many different connections between nodes of the two clusters are created. Figure 4.47 shows during what time the service directories are registered with each other, indicating when the services of the other cluster can actually be used by the first cluster. If the adjacent cluster information is not updated, the registrations will time out as seen on the blue curves to the right.

Figure 4.48 shows the development of leaf node registrations at the service directory of the clusters during the initialization phase of the protocol.

There is always a discrepancy between the number of actually available nodes and the ones registered at the service directory. Only the information actually available at the service directory is accessible to queries. With increasing stability, the numbers of registered and available nodes converge.

Complex Event Processing
When the WSNs are deployed in certain application settings, the inherent unreliability of sensor readings have to be taken into account and any kind of reasoning needs to carefully model the noise. The goal of *probabilistic complex event processing* (PCEP) is to build an infrastructure that can process a large amount of imprecise data and automatically infer and reason about the probabilities of a triggered event, using a principled probabilistic model for the underlying sensor data (see [93, 94] for further details).

To incorporate the uncertainty factor and make finer decisions on the event occurrence, a *Bayesian network* can be employed to directly model and exploit the correlations across the different sensors and the definition of a complex event language, which allows the users/applications for creating hierarchies of higher level events. A general architecture for probabilistic processing is shown in Figure 4.49.

The general architecture is composed of the following four basic elements: an *application layer* interface, a *sensor layer* interface, an *event processing engine* and a *probabilistic inference* engine.

Bayesian network techniques offer the possibility to model in the same time, the noise in the data coming from the sensors, and the possible correlations between the different sensors. For sensors and more specifically environmental sensors, the data is always imprecise and this imprecision has to be modelled and integrated in the reasoning process. Furthermore, between different sensors, some correlations can exist, for example between two temperature sensors, the correlation is actually linked to the distance between them. Incorporating this correlation in the process can on the one hand, accelerate the processing time and on the other hand increase the reliability of the results.

Bayesian networks (which belong to probabilistic graphical models) are graphs, in which the nodes represent random variables, and the (lack of) arcs represent the conditional independence assumptions [95].

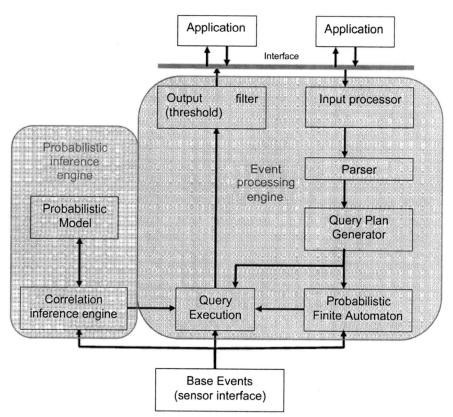

Figure 4.49 General architecture for the probabilistic processing of events [90].

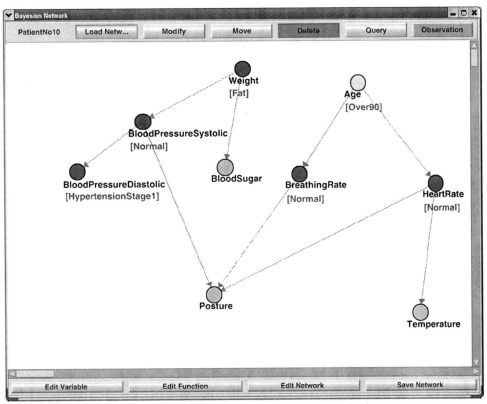

Figure 4.50   Screenshot of a Bayesian network associated to a patient [90].

Figure 4.50 shows the Bayesian network modelling of a WSN used in a medical scenario. The purpose is to determine the dangerous changes in the condition of patients in a hospital. For each patient, two variables are defined, namely his age and weight, and 4 parameters are continuously monitored: the blood pressure systolic, the blood pressure diastolic, the heart rate, and the breathing rate or the sugar rate depending on the patient's diagnosis.

The two variables are chosen arbitrary and the 4 parameters are Gaussian random variable. For each of these 6 variables, a categorization is defined to transform the continuous values to discrete ones, in order to allow the processing with Bayesian networks.

In addition to the graph structure, it is necessary to specify the parameters of the model. For a directed model, the *conditional probability distribution* (CPD) at each node must also be specified. If the variables are discrete, this can be represented as a table (CPT), which lists the probability that the child node takes on each of its different values for each combination of values of its parents.

Finally, a localization part can be defined for each patient as shown in Figure 4.51, which is a Gaussian 2 dimensions variable, and the posture of the patient is also monitored.

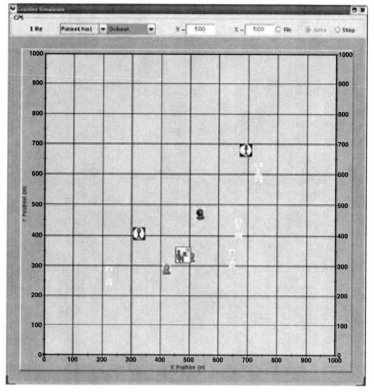

Figure 4.51   Screenshot of localization and posture in a 10 patient scenario [90].

4.3   Data Aggregation and Fusion

4.3.1   Resource Management Through Network Traffic Management

A resource management mechanism would be providing an efficient framework for the distribution of traffic in the network. In this framework the distribution of traffic on different paths is controlled to synchronise the delay of data coming from different paths towards the fusion node. To be able to do so, an *intelligent synchronized* (ISN) data aggregation protocol was proposed [90], which used the delay on the different paths as the criterion for selecting a best path, on which the data can arrive. The objective of the protocol is to design an optimal randomized data aggregation policy, which can synchronize and optimize the time of data arrival at the fusion node. The functional blocks of the ISN are shown in Figure 4.52.

The process of the aggregation of data from different nodes into a fusion node is shown in Figure 4.53.

For simplicity it can be assumed that there is one fusion node, which fuses the data of sensors from different paths. It can be seen that the data of the sensors is sent towards the fusion node on the paths with different resources. The distribution of the traffic along the paths toward the fusion node can be represented by the vector in (4.20) [89]:

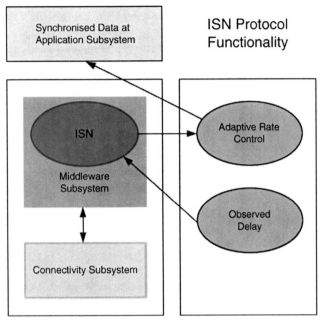

Figure 4.52 Protocol functionality of the ISN [90].

$$\vec{\lambda} = (\lambda_1, \lambda_2, \lambda_3, .., \lambda_i ..., \lambda_n) \tag{4.20}$$

where the λ_i is a traffic which is received along a particular path i.

Similarly, the distribution of the resources along the paths toward the fusion node can be given by the vector:

$$\vec{\mu} = (\mu_1, \mu_2, \mu_3, .., \mu_i ..., \mu_n) \tag{4.21}$$

where μ_i is the resource along a particular path i.

The vector delay:

$$\vec{D} = (D_1, D_2, D_3, .., D_i ..., D_n) \tag{4.22}$$

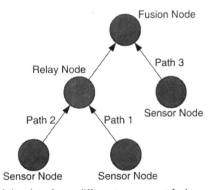

Figure 4.53 Aggregation of the data from different sensors at fusion node [90].

gives the delay values obtained on each of the n paths when the distribution of the traffic along the paths is $\overrightarrow{\lambda}$ and the and the distribution of the resources along the paths is $\overrightarrow{\mu}$.

If the total traffic that is to be received by the fusion node from all the paths is λ, then the fraction of the traffic for the particular path i, can be defined as follows:

$$P_i = \frac{\lambda_i}{\lambda} \tag{4.23}$$

Therefore, the corresponding value of the average or expected delay is obtained from the dot products of the vectors as follows:

$$(D.P)_\lambda = \sum_{i=1}^{n} D_i(\lambda P_i)P_i \tag{4.24}$$

where the subscript λ on the left hand side denotes the dependency of the average delay on the total traffic, λ that is carried over all of the n paths. The synchronized data aggregation probability, P^s, is the value of P for which D_i is equal along all the paths towards the fusion node. Similarly, λ^s is the corresponding value of λ that synchronizes all the D_i.

The value of λ_i^s is obtained as follows:

$$\lambda_i^s = \mu_i - \frac{\mu - \lambda}{n} \tag{4.25}$$

where $\mu \sum_{i=1}^{n} \mu_i$ and $\lambda \sum_{i=1}^{n} \lambda_i$.

The expected value of delay in the synchronized situation can be calculated as follows:

$$(D.P)_{\lambda_s} = \sum_{i=1}^{n} D_i^s P_i^s = \frac{n}{\mu - \lambda} \tag{4.26}$$

The ISN protocol can be described as in Figure 4.54.

The ISN protocol was tested on the two paths with different resources. In the particular example shown in Figure 4.55, the fusion node receives the total traffic of 25 kbytes/s on the two paths of two hops and three hops length.

When the value of the delay for any particular path increases, then the fraction of the traffic for that path decreases. The algorithm will forward the fraction of the traffic from the path with less resources and more delay towards the best path. It can be seen in the protocol description that the change of traffic is different for the paths with different values of resources. Figure 4.55 shows that the introduced policy synchronizes the delay of two paths from two source nodes. After several

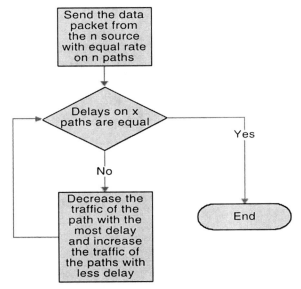

Figure 4.54   ISN protocol description [90].

iterations, the intersection of the two graphs is the point when the delays of the two paths are equal. This is shown in Figure 4.56.

The ISN protocol provides a solution, in which the delay on each path is synchronized and the expected average delay for the data coming from different paths is close to the solution for the optimal traffic distribution.

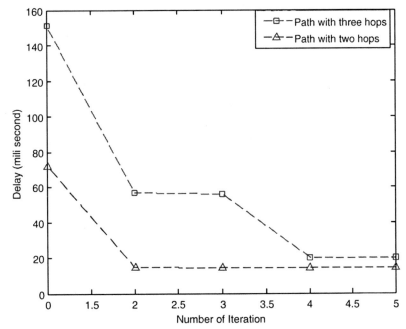

Figure 4.55   The intersection of two curves of delay after 5 iterations is the synchronized point [90].

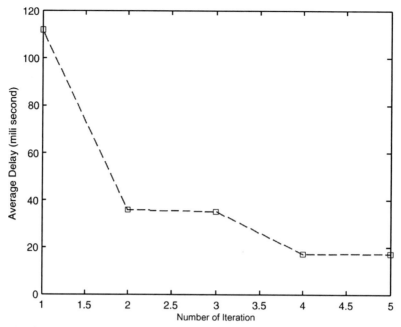

Figure 4.56 The average delay after 5 iterations, which is optimal [90].

4.3.2 Distributed Data Aggregation

Densely deployed WSNs allow environmental monitoring at extremely high spatial and temporal resolutions [90]. However, extracting the raw data from such networks can have problems, (e.g., batteries may get drained rapidly due to the excessive operation of the transceiver or data quality may deteriorate due to dropped packets caused by a network congestion. Smart farm and smart factory applications [96] are examples, in which the distributed data aggregation is essential for an energy-efficient performance.

To solve the above problems, [4] exploited the high degree of *spatial correlation* that exists between the sensor readings of adjacent nodes in a densely deployed network [90]. Thus, instead of every node transmitting individual readings, a subset of nodes, referred to as *correlating* nodes, transmits the messages representative of all the remaining nodes at any given point in time.

Every correlating node initially transmits information to the sink, hereby indicating the correlation of its readings with its adjacent neighbors. Subsequently, it continues transmitting its own readings until a change in the correlation is detected; in which case it transmits an updated correlation message. The sink then estimates the readings of the adjacent neighbors of the correlating node by combining the current readings of the correlating node with the previously transmitted correlation information. Based on a completely distributed and self-organizing scheduling algorithm that (1) prevents two adjacent nodes acting as correlating nodes simultaneously, and (2) increases the robustness and accuracy of the readings by giving every node a chance at some point to act as a correlating node, it is ensured that no node is always represented only by estimated readings.

The primary objective of the *distributed and self-organizing scheduling algorithm* (DOSA), is to help decide when a particular node should act as a correlating node and thus represent the sensor readings of its first order neighbors. During the schedule of the correlating node, the node initially transmits the correlation information to the sink node followed by its own sensor readings. All the first order neighbors do not transmit their sensor readings to the sink during this period. Since DOSA is intended to solve a scheduling problem, a *distributed graph coloring* algorithm is used to assign the schedules to individual nodes. From a graph theoretic point of view, because no two adjacent nodes can act as correlating nodes simultaneously, all the nodes chosen by DOSA to be correlating nodes need to form an independent set. Additionally, the correlating nodes for a particular instant of time need to form a dominating set since every noncorrelating node must be joined to at least one correlating node by some edge.

The subset of nodes that is both independent and dominating is known as a *maximal independent* set. A *maximal independent* set cannot be extended further by the addition of any other nodes from the graph.

To hasten the rate of assigning schedules to the nodes, DOSA utilizes the information provided by the underlying MAC protocol, a *lightweight MAC* (LMAC) [97, 98] (see also Chapter 2 of this book). While there have been many MAC protocols designed for sensor networks, none of the protocols provide neighborhood information the way LMAC does.

Additionally, the LMAC is a TDMA-based MAC, which is an added advantage as it automatically provides a sense of time, which is beneficial to DOSA. In addition, LMAC ensures that two nodes that are at least 3 hops away from each other can reuse the same time slot. It must be noted that the DOSA algorithm can run on top of other MAC protocols as well, although this would require an addition layer to keep a track of the immediate topology information [90].

Instead of coloring all the nodes from scratch, DOSA meets its requirements by building up on the colors already assigned by LMAC. An added advantage of this form of *cross-layer optimization* is that a smaller number of messages need to be transmitted for all the schedules to be assigned properly. Further, the dependence of DOSA on LMAC makes it more reactive to changes in topology as any changes in neighborhood detected by the LMAC are immediately filtered to DOSA.

Every color owned by a node represents a particular frame of time, during which a node is required to act as a correlating node. In conventional graph coloring approaches, colors are assigned to vertices such that adjacent vertices are assigned different colors and the number of colors used is minimized. DOSA allows for owning of multiple colors, (i.e., a node can have multiple schedules). Moreover, the number of colors used in DOSA is fixed and is equal to the number of slots that are assigned to an LMAC frame.

4.3.2.1   Constraints

The following two constraints must be met when two nodes u and v are adjacent to each other [90]:

- *Constraint 1*: $C_v \cap C_u = \varnothing$

In other words, two adjacent nodes cannot own the same colors. This is because two adjacent nodes should not be assigned as correlating nodes in the same time instant.

- *Constraint 2*: $C_{\Gamma(v)} = K$

All colors should be present within the one-hop neighborhood of node v, (i.e., if a node v does not own a particular color itself, the color must be present in one of its neighboring nodes that are at one-hop away. This ensures that the reading for every node will be represented at the sink node for every time instant, either *directly* or through a *correlated* reading.

Lemma 1. The combination of *constraints* 1 and 2 ensures that at any time slot, c_i, all nodes owning the color c_i, which corresponds to that time slot, form a maximal independent set on G.

Proof: At any time instant according to *Constraint* 1, two adjacent nodes will never own the color c_i, thus resulting in an independent set I. *Constraint* 2 ensures that in the closed neighborhood of every node $v \cdot V$, where V is an element of the graph, every color is present. This clearly results in a maximal independent set.

4.3.2.2 Operation

DOSA uses a *greedy* approach to assign colors to nodes. Coloring is performed using two types of colors: *LMAC colors* and *DOSA colors*. LMAC colors refer to the colors that have been assigned by the LMAC, due to the slot assignment. The DOSA colors refer to the additional colors that are assigned by DOSA to ensure that *constraints* 1 and 2 are met. This occurs after the LMAC colors have been assigned. DOSA does not have any control over the LMAC color of a node as it depends purely on the slot assignment performed by LMAC. In fact, such control is also not required. Therefore, in the following, DOSA colors are simply colors unless otherwise indicated.

The colors are acquired based on a calculated priority. A node computes its priority within its one-hop neighborhood based on its degree and the node ID. The higher the node's degree, the higher is its priority. If two neighboring nodes have the same degree, the priority is calculated based on the unique node ID; the node with the larger node ID will have the higher priority.

Once all nodes have acquired their LMAC slots, a *BeginSecondPhase* message is injected into the network through the sink node, requesting the nodes to begin the DOSA coloring phase. At this stage, every node receiving the *BeginSecond-Phase* message only has an LMAC color and does not usually satisfy the constraints. Thus, these nodes mark themselves as *Unsatisfied*. A node only attains the *Satisfied* status, when it satisfies the two constraints. Upon receiving the *BeginSecondPhase* message, a node broadcasts the *NodeStatus* message. This message contains information about the node's status (i.e., *Satisfied/Unsatisfied*) and the list of colors owned. The *ColorsOwned* field is a string of |K| bits where every colour owned by a node is marked with a '1'. The rest of the bits are marked with a '0'. Initially, a node only marks its own LMAC color as '1' due to the initial LMAC slot assignment. A neighboring node that receives the *NodeStatus* message

Algorithm 1 DOSA - Normal Initialization

Input: NodeStatusMSG(SatisfiedStatus(TRUE/FALSE), ColoursOwned)
Output: NodeStatusMSG(SatisfiedStatus(TRUE), ColoursOwned)/ NIL

1: Update(LocalInfoTable, v)
2: if LocalInfoTable contains entries from ALL adjacent nodes then
3: if SatisfiedStatus(v)=FALSE then
4: Compute Priority(v)
5: if Priority(v)=Highest then
6: $C_v \leftarrow K \backslash C_{\Gamma(v)}$
7: ColorsOwned $\leftarrow C_v$
8: SatisfiedStatus $\leftarrow$ TRUE
9: Update(LocalInfoTable, v)
10: Broadcast NodeStatusMSG(Degree, SatisfiedStatus, ColoursOwned)
11: end if
12: end if
13: end if

Figure 4.57   Steps of DOSA algorithm 1 [90].

then performs coloring using DOSA according to algorithm 1, which is shown in Figure 4.57.

Upon receiving a *NodeStatus* message, a node first updates its *LocalInfoTable* (line 1). This table stores all the information contained in the *NodeStatus* messages that are received from all the adjacent nodes. Once a node receives *NodeStatus* messages from all its immediate neighbors (line 2), and if its status is *Unsatisfied* (line 3), the node proceeds to compute its priority. Priority computes the priority of a node only among its unsatisfied neighbors (line 4), (i.e., as the time progresses and more nodes attain the *Satisfied* status, *Priority* needs to consider a smaller number of neighboring nodes. The highest priority is given to the node with the largest degree among its adjacent *Unsatisfied* neighbors. If more than one node has the same degree, then the highest priority is given to the *Unsatisfied* node with the largest NodeID.

The node that has the highest priority among all its immediate unsatisfied neighbors, acquires all the colors that are not owned by any of its adjacent neighbors (line 7). As the node has then satisfied both constraints of DOSA, it switches to the *Satisfied* state, updates its own *LocalInfoTable* and informs all its neighbors through a broadcast operation (lines 8–10). This technique corresponds to a highest degree greedy approach.

When a node dies, DOSA executes one fixed set of steps to ensure that the scheduling scheme stabilizes within a finite time. In the node addition operation however, the set of steps taken by DOSA depends on the events that occur when a new node v is added to the network. For example, the node v may detect an LMAC collision or may cause colliding or missing colors in neighboring nodes or may even cause a combination of these events. Different permutations and combinations of these events can cause the network to react in a multitude of ways, which makes it impractical to analyze the performance bounds of every particular sequence of events that causes the network to react in a certain manner. Instead, all the permutations and combinations of events can be categorized in terms of how far the network disturbance propagates when a node v is added to the network. For example, depending on the combination of events, the nodes that are either 2 or 3 hops away from node v may become *Unsatisfied*. Further detailed descriptions and performance results for DOSA can be found in [99].

4.4   Adaptive Security Framework

The diversity of the application spaces leads to different scalability settings based on the various capabilities of the sensor nodes, on the different number of nodes, and on their topologies. This imposes the need for security, privacy, and trust policies and anonymity preserving mechanisms which are scalable and, which smartly adapt to the context of each communication in order to preserve the energy and to support the end user.

4.4.1   Requirements for Security, Privacy, and Trust

Figure 4.58 shows a security framework developed in [4] for WSNs.

It includes a cross-layer security manager, which controls all issues related to data security, privacy, and trust at different levels and covers the following general security requirements through security protocols and mechanisms:

- *Confidentiality* and *secrecy* of the collected and communicated data. Confidentiality entails ensuring that the information is accessible only to those authorized to have access, and providing the assurance that the information processed by a system is protected against intentional or unintentional unauthorised access by individuals, processes, or devices.

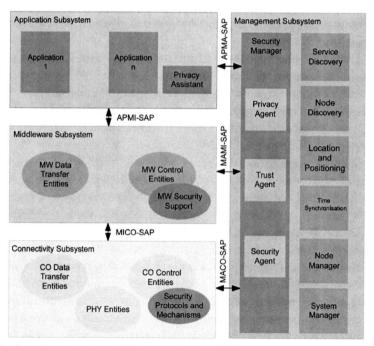

Figure 4.58   Security services in a WSN protocol stack [100].

- *Integrity* of the data that is sensed, processed, or aggregated. Data integrity is a measure of its reliability. It entails protection against unauthorized data modification, insertion, substitution, or deletion. It includes providing the means for assuring data freshness.
- *Authentication* of the originator and recipient of messages. Authentication entails providing the means for verification of the network identities, and is a prerequisite for covering the other security requirements.
- *Authorization* for accessing data or network resources. It entails providing the means to control access and approve actions by the authenticated entities.

The *security protocols* and *mechanisms* component is the most basic for the framework in Figure 4.58, because it is the only one required for all types of nodes in the network. Through this component, the *connectivity subsystem* can enforce the application of security mechanisms for every message exchange in a way that is transparent to the layers above it. The types of mechanisms that are applied depend on the preconfiguration of the node and/or the input and updates it receives from the *security manager* in the management subsystem.

Preserving the privacy of the individuals, acting within an information system generally entails keeping their personal information confidential and accessible only to authorized parties. For sensor networks, ensuring the confidentiality of the data does not suffice. For example, confidentiality of the messages content would not protect from tracking the relative location of a BSN. The additional requirements that are set are related to privacy and can be summarized as follows:

- Provision of mechanisms for controlled data disclosure based on rules and policies. Those mechanisms should operate according to the data sensitivity level and the data abstraction level, and protect from information induction through sensed data correlation.
- Protection of the identity of the nodes related to users through anonymity and pseudonymity mechanisms.
- Provision of the user's notice and choice mechanisms to be applied when a user-related device has user interaction capabilities. The responsibility of those mechanisms is to provide awareness to the users that data for them or the environment surrounding them is being collected. They also have to empower the users to control, by making informed decisions, what personal data will be disclosed and which of the network pseudonyms representing them should be used.

Anonymity mechanisms can ensure that a user may use a resource or a service without being distinguished from other users and without disclosing his identity to third parties [101]. In order for anonymity to be achieved, the data communicated in the sensor network needs either to be depersonalized or pseudonymized with respect to the identity of the sender, in cases that a legitimate network service requires user identification. *Pseudonymity* is the use of pseudonyms, such as IDs [102, 103].

In a WSN, the sensor nodes are resource-constrained and the cost to compute a new pseudonym is high, in comparison with the standard approach when the pseudonyms are changed periodically, it is of importance to identify the suitable moment when a pseudonym must be changed. Thus the node must have a mechanism to evaluate the current context in order to decide when it is the most suitable time to change the pseudonym, in order to avoid energy waste.

Finally, there are a number of architectural design requirements related to the integration of ad hoc networks with other systems (e.g., next generation cellular). The main ones can be summarized as follows:

- *Flexibility and scalability*—it should be possible to provide varying levels of security, privacy, and trust functionalities, corresponding to different node architectures, hardware limitations, user requirements, and application spaces.
- *Adaptability*—the security protocols and primitives that are used for each communication after the network deployment are selected according to the context of the communication, taking into account the trade-off among device constraints, change in context and different user's preferences.

4.4.1.1   General Architecture

A generic adaptive security framework that fulfils the above-identified requirements is shown in Figure 4.59.

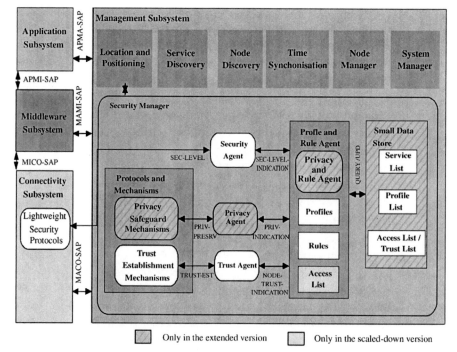

Figure 4.59   Generic adaptive security framework [100].

The *security manager (SecM)* is a logical entity, which is responsible for defining the network level of security, privacy, and trust and its decisions are based on the (context) information provided by the other management entities in the *management subsystem* (for example *node discovery*, *service discovery*, and *location positioning*), for which entities the information is already provided in the *management subsystem*. The SecM performs an adaptive secure context-aware management, which goal is to ensure the context-aware secure interactions [100].

The provided levels of security, privacy, and trust are defined as a compromise for the security, privacy, and trust policies of the network, the user preferences; and the device capability.

The *security protocol* is functionality inside the SecM. The *security protocol* functionality executes all the possible security protocols, which can be used in the system to provide the various levels of confidentiality, integrity protection, and freshness. The security protocol block has all the security protocols, irrespective of the *security level* they belong to. The security protocol gets the information about the required security level from the *SecA* and provides the appropriate protocol for the connection to the *connectivity subsystem*.

The *privacy protection* mechanisms interfere with the data to be transferred, either by pseudonymising it or by allowing/forbidding its disclosure.

The *trust establishment mechanisms* assign the trust levels to the unknown devices that the node is not preconfigured to trust. Those mechanisms include identity verification, trust transitivity, and protocols for security policies announcement, validation and negotiation. In the case of devices having a GUI, the user interference may be used for user approvals. The result of the *trust establishment* procedure is the trust level assigned to the other node together with its security profile to be used for determining the security level that will be assigned for subsequent communications with it. A high-level description of how the security level is determined is shown in Figure 4.60.

The interaction between the different components of the adaptive security framework are shown in Figure 4.61.

The SERV-QUERY (parameters) is a request, sent by the *profiles and rules agent* (PrfRlsA) to the *service list*, regarding the service. The PrfRlsA can request, with this same primitive but using different parameters, to either just get the *security level* or change the security level associated to the service. SERV-UPD is the response to the SERV-QUERY. It can carry either the security level information or just the notification that the security level has been changed.

The TRUST-QUERY (parameters) primitive is a request, sent by the PrfRlsA to the *access list/trust list*. Similar to the SERV-QUERY (parameters), the PrfRlsA can request either getting the *trust level* or changing the trust level associated to the node ID. TRUST-UPD is the response to the TRUST-QUERY. It can carry either the trust level information or the notification that the trust level has been changed in the access list/trust list.

The PROF-QUERY (parameters) is issued towards the *profile list*. The *profile database* provides the profile for the corresponding node/user through the PROF-UPD primitive.

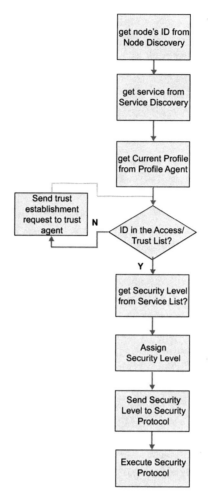

Figure 4.60 Steps in determining the security levels [100].

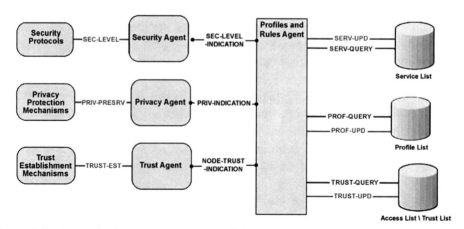

Figure 4.61 Interaction between components of the adaptive security framework [100].

The primitive SEC-LEVEL-INDICATION sends to the SecA all the information that the PrfRlsA gets from the *list and the rule module* and is relevant to determine the security level.

The primitive SEC-LEVEL sends the security level for the corresponding node/user to the security protocol entity.

The primitive PRIV-INDICATION allows for the exchange of privacy information between the *privacy agent* (PrvA) and the PrfRlsA.

The primitive PRIV-PRESRV sends the privacy (and pseudonym) information to the *privacy protection mechanism* entity.

The NODE-TRUST-INDICATION primitive allows for the exchange of trust information between the *trust agent* (TrsA) and PrfRlsA.

The primitive TRUST-EST sends the trust information to the *trust establishment mechanisms*. The trust establishment procedure is initiated by the trust agent using the TRUST-EST primitive. It is performed by the *trust agents* of the communicating nodes using the node centric data transfer primitives of the connectivity subsystem. The overall process is shown Figure 4.62.

For ease of description, the process in Figure 4.62 assumes a "waterfall" engineering approach, where each step of the process is done once and in the order shown, prior to the deployment. Other approaches are equally valid, and the process described here could be easily modified to fit them. In particular, a more

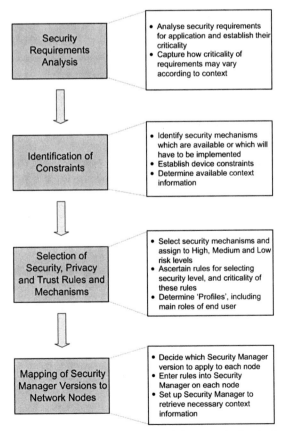

Figure 4.62 Process of tailoring the security framework for a particular deployment [100].

iterative approach, where some or all of the steps are repeated, perhaps after the deployment and during operation. For example, an end user may be able to modify the profiles and the rules according to their usage of the system, such as to remove a profile associated with a role that they never use [100].

4.4.1.2 Lightweight Configuration

The security framework is designed to be *reconfigurable*, in order to provide the most suitable levels of security, trust, and privacy functionality for different nodes. The configuration guidelines serve to assist in the specification of the most appropriate version of the framework and of its configurable components for each network node [100].

A lightweight configuration is shown in Figure 4.63.

The specific issues that need to be addressed for the configuration of each node before the deployment are the following:

1. The version of the security manager that it should be equipped with. The scaled down and the extended version are shown in Figure 4.63.
2. The set of mechanisms that the configurable components, shown in Figure 4.63, will be equipped with:
   (a) The *lightweight security protocols* component may support only a subset of the security levels (*low*, *medium*, *high*);
   (b) The *trust establishment mechanisms component* may be equipped with a subset of the mechanisms used during the trust establishment process (certificate validation, evaluation of recommendations)
3. The *profiles and rules* that the *security and privacy agents* use for providing adaptability after the network deployment.
4. The trust establishment parameters of the *access list* module, including the information for the third parties that each node is preconfigured to trust.

The specific security protocols and mechanisms that the *lightweight security protocols* component of all nodes of deployment will be equipped with, also needs to be decided on. The set of protocols and mechanisms that correspond to each security level may vary according to the criticality of each security service to the scenario.

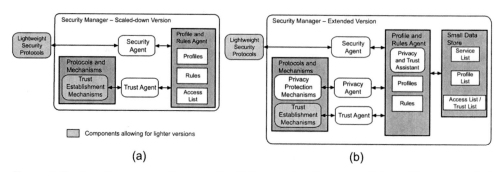

(a)   (b)

Figure 4.63 Security components supporting lighter versions in the scaled-down (a) and extended (b) versions of the framework [100].

The configuration decisions that have to be made for each node and the parameters that affects these, are shown in Figure 4.64.

The numbers surrounding the alternatives for the configuration decisions are used to specify the number of alternatives that can be selected. The security, trust, and privacy requirements for each scenario, context building block, network type, and communications scope are the critical parameters for those decisions.

Table 4.3 shows the guide for using the configuration parameters in order to reach to the configuration decisions for each node in the network.

The first step of the configuration is to characterize the capabilities and the role of the node in the network, in order to evaluate if it is possible and necessary for it to be equipped with the extended version of the security manager. Provided that a security requirements analysis has been made, the next step is to identify, in which category the node belongs to, following the hierarchy Scenario → context building block → network type (environmental/body WSN) → scopes of communication. The identification of this category is necessary in order to retrieve the security requirements that apply to this particular node. If a node belongs to more than one category, the strictest value is selected for each security requirement (Low → Medium → High, No → Yes). Table 4.3 is then used for the configuration decisions on the security manager version, the supported security levels, and the trust establishment mechanisms.

While the supported security levels and the trust establishment mechanisms depend only on the security requirements of each node, Table 4.3 shows that the extended version of the framework applies to nodes that have sufficient energy, computation, and communication capabilities; are gateways, sink nodes, or (some) relay nodes; their communications scope extends beyond the sensor network, they shall communicate with nodes of trusted subscribed networks or unknown networks; are responsible for performing a controlled data disclosure and identity management operations; provide an interface to the end users to fulfil notice and choice requirements.

The security mechanisms can be lightweight in two different ways. Firstly, the adaptability enables the selection of the "lightest" set of algorithms and the parameters for the current context, in order to avoid too strong security and the consequent wasted resources when not needed. Secondly, the "lightest" algorithms and implementations are used for the equivalent levels of security.

In terms of adaptability, the 'security level' determines the mechanisms and protocols that are used to provide the authentication, encryption, message freshness and integrity for each request. The security level may be *high, medium,* or *low,* and is re-evaluated whenever there is a change in the network state, the device state,

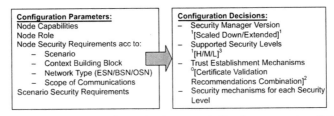

Figure 4.64   Security framework configuration issues and affecting parameters [100].

Table 4.3 Security Framework Configuration Guide [100]

Configuration Parameter		Security Manager Version			Supported Security Levels			Trust Establishment Mechanisms	
		Scaled Down	Extended		H	M	L	Cert. Valid.	Recomm. Comb.
Node Capabilities									
End sensor node, constrained energy, computation, communication capabilities		+	+	Powerful node, regular maintenance, User interface					
Node Role									
Source node / Forwarding node		+	+	Sink node / Gateway / Relay node					
Node Security Requirements									
General security requirements									
Confidentiality	W						+		
	M					+	+		
	S				+	+	+		
Integrity – Freshness	W						+		
	M					+	+		
	S				+	+	+		
Authentication	W						+		
	M					+	+		
	S				+	+	+		
Trust requirements									
Adaptive trust negotiation	Y W					+			+
	M					+		+	+
	S					+		+	+
	N							-	-
Trust revocation	Y					+			+
	N								
Privacy requirements									
Controlled data disclosure	Y		+						
	N	+							
Identity management	W	+							
	M		+						
	S		+						
Notice and choice	Y W	+							
	M		+						
	S		+						
	N	+							

the surrounding context or when the user requests another level of data protection than the current one. The most appropriate current security level is determined as a function of the trade-off between the application and user requirements, policies, context, power, and computational constraints. This adaptability aims at achieving significant savings in battery power. Without this adaptability, the standard security approach often provides the highest possible level of protection to all data. By only providing the security requirements that are strictly necessary according to the application needs, environmental context, and so forth, significant security processing and communications overhead is saved.

In terms of lightweight security algorithms and implementations, what are needed are algorithms and implementations specifically tailored to the constraints of the ad hoc network environment, as well as flexibility within these to allow for different levels of security to be applied (and resources to be saved). WSNs typically require four main types of security algorithms: encryption; integrity; message freshness; and authentication. As a general principle, use of *symmetric key cryptographic* algorithms is appropriate wherever possible. From several experimental evaluations of the energy consumption of security protocols [104–106], it is known that the symmetric cryptographic operations are considerably less costly than the asymmetric cryptographic operations, and the lightweight symmetric cryptographic algorithms are considered acceptable for resource-constrained sensor nodes.

When asymmetric cryptography is required, as another general principle use of elliptic curve algorithms wherever possible is appropriate [100]. Similar evaluations to those for symmetric cryptographic algorithms show that for asymmetric operations, the elliptic curve cryptography is considerably less costly than the traditional public key cryptography [104, 106, 107]. This is both in terms of the required processing power and the size of the messages (reduced radio power).

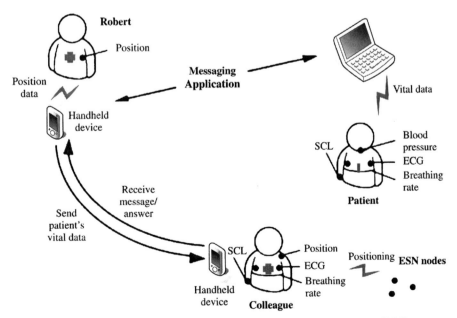

Figure 4.65   A BSN of the community application space (backup shift assistant) [100].

Table 4.4   Identified Constraints for the BSN Scenario [100]

Parameter	BSN
Network size	4
Life time	Few days to months
Node size	< 1 cm^2
Battery time	1–10 days
Deployment and maintenance	Manual deployment, easy access and placement on body, regular maintenance possible
Range	<10m (from gateway). Depends on short-range wireless access technology e.g., Bluetooth, UWB
Traffic activity pattern	Bursty when active
Mobility of sensors	Static (wrt body)
Reliability	High
Data type	Low level

Symmetric key based message authentication codes (MACs) can be used to provide the efficient integrity protection to data. A similar technique of varying the number of rounds and key length could be used to vary the level of strength in the integrity algorithm (and resources used) because the MAC size depends on the length of the blocks that the data is split to when applying the algorithm. *Hash-based MACs* could be used instead. In this case, there is no scope to vary the strength of the algorithm used. However, to provide a lower level of security, the output could be truncated from its usual size (e.g., 160 bits) to something shorter (e.g., 64 bits), which would reduce the size of the messages and save on the power consumption of the radio. As power consumption by the radio usually dominates the processing power consumption, this could be of significant benefit. This technique could also be used for the encryption based MACs. Finally, an option is simply to either use integrity protection or not, as a way of varying the security level.

An example scenario for the applicability of the security configuration is shown in Figure 4.65.

The scenario in Figure 4.65 entails the collection of both information about the vital functions and of the location, patient monitoring, and the enhanced coordination of the healthcare personnel. WSNs are used to gather and communicate context information in order for a physician (*Robert*) to be able to monitor the patient's health status and contact the next experienced colleague available in the hospital or at home if a severe patient condition comes up that he cannot deal with. Through the WSN application, the experienced colleague is informed both of the condition of the patient, and of the location and mood (e.g., stress level) of Robert, in order

Table 4.5   Available Context Data for the Patient [100]

Context Data	Sensors	Category	Bit Rate	Access
Blood pressure	Sleeve	BSN	Low [10]	User, Doctor,
Heart rate, Heart rate variability	ECG	BSN	0.3-0.5 Kbytes/s [10]	Emergency
Breathing rate	Chest strap	BSN	Low [10]	Security
Blood sugar	Glucose meter	BSN	Low	

Table 4.6   Selection of Security Services for Each Security Level [100]

Criticality for the Scenario	Security Level		
	Low	*Medium*	*High*
Authentication services			
high	Symmetric network and group keys	Symmetric link keys for pairwise authentication and μTESLA for source authentication	Elliptic curve Diffie-Hellman key agreement (ECDH), Elliptic curve digital signature algorithm (ECDSA)
Encryption			
high	No encryption	RC5 (CTR mode) 32 bits block length, 6 rounds & 24-bit key	RC5 (CTR mode) 32 bits block length, 12 rounds & 40-bit key
Integrity protection			
high	64 bit MACs CBC-MAC based on RC5	128 bit MACs CBC-MAC based on RC5	128 bit MACs CBC-MAC based on RC5
Freshness			
high	No	Relative freshness through message sequence numbers	Strong freshness through timestamps

to be fully informed when he arrives for assistance. In this scenario, BSNs are applied both on the patients, to monitor their vital functions, and on the physicians, in order to assess their availability in terms of stress level and location. The physicians communicate through mobile devices that provide them with an interface to all information received and act as the gateways of their BSNs. The BSN of the *patient* is the one selected as the example of the application of the security framework.

The first step for the configuration of the security framework for the patient's BSN is the *security requirements analysis*. This step also includes the identification of how the *change in context* affects the security requirements. The second step of the process is the identification of (device and context information) *constraints*. The constraints identified for the scenario in Figure 4.65 are summarized in Table 4.4.

There are two types of devices in the patient's BSN: the four *sensor* nodes of Table 4.4, acting as the source nodes and having constrained energy, communication, and computational capabilities, and the *gateway* node, with which the end sensor nodes communicate with, for making the information available to the hospital messaging application. The available context data is summarized in Table 4.5.

The framework configuration guidelines (see Section 4.4.1.2) are used next in order to select the security mechanisms that correspond to each security level. The selection is made according to the criticality of each security service to the scenario and the device constraints, identified in the previous steps. The scenario has high confidentiality, high integrity and freshness, and high authentication requirements.

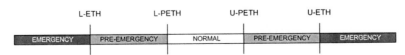

Figure 4.66   Rules for the definition of the status of the end user [100].

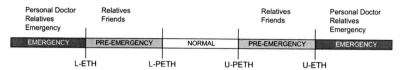

Figure 4.67   Rules for notification based on status of end user [100].

The corresponding security services for each of these security levels are summarized in Table 4.6.

This step also includes identifying the rules for deciding which security level to apply depending on the context. The framework configuration guidelines are used then to configure the security manager for each network node, (i.e., decide on a scaled-down or an extended security manager version), the supported security levels, and trust establishment mechanisms.

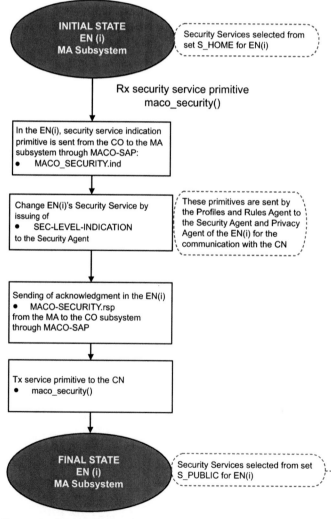

Figure 4.68   Change in the security level in the end node from a BSN based on a change in context.

Only the gateway would require the extended version of the security manager, it is the one responsible for defining the level of security and trust services for the communications with the sensor nodes on the one hand and external networks on the other. A scaled-down version would reside in the sensor nodes of the BSN, because of the very limited power, memory and computational capabilities of this node, and this version would only perform a specific security mechanism.

Among all the possible contexts, the following three context events can be considered: *normal*, *pre-emergency*, and *emergency*, which describe the health status of the patient. It is possible to outline them through *thresholds* for each context data—two thresholds are used for delineating when the patient is in an emergency case (i.e., above U-ETH or below L-ETH and four are used instead to define the *pre-emergency* case. This is shown in Figure 4.66.

In a similar way, notification rules are defined to notify a certain category of person (medical staff, family, or both) of the present situation of the user/patient.

The notification rules related to the example scenario are shown in Figure 4.67.

The change in the security level in the end node is shown in Figure 4.68.

The configuration of the rest of the nodes can be performed in a similar way.

4.4.2 Analysis and Evaluation

The security framework is evaluated against the identified requirements for security, privacy, and trust and for the given scenarios (i.e., WSNs-based). The implications that the security framework (the security, trust establishment, and privacy mechanisms) can have on the network nodes and the network operation in general, mainly in terms of resource consumption were estimated in [100]. Table 4.7 [100] summarizes the performance evaluation methodology and the main outcomes.

The results are based on a combination of experimental measurements performed by [4], as well as on several experimental evaluations that exist on the resource consumption of security mechanisms on sensor network nodes. The proposed security framework for WSNs makes use of the lightweight security mechanisms.

The main framework-wide requirement is *adaptability* and *power efficiency*, which is supported by the proposed security framework in several ways. Firstly, with the help of the security agent *security levels* are assigned. The support is provided by the security mechanisms that are applied, depending on the security needs of the communication.

Providing only the required set of security services based on current scenario and context and not wasting resources for "a full packet of security services" contributes also to the power efficiency.

Secondly, the *adaptability* is supported by the representation and establishment of various trust relationships between communicating parties. The framework includes a trust agent, responsible for establishing trust relationships and managing access lists. Finally, the adaptability is also supported by the different level of protection of the privacy of the end users acting within the network or the confidential corporate information. The privacy agent is responsible for determining if the data should be disclosed, or provided anonymously. The privacy level flags indicate

Table 4.7   Performance Evaluation Methodology and Results

	Evaluation Methodology	*Evaluation Measures and Metrics*	*Evaluation Results*
Security mechanisms	Prototype implementation on sensor testbed and analytical power consumption model	Power consumption: Benefits of the adaptability property on battery power consumption by quantifying the difference in power between having and not having the ability to adapt	Power savings from the adaptability property depend on: - The frequency of security level changes - The proportion of the total time spent in each security level during network operation
Trust establishment mechanisms	Analysis of trust establishment process and formulas for evaluating power consumption	Power consumption: Estimations based on experimental evaluations of power consumption of security primitives	Energy, computation and communication requirements: - Depend on initial parameterisation of trust establishment components - Can be optimised according to trust requirements
Privacy protection and context awareness	Simulations for varying numbers of context attributes and varying numbers of applicable rules	Response time: Influence of the complexity and granularity of the context information on the time from placing the request for data until it is filtered	Response time and memory requirements: - Can be optimized through selecting only vital rules and context attributes - Proactive approach for selection of valid rule subsets leads to smaller delays

how the user wants the data in question to be handled and revealed by the privacy agent.

The *flexibility* requirement is fulfilled by enabling the role, the capabilities, and the security needs of each node in the network to define the subset of the security manager components, which reside in the node. The extended version of the *security framework* applies to the coordinator and the gateway nodes as well as to the simple nodes without very harsh memory, battery, and computational constraints. The scaled-down version misses the *privacy agent* and the *privacy protection mechanism*, the *privacy and trust assistant*, and the *data store*. As some fundamental information from the Data Store is necessary, the simpler and essential tables containing policies, profiles, and access lists are stored in the *profiles and rules agent* as a small *access lists* component.

Except from the components that might be omitted, others allow for lighter versions to be deployed, in order to provide only a subset of the services defined. The *security protocols and mechanisms* component may support only a subset of the security levels. For highly constrained nodes with a strictly defined role, one security level may suffice for its communications with the cluster head. Moreover, the *trust establishment mechanisms* are required for nodes that, during the network lifecycle and without the reconfiguration, will need to communicate with nodes other than those they were initially configured to trust. The nodes might be equipped with a subset of the mechanisms defined, depending on their computational capabilities, their role in the network, and their communication needs.

The performance of the framework was evaluated by a very simple prototype of the framework implemented for a given scenario (i.e., *building health*) and concentrating on the ability to change the current security level of nodes in a WSN. Measurements of the power consumption of sensor nodes were then taken by running this prototype through the scenario to try and quantify the difference in energy use between having and not having the ability to adapt the security level. The implementation is shown in Figure 4.69.

The adaptable security framework was implemented for a network configuration using three Mica2 sensor nodes: two programmed as *leaf* sensor nodes and one as the *gateway* node. All three nodes were programmed with the sensor data application. The gateway was connected to a PC via the MIB510 programming board, which was connected to a serial port on the PC. The simulator application was run on the PC as well. Figure 4.69 shows the platforms, communication channels, and software applications used in the implementation.

An application on the laptop receives and displays information from the sensor nodes and configures the security level in the WSN. This application communicates with the gateway node via a serial port. The application was written in Java and based on existing TinyOS applications.

This application implemented the following functionality:

- Ability to send a request to change the security level to the attached gateway node. The sending of these requests needs to be controlled somehow, and the simplest way would be through a user interface that allows for the user to manually select the security level.

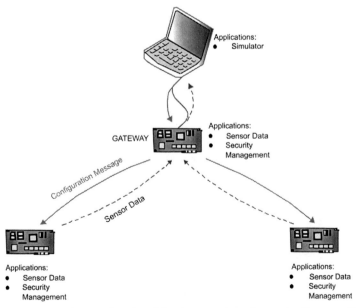

Figure 4.69 Experimental scenario-related implementation of an adaptive security framework [100].

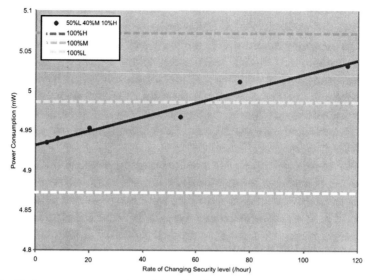

Figure 4.70   Node power consumption with and without adaptable security [100].

- Ability to read the sensor data from the attached gateway node and log it to a file.

Figure 4.70 shows the results for an experiment for the above scenario where the rate, at which the security level changed was varied, while keeping the proportion of time spent in each security level constant. The traffic matrix was identical for each experimental run.

The power consumed without the adaptive framework is also shown for comparison.

The curve labelled *100%H* shows the power consumption per node with the security level fixed at *high*. The curves labelled *100%M* and *100%L* show the power consumption for fixed security levels of *medium* and *low*, respectively. These three curves are not affected by the rate of change in the security level, and remain constant. The points labelled *50%L, 40%M, 10%H* represent the experimental values of the power consumption per node with the adaptable security framework switched on, and with the respective proportions of time spent in each security level. A linear regression curve for these points is also added.

Figure 4.70 shows that the power saving achieved by using the adaptive framework is proportional to the rate of the security level change, (i.e. to the number of configuration messages). The largest percentage power saving provided by the adaptive framework compared to a framework fixed at *high* is 2.8%. When the security level is changed fewer than 60 times per hour, the adaptive framework uses less energy than a framework that only uses the medium security level. By extrapolating the line of the best fit, it is estimated that when the rate of changing the security level is less than 160 times per hour, the adaptive framework provides power savings compared to a framework that only uses the *high* security level.

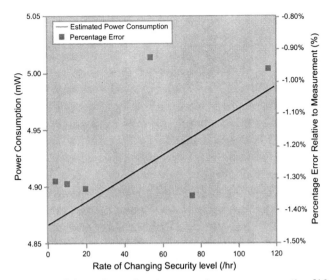

Figure 4.71 Comparison of the estimated and measured power consumption [100].

Figure 4.71 shows a comparison of the estimated and measured power consumption.

The curve represents the estimated variation in power consumption with the rate of changing the level of security in the adaptive security framework. The points labelled *percentage error* refer to the estimated power consumption per node with the adaptable security framework switched on, and relative to the one measured.

Figure 4.72 shows the estimated power consumption per 1000 sensor data messages for different WSN scenarios with the adaptive framework.

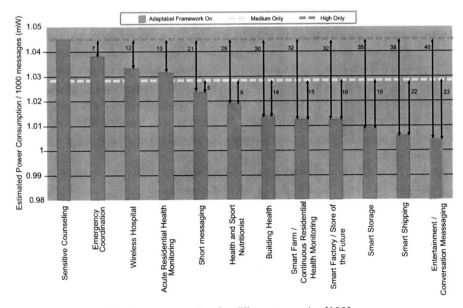

Figure 4.72 Estimated power consumption for different scenarios [100].

Figure 4.72 shows that the *sensitive counselling* scenario has the highest power consumption. This is because this scenario only uses the *high* security level and, therefore, does not benefit from using the adaptable security framework. The adaptable framework provides limited power savings for the *emergency coordination, wireless hospital* and *acute residential health monitoring* scenarios. These three scenarios only use the medium and high security levels and so the reduction in power provided by the adaptable framework is smaller. The adaptable framework provides power savings for the *wireless hospital* and *acute residential health monitoring* scenario, while the rate of changing the security level is less than 12 times per 1000 sensor data messages.

Using an adaptable framework provides the greatest power saving for scenarios that use the *low* security level most of the time, such as the *entertainment* scenario. This scenario uses the *low* security level 80% of the time with a tenth of the time spent in both *medium* and *high* security levels. The adaptive framework will provide a power saving for the *entertainment* scenario as long as the rate of changing security level is less than 40 times per 1000 sensor data messages or once every 25 messages. The maximum power saving achieved for this scenario is 4% compared to a framework fixed at *high*. While the adaptive security framework appears to benefit the *entertainment* scenario, it should be noted that the sensor nodes are also performing other functionality that may have a greater effect on the power consumed by the node and, therefore, affect the benefits of the framework to this scenario. This may also be true for other scenarios. It should also be noted that for the *entertainment* scenario, it is arguable as to whether saving battery power is of any importance, as it is relatively easy to recharge or replace batteries.

The values shown do not include the power consumed in security level changes. The numbers indicate the number of configuration messages that can be sent per 1000 sensor data messages before the power consumed using the adaptive security framework exceeds that without the framework. The curves labelled *medium only* and *high only* show the power consumption for fixed security levels of *medium* and *high*, respectively.

The performance evaluation for the proposed *flexible trust establishment* suggests that the computational, communication, and energy requirements of the trust establishment procedure depend on the initial configuration of the trust establishment components of the nodes. The initial configuration is the one that determines if the certificate validations will be performed, how many messages exchanges will be required, and at what extend the resource consuming operations will be performed.

The initial parameterization is then based on the predeployment knowledge of the network topology and the information flows, which make it possible to optimize the use of resources according to each node's trust requirements.

Although the cooperative trust establishment appears less resource consuming than the hierarchical approach, priority is given to the second option. The main reason is that the number of recommenders depends on the deployment and may not always be restricted.

Moreover, hierarchical trust establishment involves only the two interested parties and does not consume the resources of other network nodes.

The preliminary evaluation of the context-aware privacy protection mechanisms and the complexity of the context assessment have been done for 50, 100, 150 and 200 applicable rules and 3, 6, and 10 number of context attributes [100]. Since the rules were read from the data store or the valid subset proactively stored in the memory, there is no significant difference in the ratio read from datastore vs. read from memory with 6 and 10 context instances for over 120 applicable rules. The memory usage for compressed and uncompressed implementation of the rule agent shows that the gain is bigger with higher number of context attributes—70% with 10 context attributes. The proposed model of rules and the proactive approach for selection of a valid rule subset lead to a delay much less than 1 sec (where 1 sec is a limit, which sometimes matters to users). Additional performance parameters, such as for example memory cost and influence of how often the context changes, have to be further considered [100].

Table 4.8 summarizes the recommendations based on the performed by [4] on how to address the security, privacy, and trust in a WSN that were described above.

The diverse security needs of WSN scenarios can be satisfied by the definition of an adaptive security framework, able to make use of suitable lightweight security mechanisms to continually select and implement the most appropriate level of security, privacy, and trust according to the current context. The framework provides many options designed to make it suitable for numerous different scenarios and to provide adaptability within these scenarios.

The adaptability provided by the security framework aims to achieve significant savings in resources, and in particular battery power. Without this adaptability, the standard security approach often provides the highest possible level of protec-

Table 4.8 Overview of the Proposal for e-SENSE Security Framework [100]

	Mechanisms	*Flexibility and Configurability*	*Adaptability*
Security services	Various sets of mechanisms for authentication, encryption, integrity and freshness (Table 8.3)	For each node, selection of sets of mechanisms for the security levels according to the scenario and the node security requirements (Section 7.5 Configuration Guidelines)	Selection of appropriate security level for each communication according to node profile, the applicable rules and the current context information (Figure 8.7)
Privacy services	Mechanisms for privacy protection and controlled information disclosure according to user/application/service profile and current context (Section 8.3)	For each node, privacy protection components are included according to node privacy requirements (Table 7.1)	Controlled information disclosure decisions depend on the node profile, the applicable rules and the current context information (Section 8.5)
Trust services	Both hierarchical and cooperative trust establishment mechanisms, combined on common evaluation metrics (Table 8.5)	For each node, selection of supported trust establishment mechanisms according to node trust requirements (Table 7.1)	Mechanism to be applied depends on node preconfiguration and available evidence from trust target or neighboring nodes (Figure 9.7)

tion to all data. By only providing the security requirements that are strictly necessary according to the application needs, environmental context, etc., significant security processing and communications overhead can be saved. The framework enables self-reliance and minimizes the need for maintenance by providing for self-configuration of nodes according to their individual context and defined policy. In addition, the use of policies means that changes of requirements can be met rapidly without affecting implementations on the nodes (policies define "what" is needed and not "how" it is achieved, which is up to the nodes themselves). The actual level of these benefits is best validated in practice, as well as the effect of the usual sensor network constraints [100].

4.5   Conclusions

This chapter described advances in the state of the art in the area of context-aware and secure ad hoc communication networks. In particular, focus was on frameworks and mechanisms for context-aware service discovery and lightweight security for MANETs, PNs, and WSNs.

The definition of context varies depending on the type of network and scenario. Some scenarios distinguish between user-specific information and context (e.g., PNs and PN-Fs), treating context as environmental information (e.g., WSN), some of the platforms make no such distinction. For example, The IST project DAIDALOS distinguished between the users' personal attributes and preferences, but all other information such as the location and activity of the user is treated as context. The essential difference is that context information is transient and changes over time whereas user information has a longer life, being built up and inferred from.

The use of ontologies for information, knowledge and context representation and management is one common factor.

In all cases, the security needs of the scenarios can be grouped according o the type of application spaces.

The scenarios might have widely diverging security requirements and in some of them these requirements may change rapidly according to the changes in context, such as type of information generated by the nodes, the location, the current service, user preferences, and so forth. An adaptive security framework can help in dynamically managing security, privacy, and trust to always be able to apply the correct mechanisms for the current situation, and in addition, could significantly save on resources, such as battery power.

Identity management in ad hoc networks can be based on the concept of VIDs. A VID is a way for the user to group the policies that govern, which information is disclosed from the personal network to the outside, and which access rights apply. A careful establishment of policies, therefore, enables the user to exercise control, over which information is disclosed to outsiders and thereby enforce, to an extent, some privacy goals (e.g., unlinkability of two or more VIDs).

The user profile structure is essential to the embedding of user privacy policies within the profile entries themselves, ensuring a smooth adaptation of the level of user preferences and disclosures to the user's needs. The need for different VIDs of the user in order to protect his/her anonymity was adopted by the FP6 IST project MAGNET Beyond in the form of user data sets that cover a user's real and anonymous identities. The VID concept was originally proposed by the FP6 IST project DAIDALOS, for the selection of a user-oriented service within a scenario integrating MANETs with cellular networks.

Security spans a whole range of issues, from the mutual authentication for network and service access to the link layer protection based on channel bindings. An identity management and anonymity infrastructure based on VID can be supported through different security frameworks.

The FP6 IST project e-SENSE proposed a novel context-aware service discovery approach to optimize the efficiency of the discovery in a WSN. The approach exploited ontological reasoning for a semantic-based service matching, integrating cost metrics in service descriptions, and optimizing the discovery process for low latency in order to exploit short-lived network access.

Context-aware access to services implies that the services should not only be accessible everywhere, but also that they should be adaptable to the user's preferences and networking context. The innovations of the FP6 IST projects focused on the mechanisms for collecting and processing of the relevant information from the network infrastructure and the services themselves and their use. The information can be used for the provisioning of services as well as to assist the user in the selection or customization of the service. Due to the continuous changes of the user's context, which may be rapid and unexpected, a service may have to reconfigure itself frequently. The service adaptation can be a source of information for personalization and adaptation to user behavioural patterns.

Context information can affect not only services in use, but also can influence the process of discovery, selection, and composition of services. The described in this chapter architectures integrally support discovery and (re)composition.

With the trend towards pervasive computing environments and the Internet of the future, context-awareness and security gain new significance and new challenges that must be overcome. Enabling techniques, such as self-monitoring, self-adaptation, and self-management become key into providing the future user with reliable and fulfilling communication and services.

References

[1]   FP6 IST Projects in *Mobile and Wireless*, at http://cordis.europa.eu/ist/ct/proclu/p/mob-wireless.htm.

[2]   FP6 IST projects in *Broadband for All*, at http://cordis.europa.eu/ist/ct/proclu/p/broadband.htm.

[3]   FP6 IST Project CRUISE, at http://cordis.europa.eu/ist/ct/proclu/p/broadband.htm.

[4]   FP6 IST Project E-SENSE, at http://cordis.europa.eu/ist/ct/proclu/p/mob-wireless.htm.

[5]   FP6 IST Projects MAGNET and MAGNET Beyond, at http://cordis.europa.eu/ist/ct/proclu/p/mob-wireless.htm.

[6]   FP6 IST Project ORACLE, at http://cordis.europa.eu/ist/ct/proclu/p/mob-wireless.htm.

[7]   FP6 IST Project PULSERS, at http://cordis.europa.eu/ist/ct/proclu/p/mob-wireless.htm.

[8]   FP6 IST Project RESOLUTION, at http://cordis.europa.eu/ist/ct/proclu/p/broadband.htm.

[9]   Prasad., R., and A., Mihovska (Eds.), *New Horizons in Mobile Communications: Radio Interfaces*, Vol. 1, Norwood, MA: Artech House 2009.

[10] Prasad., R., and A., Mihovska (Eds.), *New Horizons in Mobile Communications: Networks, Services and Applications*, Vol. 2, Norwood, MA: Artech House, 2009.

[11] Prasad., R., and A., Mihovska (Eds.), *New Horizons in Mobile Communications: Reconfigurability*, Vol. 3, Norwood, MA: Artech House, 2009.

[12] Challengers. 2008.

[13] FP6 IST Project ORACLE, Deliverable 2.6, "Protocol Architecture and Sensing Parameters: Final Specifications," October 2008, http://cordis.europa.eu/ist/ct/proclu/p/mob-wireless.htm.

[14] FP6 IST Project MAGNET Beyond, Deliverable 1.5.1, "Inclusion of Organizational, Customer and Market Aspects of Business Requirements," at http://cordis.europa.eu/ist/ct/proclu/p/mob-wireless.htm.

[15] FP6 IST Project MAGNET Beyond, Deliverable 1.4.2, "Defining Usability of PN Services," December 2007, http://cordis.europa.eu/ist/ct/proclu/p/mob-wireless.htm.

[16] Mullins, R., (Ed.), "Context and Knowledge Management," *Mobile Service Platforms Cluster*, White Paper," November 2008, http://cordis.europa.eu/ist/ct/proclu/p/mob-wireless.htm.

[17] FP6 IST Project DAIDALOS and DAIDALOS II, at http://cordis.europa.eu/ist/ct/proclu/p/mob-wireless.htm.

[18] FP6 IST project MAGNET Beyond, Deliverable 1.2.3, "The Role of User Profiles in PN Services and Context Awareness," June 2008, at http://cordis.europa.eu/ist/ct/proclu/p/mob-wireless.htm.

[19] Dey, A. K., "Providing Architectural Support for Building Context-Aware Applications," PhD thesis, Georgia Institute of Technology, Atlanta, GA, November 2000.

[20] FP6 IST Project MAGNET Beyond, Deliverable 4.2.3, "Specification of User Profile, Identity, and Role Management for PNs and Integration to the PN Platform," March 2007, http://cordis.europa.eu/ist/ct/proclu/p/mob-wireless.htm.

[21] FP6 IST project DAIDALOS II, Deliverable 122, "Updated Daidalos II Global Architecture," November 2007, at http://cordis.europa.eu/ist/ct/proclu/p/mob-wireless.htm.

[22] O'Reilly, K., "What Is Web 2.0 - Design Patterns and Business Models for the Next Generation of Software," Vol. 2006, 2005.

[23] FP6 IST Project MEMBRANE, Deliverable 2.1, "Most Relevant Community/Market Needs and Target Deployment Scenarios for Wireless Backhaul Multi-element Multihop Networks," June 2006, at http://cordis.europa.eu/ist/ct/proclu/p/broadband.htm.

[24] FP6 IST Project CRUISE, Deliverable 210, "Main Research Challenges related to WSN Architecture and Topology Control," August 2006, http://cordis.europa.eu/ist/ct/proclu/p/broadband.htm.

[25] FP6 IST Project CRUISE, Deliverable 220.2, "Framework for Networks and Data Aggregation Protocols in WSN, " December 2007, at http://cordis.europa.eu/ist/ct/proclu/p/broadband.htm.

[26] Wan, C.-Y., S. B. Eisenman, and A. T. Campbell, "CODA: Congestion Detection and Avoidance in Sensor Networks," in *Proc. of ACM SenSys*, Los Angeles, CA, November 2003.

[27] Galluccio, L., A. T. Campbell, and S. Palazzo, "CONCERT: Aggregation-based Congestion Control for Sensor Networks" in *Proc. of ACM SenSys*, San Diego, CA, November 2005.

[28] Shrivastava, N., et al., "Medians and Beyond: New Aggregation Techniques for Sensor Networks," in *Proc. of ACM Sensys*, Baltimore, Maryland. November 2004.

[29] Scaglione, A., and S. Servetto, "On the Interdependence of Routing and Data Compression in Multihop Sensor Networks," in *Proc. of Mobicom*, Atlanta, GA, September 2002.

[30]   Erramilli, V., I. Matta, and A. Bestavros, "On the Interaction between Data Aggregation and Topology Control in Wireless Sensor Networks," in *Proc. of IEEE SECON*, Santa Clara, CA, October 2004.

[31]   Intanagonwiwat, C., et al., "Impact of Network Density on the Data Aggregation in Wireless Sensor Networks," in *Proc. of IEEE ICDCS*, Vienna, Austria, July 2002.

[32]   Intanagonwiwat, C., R. Govindan, and D. Estrin, "Directed Diffusion: A Scalable and Robust Communication Paradigm for Sensor Networks," in *Proc. of the Sixth Annual International Conference on Mobile Computing and Networks*, Boston, Massachusetts, August 2000.

[33]   Haas, Z. J., J. Y. Halpern, and L. Li, "Gossip-based Ad Hoc Routing," in *IEEE/ACM Transactions on Networking*, Vol. 14, No. 3, June 2006.

[34]   Luo, J., P. T. Eugster, and J. P. Hubaux, "Route Driven Gossip: Probabilistic Reliable Multicast in Ad Hoc Networks," in *Proc. of IEEE Infocom*, San Francisco, CA, April 2003.

[35]   Chandra, R., V. Ramasubramaniam, and K. Birman, "Anonymous Gossip: Improving Multicast Reliability in Mobile Ad Hoc Networks," in *Proc. of ICDCS*, Phoenix, AR, April 2001.

[36]   Boyd, S., "Analysis and Optimization of Randomized Gossip Algorithms," in *Proc. of the 43rd Conference on Decision and Control*, Paradise Islands, The Bahamas, December 2004.

[37]   Vahdat, A., and D. Becker, "Epidemic Routing for Partially-Connected Ad Hoc Networks," *Technical Report CS-200006*, Duke University, April 2000.

[38]   Subramaniam, D., P. Druschel, and J. Chen, "Ants and Reinforcement Learning: A Case Study in Routing in Dynamic Data Networks," in *Proc. of IJCAI*, Nagoya, Japan, August 1997.

[39]   Dimakis, A. G., A. D. Sarwate, and M. J. Wainwright, "Geographic Gossip: Efficient Aggregation for Sensor Networks," in *Proc. of the Fifth International Conference on Information Processing in Sensor Networks*, Nashville, Tennessee, April 2006.

[40]   Braginsky, D., and D. Estrin, "Rumor Routing for Sensor Networks," in *Proc. of WSNA*, Atlanta, GA, September 2002.

[41]   Lee, S., B. Bhattacharjee, and S. Banerjee, "Efficient Geographic Routing in Multihop Wireless Networks," in *Proc. of MobiHoc*, Urbana Champaign, IL, May 2005.

[42]   Kuhn, F., R. Wattenhofer, and S. A. Zollinger, "Worst-case Optimal and Average-Case Efficient Geometric Ad Hoc Routing," in *Proc. of Mobihoc*, Annapolis, MD, June 2003.

[43]   Vass, D., and A. Vidacs, "Distributed Data Aggregation with Geographical Routing in Wireless Sensor Networks," in *Proc. of ICPS Conf.*, Istanbul, Turkey, July 2007.

[44]   Madden, S., et al., "TAG: A Tiny Aggregation Service for Ad-Hoc Sensor Networks," in *Proc. of the 5th symposium on Operating systems design and implementation*, Boston, MA, December 2002.

[45]   http://www.microtopss.msrg.utoronto.ca.

[46]   Baldoni, R., et al., "Content-Based Publish-Subscribe over Structured overlay Networks," in *Proc. of ICSCS*, Columbus, OH, June 2005.

[47]   Wun, A., M. Petrovic, and H.-A. Jacobsen, "A System for Semantic Data Fusion in Sensor Networks," in *Proc. of DEBS*, Toronto, Canada, June 2007.

[48]   Michaelides, M. P., and C. G. Panayiotou, "Subtract on Negative Add on Positive (SNAP) Estimation Algorithm for Sensor Networks," in *Proc. of the 7th IEEE International Symposium on Signal Processing and Information Technology*, Cairo, Egypt, December 2007.

[49]   Niu, R., and P. Varshney, "Target Location Estimation in Wireless Sensor Networks using Binary Data," in *Proc. of Int. Conf. Acoust., Speech, and Signal Processing*, May 2001.

[50]   Chan, H., A. Perrig, and B. Przydatek, "SIA: Secure Information Aggregation in Sensor Networks," in *Proc. of ACM Sensys*, Los Angeles, CA, November 2003.

[51] Yang, Y., et al., "SDAP: a Secure Hop-by-Hop Data Aggregation Protocol for Sensor Networks," in *Proc. of ACM Mobihoc*, Florence, Italy, May 2006.

[52] Roy, S., S. Setia, and S. Jajodia, "Attack-Resilient Hierarchical Data Aggregation in Sensor Networks," in *Proc. of SASN*, Alexandria, VA, October 2006.

[53] IST FP6 Project E-SENSE, Deliverable D 2.2.1, "Initial e-SENSE System Architecture," November 2006, at http://cordis.europa.eu/ist/ct/proclu/p/broadband.htm.

[54] Fielding, R. T., "Architectural Styles and the Design of Network-based Software Architectures," *Ph.D. Dissertation*, Dept. Information and Computer Science, University of California, 2000.

[55] Bechhofer, S., et al., February 2004, "OWL Web Ontology Language Reference," available: http://www.w3.org/TR/owl-ref/.

[56] Prud'hommeaux, E., and A. Seaborne, " SPARQL Query Language for RDF," February 2006, at: http://www.w3.org/TR/rdf-sparql-query.

[57] FP6 IST project DAIDALOS II, Deliverable 211, "Concepts for Networks with Relation to 5 key Concepts, Especially Virtual Identities," September 2006, http://cordis.europa.eu/ist/ct/proclu/p/mob-wireless.htm.

[58] Russell, S. J., and P. Norvig, *Artificial Intelligence A Modern Approach*, Second Edition, Upper Saddle River, NJ: Pearson Education, 2003.

[59] FP6 IST Project MAGNET Beyond, Deliverable 2.2.1, "Specifications of Interfaces and Interworking between PN Networking Architecture and Service Architectures," February 2008, at http://cordis.europa.eu/ist/ct/proclu/p/broadband.htm.

[60] FP6 IST Project SPICE, Deliverable 2.2, "Semantic Publication Architecture for Service Enabler Components," April 2007, at http://cordis.europa.eu/ist/ct/proclu/p/mob-wireless.htm.

[61] Berners-Lee, T., "Enabling Standards and Technologies—Layer Cake," 2002; http://www.w3.org/2002/Talks/04-sweb/slide12-0.html.

[62] RDF Vocabulary Description Language 1.0: RDF Schema, February 2004, http://www.w3.org/TR/rdf-schema/

[63] RDF/XML Syntax Specification (Revised), W3C Recommendation, February 2004, http://www.w3.org/TR/rdf-syntax-grammar/

[64] Prasad, R., and A. Mihovska (Eds.), New Horizons in Mobile Communications: Networks, Services, and Applications, Vol. 2, Norwood, MA: Artech House, 2009.

[65] IST FP6 Project MAGNET Beyond Deliverable D 2.3.2, "PN Secure Networking Framework, Solutions, and Performance," September 2008, at http://cordis.europa.eu/ist/ct/proclu/p/broadband.htm.

[66] IST FP6 Project MAGNET Beyond Deliverable D 2.3.1, "Specification of PN Networking and Security Components," December 2007, at http://cordis.europa.eu/ist/ct/proclu/p/mob-wireless.htm.

[67] IST FP6 Project MAGNET Beyond, Deliverable 4.2.1, "State of the Art and Functional Specification of Network Level Security Architecture," September 2006, http://cordis.europa.eu/ist/ct/proclu/p/mob-wireless.htm.

[68] IST FP6 Project MAGNET Beyond, Deliverable 4.1.1, "Preliminary Secure Extended Architecture," December 2006, http://cordis.europa.eu/ist/ct/proclu/p/mob-wireless.htm.

[69] Winer, D., "XML-RPC Specification," http://www.xmlrpc.com/spec, June 1999.

[70] FP6 IST Project MAGNET, Deliverable 2.2.3, "Resource and Context Discovery System Specification, " December 2005, at http://cordis.europa.eu/ist/ct/proclu/p/mob-wireless.htm.

[71] FP6 IST Project MAGNET Deliverable 2.4.3, "Refined Architectures and Protocols for PN Ad-hoc Self-configuration, Interworking, Routing, and Mobility Management," December 2005, http://cordis.europa.eu/ist/ct/proclu/p/mob-wireless.htm.

[72] Lanza, J., L. Sanchez, and L. Muñoz, "Cluster Head Selection and Maintenance over Heterogeneous Mobile WPANs: An Experimental Approach," in *Proc. of WPMC'06*, San Diego, CA, September 2006.

[73]   Olsen, R. L., H. P. Schwefel, and M. B. Hansen, "Quantitative Analysis of Access Strate-
       gies to Remote Information in Network Services," in *Proc. of IEEE GLOBECOM*, San
       Fransisco, CA, November–December 2006.

[74]   Olsen, R. L., H. P. Schwefel, and M. Bauer, "Influence of Unreliable Information on Con-
       text Aware Service Discovery," in *Proc. of the Third Workshop on Context Aware Proac-
       tive Systems*, Guildford, the United Kingdom, June 2007.

[75]   Olsen, R. L., "Enhancement of Wide-Area Service Discovery using Dynamic Context In-
       formation," *Ph.D. dissertation thesis*, Aalborg University, January 2008, ISBN: 87-92078-
       37-0.

[76]   Schwefel, H. P., M. B. Hansen, and R. L. Olsen, "Adaptive Caching strategies for Context
       Management systems," in *Proc. of PIMRC*, Athens, Greece, September 2007.

[77]   IST FP6 Project E-SENSE, Deliverable D 4.1.1, "Distributed Services Concept," December
       2006, http://cordis.europa.eu/ist/ct/proclu/p/mob-wireless.htm.

[78]   ZigBee Alliance ZigBee Specification, Version 1.0, June, 2005, at www.zigbee.org.

[79]   Getting, I., "The Global Positioning System," in *IEEE Spectrum*, Vol. 30, No. 12, Decem-
       ber 1993, pp. 36–47.

[80]   Elson, J., L. Girod, and D. Estrin, "Fine-Grained Network Time Synchronization using
       Reference Broadcasts," in *Proc. of Fifth Symposium on Operating Systems Design and
       Implementation (OSDI)*, 2002, Vol. 36, pp. 147–163.

[81]   Ganeriwal, S., R. Kumar, and M. B. Srivastava, "Timingsync Protocol for Sensor Net-
       works," in *Proc. of ACM SenSys*, 2003.

[82]   Maroti, M., et al., "The Flooding Time Synchronization Protocol," in *Proc. of ACM Sen-
       Sys*, Baltimore, MD, November 2004, pp. 39–49.

[83]   Zhu, F., M. W. Mutka, and L. M. Ni, "Service Discovery in Pervasive Computing," in *IEEE
       Pervasive Computing*, October 2005, pp. 81–90.

[84]   Bluetooth SIG, "Specification of the Bluetooth System," February 2003.

[85]   Sun Microsystems, "Jini Technology Core Platform Specification," Version 2.0, June 2003.

[86]   UPnP Forum, "UPnP Device Architecture 1.0," December 2003.

[87]   Guttman, E., et al., "Service Location Protocol," Version 2, IETF RFC 2608, June 1999.

[88]   Salutation Consortium, "Salutation Architecture Specification," 1999.

[89]   IST FP6 Project E-SENSE, Deliverable D 4.2.1, "Distributed Data Processing Concept,"
       http://cordis.europa.eu/ist/ct/proclu/p/mob-wireless.htm.

[90]   IST FP6 Project E-SENSE, Deliverable D 4.2.2, "Distributed Data Processing Framework,"
       December 2007, http://cordis.europa.eu/ist/ct/proclu/p/mob-wireless.htm.

[91]   Lester, J., B. Hannaford, and G. Borriello, " 'Are You with Me?'—Using Accelerometers to
       Determine If Two Devices Are Carried by the Same Person," In *Proc. of the 2nd Interna-
       tional Conference on Pervasive Computing*, 2004, pp. 33–50.

[92]   Marin-Perianu, R. S., et al., "A Context-Aware Method for Spontaneous Clustering of
       Dynamic Wireless Sensor Nodes," *Technical Report TR-CTIT-07-66*, Pervasive Systems
       Group, University of Twente, 2007.

[93]   Wang, D. Z. et al., "Probabilistic Complex Event Triggering," This paper was produced as
       a partial fulfilment of the project requirements for the courses: CS262A, "Advanced Topics
       in Computer Systems" and CS281A, "Statistical Learning Theory—Graphical Models"
       from the University of California at Berkeley.

[94]   Rizvi, S., "Complex Event Processing Beyond Active Databases: Streams and Uncertain-
       ties," Technical Report No. UCB/EECS-2005-26 from the University of California at
       Berkeley, 2005.

[95]   Heckerman, D., "A Tutorial on Learning With Bayesian Networks," Technical Report No.
       MSR-TR-95-06 from Microsoft Research Advanced Technology Division, 1996.

[96]   IST Project e-SENSE, Deliverable D1.2.1, "Scenarios and Audio Visual Concepts," Septem-
       ber 2006, at http://cordis.europa.eu/ist/ct/proclu/p/mob-wireless.htm.

[97]   Van Hoesel, L. F. W., and P. J. M. Havinga, "A Lightweight Medium Access Protocol (LMAC) for Wireless Sensor Networks: Reducing Preamble Transmissions and Transceiver State Switches," In *Proc. of INSS*, Tokyo, Japan, June 2004.

[98]   Van Hoesel, L. F. W., and P. J. M. Havinga, "Design Aspects of an Energy-Efficient, Lightweight Medium Access Control Protocol for Wireless Sensor Networks," Technical Report TR-CTIT-06-47, Enschede, July 2006.

[99]   Chatterjea, S., et al., "A Distributed and Self-Organizing Scheduling Algorithm for Energy-Efficient Data Aggregation in Wireless Sensor Networks," Technical Report TR-CTIT-07-10, Enschede, February 2007.

[100] IST FP6 Project E-SENSE, Deliverable D 2.3.1, "e-SENSE Security, Trust, and Privacy Framework," December 2007, http://cordis.europa.eu/ist/ct/proclu/p/mob-wireless.htm.

[101] Beresford, A. R., and F. Stajano, "Location Privacy in Pervasive Computing," in *IEEE Pervasive Computing*, Vol. 2, Issue 1, January 2003, pp. 46–55.

[102] Pfitzmann, A., and M. Hansen, "Anonymity, Unlinkability, Unobservability, Pseudonymity, and Identity Management—A Consolidated Proposal for Terminology," Version v0.27, February 2006.

[103] Gruteser, M., and B. Hoh, "On the Anonymity of Periodic Location Samples," in *Proc. of Conference on Security in Pervasive Computing*, 2005.

[104] Wang, Y., B. Ramamurthy, and X. Zou, "The Performance of Elliptic Curve Based Group Diffie-Hellman Protocols for Secure Group Communication over Ad Hoc Networks," in *Proc. of the IEEE International Conference on Communications* (ICC), 2006, pp. 2243–2248.

[105] Guimaraes, G., et al., "Evaluation of Security Mechanisms in Wireless Sensor Networks," in *Proc. of Systems Communications*, 2005, pp. 428–433.

[106] Wander, A., et al., "Energy Analysis of Public-Key Cryptography for Wireless Sensor Networks," in *Proc. of the Third IEEE International Conference on Pervasive Computing and Communications*, 2005, pp. 324–328.

[107] Arazi, B., et al., "Revisiting Public-Key Cryptography for Wireless Sensor Networks," in *IEEE Computer*, Vol. 38, 2005, pp. 103–105.

CHAPTER 5

Network and Enabling Architectures

This chapter describes network architectures proposed and designed in the scope of the European Framework Program Six (FP6) Information Society and Technologies (IST) projects [1, 2] for the support of device and user communications in an ad hoc stand-alone or integrated scenario. A number of projects provided significant contributions to the advancement of the ad hoc network concepts and standards in various scenarios. The network architectures and principles described in this chapter continue on the basis of the innovative concepts described in the previous chapters of this book to complete the overall picture of the structure, requirements, challenges, and methodologies that have had a direct impact on the development of standards and similar innovations in the area of ad hoc networks.

The work of the FP6 IST projects resulted in defining both the architectures for evolvable multiaccess communication networks and the migration packages leading to their stepwise implementation [3]. Some of the novel concepts developed were related to network context gathering, management and delivery of network state information to service platforms (e.g., the projects MAGNET and MAGNET Beyond [4], DAIDALOS I and II [5], Ambient Networks [6], e-SENSE [7]); to a new class of dynamic and autonomous networked systems, foreseen with optimized solutions and striking a balance between ease of use, trustworthiness, and flexibility [4]; to the development of self-adaptive and service-aware distributed architectures towards next generation multi-access networks (e.g., the project MEMBRANE [8], the project FIREWORKS [4, 7, 9], the project E2R and E2R II [10]), to generic and flexible architectures for support of interworking within a multihop communication system and with other systems.

This chapter is organized as follows. Section 5.1 introduces into the main challenges and requirements for ad hoc network architectures. Section 5.2 describes the major challenges and innovations in the architecture design in support of PNs and PN-Fs. Section 5.3 describes the challenges and proposed innovations for the realization of enabling architectures for the support of WSNs. Section 5.4 describes a device protocol architecture for opportunistic networks. Section 5.5 is focused on the introduction of positioning in the network and terminal architecture as a service focused on the provision of end-to-end user experience in the context of composite network environments. Section 5.6 describes some testbed realizations for the testing of innovations for WSNs. The integration of remotely located and separately developed testbeds is also discussed as an important aspect of future emerging technologies and the coverage of their multidisciplinary aspects. Section 5.7 concludes the chapter.

5.1 Introduction

Research in the evolution of network architectures is proposing interesting and complex solutions. The role of the network and enabling architectures is to provide the backbone framework for the running of the various mechanisms, protocols and interactions required for the successful operation of a given system. For example, packet scheduling, power control, and routing are commonly used techniques to enhance and achieve target *quality of service* (QoS) in wireless networks. However, these control mechanisms and protocols are often designed separately without the careful consideration of their inter-dependency. A properly designed architecture supporting the joint operation of these mechanisms can benefit the performance of the ad hoc wireless applications (see Chapter 3 of this book).

The way a given architecture is designed is depending on the scenarios and requirements, and the objectives of the mechanisms to be supported by it. In the context of next generation communications, some common requirements that can be identified are the need for flexibility, adaptivity, scalability, and a generic nature. Other requirements that follow for these main ones are self-organization, self-management, and reconfigurability, and these last ones are more scenario dependent.

For example, in a multihop cellular communication system enhanced by relays and foreseen with different radio transmission modes in support of ubiquity, would require a cooperation supporting architecture for radio resource management (RRM) targeting the interworking within the system and with other systems [12]. Flexibility would allow for the support of various scenarios based on enhanced logical functionalities, while keeping the number of physical entities to a minimum to avoid additional signaling overhead. In addition, the architecture must incorporate the relay-communication and RRM functionalities.

In another example, within a network of collaborative user terminals utilizing distributed sensing and decision-making, the communication protocols required will be significantly more complex than those encountered with conventional wireless terminals [13]. A supporting protocol architecture should ensure the functionalities of control and management of the protocols, robustness of signaling via the opportunistic channels, and scalability in terms of protocol overhead and message response times. Both, fast transfers within the geographical or network-topological vicinity as well as assured network-wide end-to-end communication of context information must be possible to minimize the network signaling load (i.e., to disseminate information only if and when relevant). To enable the forwarding of the sensor information, a common encoding of identification elements or of complete *protocol data units* (PDUs) for link and network layer will greatly simplify the protocol architecture and will help to avoid dedicated routing and gateway functions that must otherwise be colocated to each terminal.

As pointed out in Chapter 4 of this book, the need (originally from network considerations) to bring together mobility, authorization, authentication, and accounting; charging, auditing; and QoS combined with the need to manage the resources from the network side and the overall security concerns is relevant beyond the pure network and service levels as it directly concerns the overall experience of the user [14]. [5] proposed the architectural concept of joint *mobility, AAA auditing and charging (A4C), resource management, QoS, and security* (MARQS). Another issue is to find a way to get a user to access services, networks and content in a simple way, while keeping things secure, and maintaining the user privacy, and one approach was introduced by the concept of virtual identities (see Chapter 4 of this book). The concept of VID cuts across layers and allows for access (and if needed reachability) independently of a particular device [14]. Because there is a range of providers with their own data, context or personalization information and access restrictions, a more universal access by a user can only work if some of these information items can be combined. A part of this can be solved by the VID giving a view of multiple providers, but this needs to be complemented by the sharing of data and the provision of access rights to users between providers. The concept of *federation* is proposed in [5]. Such a *federation* must work in a range of cases from extreme trust to very limited and managed trust between providers, and must not only work horizontally (i.e., between network operators), but vertically (e.g., between network operators and content providers and in principle with other forms of business). The *federation* also supports cooperation between competing parties required to achieve the final service provisions, with all or at least most parties having a net gain through cooperation.

5.1.1   Seamless Integration of Broadcast

A managed approach to support a controlled and chargeable access far beyond services, and also addressing single devices is also required in the context of pervasive systems. Integration architectures for seamless integration of various technologies such as *digital video broadcasting* (DVB) with wireless and fixed access technologies, such as the 802.x series, should be targeting both the technology as well as the service level. This separation of concerns allows for the optimizations across the whole bandwidth of offerings, from network via service to content [14].

Somewhere in-between the technologies and the (user) services are the network services, such as mobility management, QoS provisioning, charging, and accounting or security mechanisms. The need for some major improvements in order to support broadcast. While, traditionally, the broadcast networks have been used to distribute broadcast services, the idea was to use the broadcast networks for any type of service, and provide broadcast services over any type of network. This is referred to as *seamless integration of broadcast* [14].

To be able to dynamically select the transmission media (unicast or broadcast technologies) to reach the user, and to optimally select the transmission mode (unicast, multicast, or broadcast) according to the service, the broadband distribution of multimedia content to huge groups of users, as well as personalized interactive broadband multimedia sessions must be supported.

Over broadcast technologies such as DVB, some service information may be multicasted on a regular basis (on a defined channel), enabling in particular receive-only terminals to provide the service information to the user. *Reactive* mechanisms, where services are announced as a reaction to a service query, are required since the number of possibly available services can be infinite [15, 16].

The requirement that all traffic (including live TV broadcast and carousel service) is based on the *Internet Protocol* (IP) requires IP multicast support for all networking technologies [17]. In order to provide for scalable and efficient transmission mechanisms, the *dynamic switching* between unicast and multicast delivery should be part of the network design [14]. The dynamic switching is based on a set of related parameters and should not only consider the number of receivers on a specific part of the network, but also the network capabilities (e.g., the drawbacks of multicast distribution over WLAN), the service requirements, and the user (group) preferences.

The integration of a *multimedia broadcast/multicast service* (MBMS) subsystem in the radio environment of a non-broadcast-dedicated technology (e.g., TDMA) provides the advantage of resource sharing. MBMS is aimed at delivering identical multimedia contents either to all, or to a selected group of MTs. An early implementation of MBMS was brought to the market with the Release 6 of the *Universal Mobile Telecommunication System* (UMTS) [18].

However, this imposes requirements on the resource and QoS management architectures. For example, some prioritization of services must be performed based on the *mobile terminal* (MT) capabilities, between the MBMS and non-MBMS bearer services. Because the radio channels must be established dynamically, the multicast service activation requires that the entities controlling the activation of the MBMS bearer interface with the upper layer entities at the service level. This impacts the full flow for resource activation, from the QoS management entity (e.g., broker) to the base station [16].

A QoS architecture that supports QoS-enabled multicast, including service differentiation over unidirectional links, was proposed by [5]. A *performance manager* decides that a service or flow is worthy to be moved to multicast or broadcast technologies, to support the numerous users that might be requesting or receiving the same content. The reverse functionality, from multicast or broadcast to unicast is another interesting research topic. In general, given the differences in technology support between the broadcast/unicast media/technology, the integration of these features in a consistent way is very challenging.

In a multihop cellular communication system that supports relaying, the MBMS provision must be organized differently as compared to the UMTS case [19]. The placement of additional relay nodes (RNs) in a *relay enhanced cell* (REC) makes it possible to chose those of them, which should assist the process of broadcasting or multicasting. In other words, depending on the network load and the willingness of the users to subscribe to MBMS it might be not necessary to exploit all the RNs. Apart from the technical issues, it is also crucial to focus on the possible business models. Since UMTS does not offer feedback with respect to MBMS, it is neither possible to measure the audience response, nor confirm the reception of a message. It is worth considering whether the system should additionally provide the option of receiving feedback from UTs [19].

Figure 5.1 shows a scenario that uses advanced MBMS with feedback.

Due to the existence of REC, the BS can perform broadcasting and multicasting either in a *blind* or an *advanced* way. Because the distinct regions of the REC are served by different RNs, and the central region—by the BS itself, it is possible to identify those regions, to which the multimedia contents should be really delivered. It is rather unlikely that there is a need for such an operation in the case of broadcasting, however, it is probable that a multicast group will not be spread throughout the whole REC. As a result, such an *advanced contents* delivery can increase the throughputs and decrease the latencies.

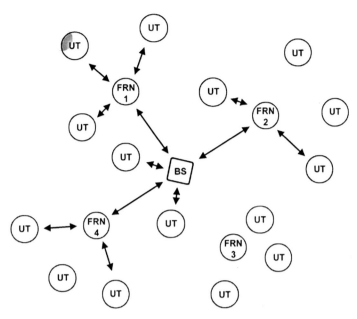

Figure 5.1   Advanced MBMS with feedback [19].

In the case of the *blind* approach, no attention is paid to whether there are regions in the REC, where users are unwilling to accept multicast traffic or not. Consequently, the contents are delivered to them either directly or via an intermediary RN. Such an approach would in essence emulate a kind of a legacy cell, which is not relay enhanced and, therefore, not optimized for REC multihop radio systems [11]. The idea behind the advanced version is to deliver multicast traffic to those regions only, where the UTs belonging to the relevant multicast group are located. This means that in the situation shown in Figure 5.1, the dedicated link between BS and RN 3 will be less occupied and, therefore, able to carry other data.

Service providers are usually not very willing to cover the costs of broadcasting if the so-called audience measurements are not feasible [19]. If this is the case, the broadcasting can function as the means of encouraging users to subscribe to multicast services. Consequently, by analyzing whether the number of subscriptions is growing or going down, a given service provider might be able to estimate the impact of their advertisements.

The availability of feedback would make MBMS more attractive to a wide range of service providers. This way broadcast advertisements would be delivered to the users not only by those providers, who expect revenue in the form of an increased subscription to the multimedia contents they can offer, but also other companies, eager to advertise their products provided they know how many potential clients have received their message.

Multicast has gain attention recently with the emerging of new application scenarios where networking among wireless sensors benefits the collecting and distributing of information [20]. Examples of such scenarios include the data collection with mobile sensors during the tracking of firefighters in burning buildings, disaster rescue operations, and health care applications including body sensor networks (BSNs).

Multicast algorithms for WSNs can basically be divided into *tree-based* and *mesh-based* according to how packets are routed through the network. In general, tree-based approaches suit better in static environments where mesh-based protocol would incur an excessive control overhead. On the other hand, meshes of forwarding paths provide resiliency against mobility. Thus, multicast integration in WSNs is closely related to *topology control*. The goal of topology control is to establish and maintain the graph that represents the communication links between the network nodes, such that the graph has some desired properties. For example, the resulting network should provide a sufficient level of connectivity without sacrificing in energy consumption, communication capacity, or sensing coverage. The topology control for WSNs needs to achieve an optimal balance between the following important characteristics of WSNs:

- Energy efficiency;
- Connectivity (including fault tolerance and robustness to mobility);
- Communication capacity (network throughput, delay distribution, fairness);
- Sensing coverage.

These are partially contradicting goals. For example, energy conservation and the increase of communication capacity by way of reducing the interference lead to fragility of the network connectivity in the short-term [20]. On the other hand, energy conservation increases the network lifetime and thus maintains connectivity in the long-term. Also, when the sensing radius determines the deployment pattern, the resulting topology does not necessarily form an optimal communication graph.

5.1.2 Mobility Support

Depending on the objectives of the system, mobility support will impose different requirements. For example, a *mobile broadband wireless access* (MBWA) system concept using relays, should be capable of providing cost-effective and reliable *triple play* services [e.g., high-speed Internet, telephony (VoIP) and television (video on demand or regular broadcasts], in several operational scenarios [21]. This means that the system should be able to fulfill the user requirements imposed by the offered services under various conditions in terms of user density and traffic demand, the propagation conditions, and the user mobility. The performance and the configuration of the system are likely to vary across different environments and there should not be any inherent limitations in the system operation for a given circumstance (e.g., high velocity trains). At the same time, the deployment cost must be as low as possible in order to allow profitability.

The system performance is ultimately assessed by the quality of a specific service that is delivered to the users and the associated deployment cost. Mobility management techniques support the user mobility, including the traffic balancing, which is essential for a network to use efficiently the resources of the system and ultimately translates into QoS. In a multihop communication system, where different radio modes will be operating together [11], the traffic balancing between modes is a very important mechanism to keep the network operating in a normal state serving the users with highest possible QoS. The intramode and intermode handover are the key mechanisms to support the traffic balancing inside the system [22]. The most essential trigger for handover is the signal strength (like the handovers in the currently deployed networks), but also on load and service based information, mobility (velocity), user's subscription profile, user's preferences, terminal capabilities, user's environment (indoor, outdoor, etc.), location, interference, and statistical information. When the system supports different radio modes, it implies a diversified set of deployment scenarios, then the mobility and the location of the users will play a very important role in the handover process. The requirements on a mobility architecture then will be related to the location of the functionalities in order to support the user mobility, with a focus on ensuring low delays and jitter, seamless roaming, low signaling overhead, and so forth. It should further be considered whether the mobility control is best performed in a centralized or a distributed approach.

In another scenario, involving a relay-enhanced broadband system, the mobility issues should remain as compatible as possible with 802.16e. Considering, that the scenario can involve static RNs, the only mobile entities are the subscriber stations (SS), which should comply with already defined functionalities [21]. Upon considering mobile RN stations, critical dynamics are introduced and an appropriate functionality is needed to cope with them. Some of these could be the support of mobile RN handover, handover decision for subordinates stations, mobile RN scanning, which would help a mobile RN to scan for candidate stations. A mandatory functionality for the multihop BS of this category would be the *network topology advertisement*, which was introduced in 802.16e and involves the broadcast of topology information by the BS [21].

Mobility between different local mobility domains can be handled by a global mobility protocol [23]. This approach results in separating two mobility management domains: local and global, which are managed independently. The independence, however, is not available in the traditional hierarchical mobility management approaches and requires a novel one that can provide for seamless handovers in a multitechnology environment, not only between the different access networks but also between: infrastructure, MANETs and moving networks (NEMOs).

Multihop positioning applied to MANETs, allows for determining the locations in a distributed fashion wirelessly through the hybrid use of deployed positioning technology and distance. The accuracy of the positioning is proportional to the density of both the MANET and the static nodes.

In a wireless sensor network (WSN) where a large number of sensor nodes cooperate among themselves to monitor an area, mobility will be related to environmental influences (e.g., wind or water) and to physical mobility between the sensor nodes (e.g., attached to or carried by mobile entities). Further, the sensors would possess automotive capabilities. Sensor networks and their applications are subject to constraints such as limited processing, storage, communication capabilities, and limited power supplies.

A number of approaches exploiting mobility for data collection in WSNs have been proposed in recent years [24–26]. These approaches can be categorized with respect to the properties of sink mobility as well as the wireless communication methods for data transfer.

In an integrated sensor and other system environments (e.g., cellular, wireless local area networks (WLAN), where the data is sensed by sensors and then transported towards the data storage systems and applications servers, the user terminals (UTs) of these types of networks might act as mobile gateways collecting the data from the WSN and forwarding it.

This scenario can be also considered as an extension of the traditional WSN scenario where a sink collects information from the sensor nodes distributed in the environment, through wireless links. In this case, there are multiple sinks, and their locations are not known. Moreover, the sinks are in fact gateways forwarding the information collected to the higher levels through heterogeneous wireless interfaces characterized by different parameters in terms of transmit power, capacities. The architectural challenges that arise from such a scenario are related to heterogeneity (different radio communication techniques are involved at the various interfaces); the air interfaces and devices implement different communication paradigms (e.g., mesh or flat topologies) at the different levels; mobility between different levels of the system, and that mobile gateways are devices carried by people, such as laptops or cellular phones, and, therefore, are not specifically deployed for the aim of collecting data from the environment [20].

An example is given in Figure 5.2.

The area is covered by some indoor UMTS stations, and the employees working in the offices carry UMTS mobile devices, also equipped with ZigBee air interfaces. These MTs can interact with the ZigBee-enabled small devices distributed over the corridors, and inside the offices. Such devices might provide localization and logistic information, and are also able to detect the presence in their immediate neighborhood of objects, such as laptops, printers, pieces of equipment, etc., which have ZigBee-enabled, low cost devices that communicate with the nodes distributed in the environment. In this scenario, every employee can scan the environment to get the information on the localization of the movable objects. This can also be done

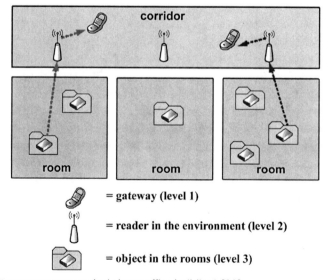

Figure 5.2   Heterogeneous scenario (a large office building) [20].

through Web services implemented in the intranet serving the building: the user sitting in his/her office will get the requested information through a sequence of links from the lower level (the objects) to the upper (the access ports, bringing the information to the infrastructure and the Internet or intranet). In order to monitor the objects, it is important to guarantee that the network connectivity remains above a specific threshold.

5.1.3   QoS Support

The requirements for mobility support have an influence on the design process of every element of the system. In essence, a properly functioning mobility support is the first step towards QoS provisioning in future wireless environments. Other aspects of importance are ensuring the service creation and delivery in interdomain communications scenarios. This would include the negotiation of *service level agreements* (SLAs), interworking of infrastructures, and legal aspects of service provisioning.

A common issue, when discussing a QoS-enabled architecture behavior, is the identification of the collection of packets, to which a specific behavior applies. This is usually known as a *session*, and may not always be easily identified (e.g., the set of packets under scope of a *Session Initiating Protocol*—SIP session may require deep packet inspection) [23].

Another characteristic of a session is that it implies a relationship between some network nodes. For example—a SIP session will involve a number of network nodes (SIP clients, SIP proxy) that are involved in the exchange of packets. The scope of what a *session* is varies with the particular subject of interest. The following types of sessions can be identified:

- *Network access session*: a session between a device (a mobile terminal) and a network point of attachment. Typically, an access session involves authentication for network access.
- *Transport session*: a transport session connects node in the network that can exchange QoS and mobility management packets. Multiple transport sessions can exist within the same network access session.
- *Application session*: runs on top of the (multiple) transport sessions. It exchanges application-specific packets among the distributed application parts.
- *Pervasive session*: a session that is directly mapped onto user goals and intentions. It can be context-aware and personalizable and will control the overall coordination of the multiple application sessions that might interact with each other based on the user context.

A QoS architecture must implement functionalities for admission and congestion control (at the network layer), for resource allocation (at MAC layer), for service discovery and context-awareness (at service and application layers). Cross-layer approaches, therefore, benefit the achievable performance of QoS architectures (see Chapters 2 and 3 of this book). Policy-based management architectures are another approach to achieving QoS in future networks.

QoS policies allow the operator for controlling dynamically the negotiated QoS for the mobile users [27]. Policies are used to enhance the negotiation mechanisms. The following are examples of QoS policy for mobile multicast are [28]:

- Regulation of access to the QoS considering a layered multicast QoS model and user profiles;
- Enhancement of the QoS for particular applications based on a restriction of resource usage at a certain time;
- Optimized handover decisions using specific criteria for the selection of appropriate access network to reduce costs and avoid an overload;
- Usage of specific routes for specific multicast groups, services, and applications;
- Multicast service redirection, when in case of a failure of access router specific multicast access router should be used to continue the multicast services.

QoS in a pervasive environment context means support of adaptation to changing contexts, movement, and user requests across personal and embedded devices. Pervasive computing promotes the idea that embedding computation into the environment would enable people to move around and interact with computers more naturally than they currently do [28]. Pervasive computing is about supporting the user in the access to services no matter where the user is and providing an improved user experience by taking into account the context-aware user preferences.

In a multidomain heterogeneous environment [5], in order to guarantee end-to-end QoS, signaling must flow between the domains. As a result of the inherent restrictions of the federation level, signaling must pass between domains through a federated central entity that can assume the role of a proxy for QoS purposes. An example of interdomain QoS signaling is shown in Figure 5.3.

When establishing an interdomain session in this scenario, the *access network* (AN) QoS brokers (ANQoSBrs) communicate directly as if it is an intradomain session establishment. This communication can be to exchange information about QoS resources request/release, QoS rereservation (allocated resources change for ongoing session), QoS resources availability request, and QoS signaling authentication and security. The *core network* (CN) QoSBrs for this scenario are not involved in the signaling path for the session establishment; they only manage the virtual tunnel that connects the two involved ANs as if they were same domain ANs. The

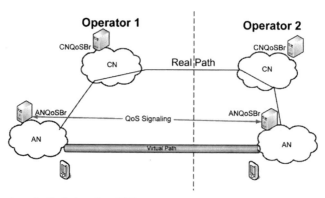

Figure 5.3 Interdomain QoS signaling [28].

administrative domains have a full information about each other's topology and the user profiles are fully disclosed between the domains. Because of the topology knowledge, all the network entities may communicate directly.

MANETs will be an integral part of next generation networks A MANET is a collection of autonomous *mobile nodes* (MN) that communicate using wireless links without the support from any pre-existing infrastructure network. For their integration into next generation networks, Internet connectivity is required for, which then can extend the range of hotspots by providing multihop connectivity from the MNs towards the Internet through one or more gateway nodes by utilizing the packet forwarding capabilities of the intermediate nodes via the multihop paths [28]. Therefore, a QoS supporting architecture should implement protocols and mechanisms for connecting the MANETs to the public Internet.

SIP is a signaling, presence, and instant messaging protocol developed to set up, modify, and tear down multimedia sessions, request, and deliver presence and instant messages over the Internet [29]. SIP is a *client/server*-based protocol, and a *user agent server* (UAS) needs to be reachable on the network so that SIP systems can work properly. As MANETs are dynamic networks formed by peer nodes, a standard SIP architecture is clearly not applicable to them as registrars and proxies are fixed, static, and centralized entities. Therefore, the SIP protocol cannot be deployed as is in isolated MANETs. In Internet-connected MANETs, however, the end points located in the ad hoc network can reach other parties located in the Internet (and thus also SIP proxies and registrars) through the gateway nodes, but when two nodes in the MANET need to communicate via SIP, any SIP signaling will traverse the gateway, which is a severe performance limitation [28]. Therefore, alternative approaches are required. There are several studies on how SIP services can be provided in isolated MANETs. [5] provided an analysis on the alternatives for providing SIP services in Internet connected MANETs.

The support of a service discovery framework is useful in MANETs to give users the possibility to discover people, services, or devices in the network. As this approach does not consider the scope of an Internet connected MANET where a MN can contact a node outside the MANET this can be achieved by a mechanism designed to enable SIP services in this scenario.

Assuming the use of SLP as a service discovery framework, if the *callee* is located outside the MANET, the *caller* will also issues a broadcast SLP request message for service "SIP" due to the inexistence of this information in the local cache, but it will not get a reply. After a certain timeout period the caller assumes that the callee is located in the Internet, then it sends the INVITE directly to the gateway, which in turn uses the standard SIP proxy mechanism to locate the callee proxy. Figure 5.4 shows how messages are exchanged in this case.

A similar mechanism can be applied to distributed SIP and an integration with a routing protocol and the main difference resides on how the distributed registration is achieved inside the MANET.

5.1.3.1 QoS Architectures

With the aim of implementing a scalable and robust architecture, capable of opening the inherent possibilities in a wide range of wireless and fixed networking technologies

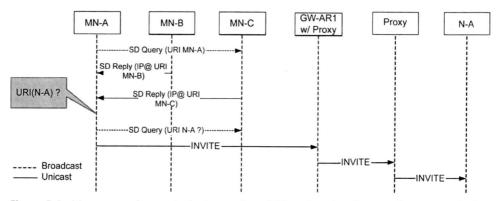

Figure 5.4 Messages exchanges in the integration of SIP and service discovery for MANETs [28].

and especially for the integration of mobility with broadcast and end to-end QoS, the QoS architecture should follow a hierarchical organization of the network and the QoS control entities [28]. Hierarchical networks have the advantage of being more scalable and to have an enhanced support for mobile networks (and MANETs), although they require a increase in the resource management control's complexity. Figure 5.5 shows two administrative domains, each of them with several ANs and one CN. *Access routers* (AR) connect the UTs to the ANs.

The CN, composed by a multiprotocol label switching (MPLS) domain, grants access between each one of the ANs that compose each domain and between the

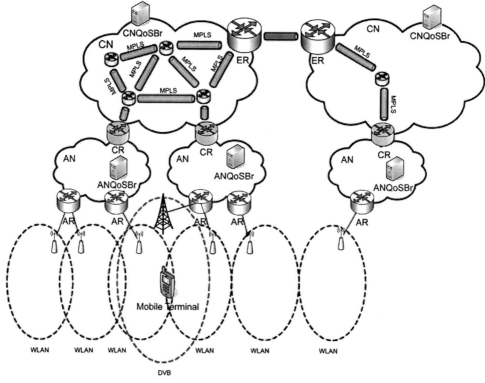

Figure 5.5 Hierarchical QoS architecture model [28].

ANs and the foreign domains, through the *egress routers* (ER). The ANQoSBrs perform both admission control and network resources management functions in the AN. The CNQoSBr manages the MPLS domain that connects each one of the domain's ANs and the connections resources allocated to the inter-domain connections. This entity, opposed to the ANQoSBr, only manages the aggregates of flows.

Regarding specifically mobility, QoS can be analyzed in the following two main directions: signaling protocols behavior in mobility environments and known implemented architectures.

The signaling protocols for QoS typically are confined to extensions to the well-known *resource reservation protocol* (RSVP) and the *new extensible IP signaling protocol* (NSIS) [30]. RSVP has many inefficiencies concerning mobility, most of which relate to the following three main factors: (i) lack of mechanisms and/or difficulty to produce advanced reservations, lack of support for operation in tunnels (e.g., as in *mobile IP*—MIP). These inefficiencies had produced the extensions to RSVP (e.g., MRSVP) or the specific extensions to mobility protocols such as MIP.

NSIS is a new framework that aims to define a generic and extensible family of protocols for in-band signaling. NSIS takes special care with mobility, which starting point is the RSVP known problems. NSIS allows for advanced reservations and operates over tunnels.

The *integrated services* protocol (IntServ)/RSVP has drawbacks for large-scale networks (e.g., operator scenarios) and is hard to interoperate with mobility, therefore, typical solutions use either a pure *differentiated services* (DiffServ) architecture or a mix IntServ/Diffserv in access and core. Hence, mobility becomes a problem of context/state transfer between the previous and next positions in the network. Interactions with QoS brokers and mobility-specific entities (e.g., *mobility anchor point*—MAP in *hierarchical MIP*—HMIP or the context transfer between adjacent ARs) are the focus of these architectures.

5.1.3.2 Policy-Based Architectures

The specification of a policy has to follow a specific policy language. The usage of *ontologies* for policy definition is widely accepted. Several solutions are based on this approach [31, 32]. Other proposals are based on formal logic approaches [33]. [5] defined a policy ontology in *Web Ontology Language* (OWL) [34]. The drawback of using XML based-representation of policies is the requirement for a dedicated *graphical user interface* (GUI) used to define the policies.

An *information model* covering the managed entities is required for a policy-based framework. Standardization of these models is required to enable the consistent exchange of information between systems provided by different vendors. The *Distributed Management Task Force* (DMTF) developed the *Common Information Model* (CIM) [35]. One of the schemes included in the CIM model contains a policy model defined in cooperation with IETF. This model named the *Policy Core Information Model* (PCIM) [36], later on evolved to PCIMe [37].

In parallel to DMTF work, the *Telemanagement Forum* (TMF) developed the *Shared Information and Data Model* (SID) [38]. One of the elements of this model is the *Directory Enabled Networking*—a new generation (DEN-ng) model, which also covers aspects related with policies. The SID information model is part of a broader concept, the *Next Generation Operations Systems and Software* (NGOSS). NGOSS defines for the service providers and their suppliers a comprehensive, integrated framework for developing, procuring, and deploying an operational and business support systems and software. This approach is more business oriented, which results in a more abstract information model. The DMTF approach is more technically oriented.

In [31], a specific information model for *multiagent* environment was designed. This information model was developed as ontology written in (OWL). [5] adapted the Common Information Model and developed the required extension schemes to integrate the CIM classes with policy ontology.

One of the most important aspects in opportunistic networks is the selection of policy/policies to be applied to a certain communication scenario and the process should be clearly identified within the decision-making framework [13]. This process involves the general understanding of all input entities such as current system contexts, profiles, and policies through the machine understandable languages. Among the universal contexts, an efficient procedure also needs to be engaged in order to narrow down the search for the suitable policy/policies and this can be introduced as a context-filtering mechanism. The decision-making is then employed to search for the appropriate policy. This crucial process should be able to refine the search and possibly transform the existing policy/policies in case no appropriated policy can be found. A policy framework for support of opportunistic communications is shown in Figure 5.6.

Any decision made within the policy framework in Figure 5.6 is based on *context*; context, in general, is any information that can be used to characterize the situation of an entity. The interpretation of this information follows rules that are captured in different types of policies, which apply in various communication situations. A *profile* is a description of configuration settings, capabilities, or resources associated with a user, application or a group of these. Profiles can be seen as a group of single context or resource information, to summarize and structure information for better and more efficient handling. Figure 5.6 shows only a schematic overview, and in a detailed architecture representation, each relation can include

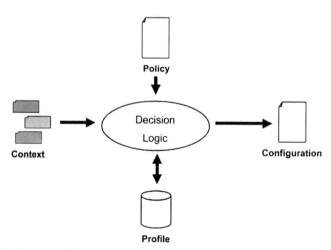

Figure 5.6 Policy framework for opportunistic communications [39].

further steps for the processing and aggregating of information. Not shown is the general *cognition loop*, the resulting configuration of networking equipment will influence the captured context and the overall system will adapt to the changing environment [39].

In a broader view, all kind of information that needs to be exchanged, including policies, profiles and configuration data can be understood as context information. This view offers to use the context management system as a basis for the policy framework. Using a context management system as a basis for the policy framework can simplify some of the tasks of the policy framework.

5.1.3.3 Context-Aware Architectures

Ambient intelligence is a key component for future beyond 3G mobile and wireless communication systems and platforms with their respective service environment [40]. The enabling technology that provides systems with information to allow for ambient intelligence requires the capturing of context that enables the convergence of information collected through many input modalities, rather than relying on the active user interactions or specialized sensor systems gathering only limited information about particular parts of a system. The richness of information that is required to fully capture the ambient intelligence demands a multitude of multisensory information. To obtain this information, potentially, a large variety in terms of their sensing capabilities as well as a large number of sensors is required. The sensors may communicate among themselves or via gateways with other systems and networks (e.g. other sensor networks, Cellular, WLAN, PAN, or the core network). The majority of the sensors in these areas will be wireless, mainly for the ease of deployment and convenience. [7] proposed a framework that enables the convergence of information and focusing on energy-efficient WSNs that are multisensory in their composition, heterogeneous in their networking, and either mobile or embedded in the environment. This encompasses any structure from single sensors to thousands of sensors collecting information about the environment, a person, or an object. The proposed framework is able to supply ambient intelligent systems with information in a transparent way hiding the underlying technologies, thus enabling simple integration of context sources and autonomous operation. The architectural vision towards ambient intelligence is shown in Figure 5.7.

Capturing, classifying, filtering, and sensing the situation and context through phenomena and signals from the physical environment will support and significantly enhance and enrich personal, family, and community focused mobile applications and services as well as enhance the wireless communication systems. [7] developed the system concept and architecture to collaboratively capture a user's context, and to pre-process rough data into meaningful data to be input into a user's (individual, family, community) profile enabling context-enabled applications and services to a user at the right time. The integrity and sensitivity of such information demands a careful consideration on security, privacy, and trust of a user. Some of these issues were already discussed in Chapter 4 of this book.

The key requirements for the architecture shown in Figure 5.7 are ultra-low-power operation (in particular, for the communications but also for the local processing of the sensor information) and multidimensional scalability with respect to

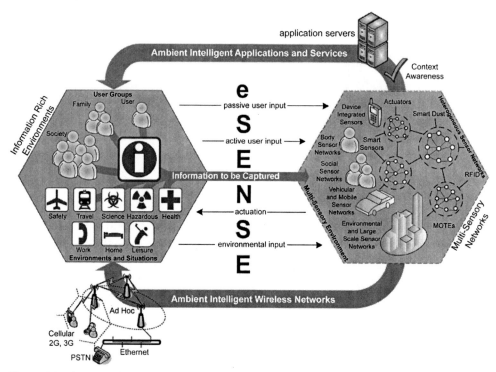

Figure 5.7 Support of ambient intelligence through context capturing [40].

mobility, number of sensors, diversity of sensor classes, sensor network types, and sensor payload types. Also, presenting the captured information to ambient intelligent systems, achieving transparency with respect to the underlying sensor systems is of importance.

5.1.4 Security, Privacy, and Trust

Before a mobile user can request any service, it must get basic network access and global IP connectivity [23]. In this process the user identifies to a network access provider to get authenticated. The user first establishes a network session, during which the MT is authenticated and authorized to access the network resources. Information servers, such as an ID repository are contacted during the session setup, using a non-authenticated network session. A security architecture in support of the above networks access must incorporate the following features:

- Security bootstrapping procedure;
- Federated authentication and authorization (AA);
- Privacy by means of pseudonimity and unlinkability protection for users.

The identity management components should be intrinsically connected to the security framework [23].

In a private personal area network (P-PAN) and personal networks (PNs), which connects the public network to the personal life involving various infrastruc-

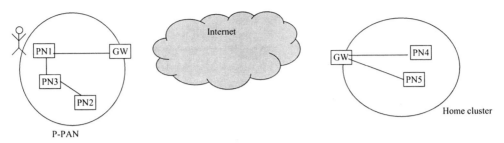

Figure 5.8   Basic PN architecture with two clusters [41].

tures and tiny devices, the security requirements become very high. A security architecture will benefit from lightweight and adaptive solutions to cover all possible scenarios [4]. A basic PN architecture is shown in Figure 5.8.

A PN is a network of personal nodes, consisting of one or more clusters. The network forms one security domain where all nodes have equal access rights to participate in the network. The specific functionality that is needed to maintain a PN from a security point of view can be divided on the following levels:

- PN management;
- PN network layer;
- PN link layer.

Additionally, a solution on service level is important. Strictly speaking, this functionality exceeds the PN (networking) layer but relates to the following:

- PN service layer;
- PN context.

Clusters can be connected via other networks by gateways. Upon configuration in an external network, the gateway will try to connect and register itself at its PN agent. The PN agent is the central registration point for the clusters and should always be reachable. The PN agent client entity is the gateway that communicates with the PN Agent (see also Chapter 4 of this book). The cluster, in which the user is present is the P-PAN. The communication with the user is maintained by the PN manager (GUI). The imprinting of devices to the PN is controlled by the user through a PN manager and handled by the *PN certification authority* (PNCA) that is present in the P-PAN. The use of the PNCA should be protected by some kind of system operator password or hardware dongle, which is needed to protect the access to the security database (i.e. secret key) of the PNCA. In addition, this PNCA maintains a PN node list with the identifiers and certificates of the imprinted devices. The PNCA also maintains a list of revoked certificates (CRL).

Any update of the CRL will be tagged with a sequence number and signed with the secret PN key. The PNCA will broadcast the CRL within the PN. It is the task of the PN agent to store the latest CRL as a reference.

As the user may change cluster and devices, the PN manager and the PNCA may appear on different devices. This issue of the PNCA resilience requires synchronization between the databases of the possible PNCA instances. The counterpart of the PNCA is the *trust establishment module*, which executes the required protocol on the new PN device and must be available in all PN devices.

This module executes the first and second phase of the imprinting procedure. The second phase is triggered when the lower layer neighborhood discovery module detects a new PN node, with which there is not yet a shared secret key. The outcome of this second phase is a shared secret key that is used to derive session keys on the lower layers. At initialization, the trust establishment module generates a node specific public/private key-pair that will be used in the PN certificate. In addition, some kind of prevention should be implemented to deny a second certification request or certificate removal by any other user than the PN user himself.

If a node has been disconnected from the PN, the trust establishment module shall check for any updates of the CRL at another node in the cluster, comparison with the CRL version number will detect the need for a CRL update. The hierarchy in CRL update is: PN agent, PN agent client (GW) and finally, any other neighboring node.

As certificates have a limited lifetime, a certificate renewal procedure needs to be implemented. This procedure should start with the detection of a "near" expiration and the alerting of the PN user in the P-PAN. If a renewal of the certificate is needed, the PN manager (i.e., user) will start the, password protected, PNCA functionality that will issue the new certificate. The expiration detection functionality could be carried out by a latent PNCA functionality but because of the PNCA resilience and the fact that the PN agent will be used for advertising the CRL, it is preferred to use the PN agent for certificate expiration detection.

A similar approach to the provision of security, privacy and trust can be applied to WSNs. This was discussed in details in Chapter 4 of this book.

5.1.5 Device Architectures

A multitude of devices is related to ad hoc networks. Depending on the scenario, the requirements to the devices will differ. Some common factors related to all mobile devices for next generation communications is the requirement for a certain degree of intelligence, whether to sense the environment, to decide on the resource allocation, or for talking to other devices. Multiradio capabilities are one strong requirement for a mobile device.

In this context, research within the FP6 IST program focused on modifications or new terminal architectures depending on the identified scenarios.

A terminal for an integrated heterogeneous scenario should be provided with the following capabilities [23]:

- Several network interfaces for mobility and multihoming;
- QoS support on heterogeneous network interfaces;
- Security and privacy-enhancing components (e.g., for use of a VID framework);
- A runtime environment for flexible service management.

A terminal in the context of opportunistic networks should be provided with cognitive radio capabilities. A common *application programming interface* (API) toward the sensing part of the terminal would allow for a broader interaction between the different types and brands of terminals. This requires the usage of well-established data formats and standards and an evolutionary approach in the protocol/interface development to increase the acceptance and lower the risk.

A terminal in the context of PNs, should be able to support a low data rates (LDR) and high data rates (HDR) WPAN-based air interfaces. For example, [4] proposed for the LDR air interface, a PHY layer using *frequency modulated ultrawide band* (FM-UWB) and a MAC layer based on IEEE 802.15.4, while for the HDR, a *multicarrier spread-spectrum* (MC-SS)-based PHY layer and a MAC layer scheme using IEEE 802.15.3 are defined. This required that a convergence layer for coexistence support had to be introduced in the mobile terminal architecture.

5.2   Networking and Enabling Architectures for Personal Networks

The PN includes the P-PAN and a (dynamic) collection of personal nodes and personal devices, organized in clusters that are attached to the P-PAN from remote locations and have an established common long-term trust relationship. This means that the PN can be seen as a *collection of clusters* that are geographically dispersed. These clusters constitute the *building blocks* of the PN.

The main idea of the PN network layer architecture is to separate the communication among the nodes of the same user from the communication of other nodes. Nodes belonging to the same owner form *clusters* of personal nodes and they can communicate with any other personal node in that cluster without using foreign nodes. In this way, the communication, routing, and other self-organizing mechanisms can be protected on a local scale. For the global scale, tunnels are established between the clusters to both accommodate and protect the intercluster communication. Figure 5.9 shows a high level abstraction of the proposed architecture.

In this architecture, the home network of a person will be one cluster, the car network another, the PAN around the person a third, and so on. Clusters work as local networks and, therefore, need their own local routing, addressing, self-configuration, and other internal mechanisms. The formation of clusters is a purely

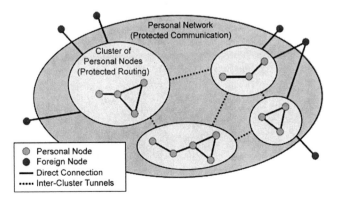

Figure 5.9   A PN network layer architecture [42].

local process and does not need any support from the infrastructure. It is based only on physical connectivity and trust. Further, the functionality for the formation and the management of the cluster can be provided either in a centralized fashion, in which a node acts as a central authority that manages the cluster, or in a distributed fashion, in which, the functionality is distributed over all personal nodes in the Cluster.

The management of the PNs can be divided into two domains; the *clusters management* (management not restricted to the network level but is used in more general sense here) and the *PN management*. The cluster management comprises all management functions, in which the actors are restricted to the devices and the nodes inside a cluster. There is no participation of *foreign nodes*. These management functions are usually performed during the cluster formation. Examples of such management functions are the *selection of a node* to perform the *master node* functions (e.g., the *service management node* in the service abstraction level), the *gateway nodes* selection (possibly through discovery), and the local service discovery.

This type of ad hoc PN either can be composed by the merging of clusters or by the access of the foreign nodes or nodes from the cluster across the ad hoc networks. Examples of the management functions performed at this level are: *master to master coordination*, *foreign node/device trust* negotiation, routing and discovery, and so forth.

The different management domains in the PN architecture are shown in Figure 5.10.

The PN establishment and maintenance can rely on the agent functionality, and then it is referred to as an *agent-based PN establishment*. This can be further classi-

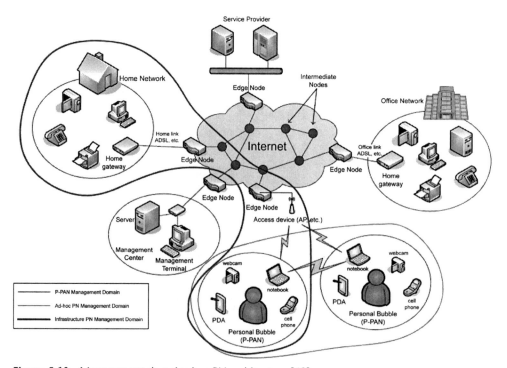

Figure 5.10   Management domains in a PN architecture [42].

fied depending on the fact the agents are located in a cluster or in the interconnecting structure. In the absence of agents in the interconnecting structure or clusters, or when an agent in a cluster needs an access to the Internet for its functionality, and the remote clusters still want to communicate, then it is an ad hoc PN establishment. For example, when the interconnecting network is ad hoc-based, the remote cluster should, hence, be found using autonomous means, hereby aided by intermediary foreign nodes. To this end an ad hoc network solution needs to be developed enabling a direct secure connection between two remote clusters.

The different types of PN establishment will lead to different architectural decisions for the PN establishment and maintenance. It must be mentioned that different types can be present within the same PN. Figure 5.11 shows an overview of the possible PN establishment scenarios.

The main building blocks that fulfill the functionalities that are supported by the personal nodes in order to form the PN are shown in Figure 5.12.

The following Sections describe the PN network components and how these can be implemented in order to solve the PN problematic.

5.2.1   PN Networking Functionalities

Figure 5.13 [42] shows the PN networking functionalities.

The *tunnel manager* service at the *gateway*, provides connectivity to the nodes and devices outside the *cluster*, and is responsible for setting up secure tunnels between the clusters of the PN. The location of the other cluster gateways is obtained via the *PN agent* client.

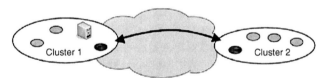

a) Agent based PN establishment with Agent capability in a Cluster

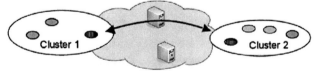

b) Agent based PN establishment with Agent capability in the Interconnecting Structure

c) Ad hoc PN establishment

Figure 5.11   Different PN establishments [42].

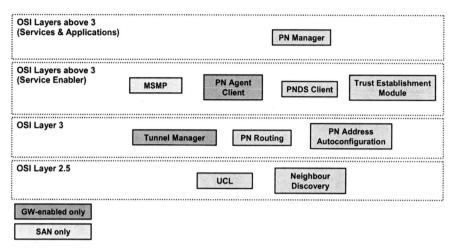

Figure 5.12 Main building blocks of personal node functionalities [43].

Another PN network service is the configuration of the nodes. A separate PN addressing space can be used for the PN communication, where every PN node can receive a unique PN IP address, thereby creating a PN overlay. Two addressing schemes can be used, namely the following: a *hierarchical* addressing scheme similar to the one for NEMO [14], or a *fixed (flat)* addressing scheme. In order to establish IP connectivity within the PN, a routing functionality is needed, which is provided by a *PN routing protocol module.*

5.2.2 PN Link Layer Functionality

The cluster formation in a PN should be *self-configuring* and *secure*. A *neighborhood discovery* module allows for this cluster formation [41]. Clusters advertize their presence, which would affect the privacy of the PN. After the detection of a PN cluster, the PN devices will be authenticated before the access to the PN is granted. The responsible for authenticating the certificate and the creation of the shared secret key is the *trust establishment* module. This shared secret key is the basis for the encryption between the two nodes. The respective session keys will be derived from this shared secret key and this is dependent on the type of the link that is used between the nodes (e.g., Bluetooth, WiFi, IPSec). This link-specific session key generation is handled by a *universal convergence layer* (UCL) [44].

Figure 5.13 Functionalities for PN networking [41].

The security and the access control is handled at the link layer. Within the PN all nodes have the same security level, so, in principle, there is no distinction between the PN nodes and the friendly nodes [41]. The friendly nodes within a PN should be avoided because the friendly nodes being part of the PN multihop structure will have access to unencrypted data. Allowing friendly nodes to access this data introduces a *security risk*. If friendly nodes are part of the architecture, one should, therefore, handle security on a *higher* level than the link level. This would correspond to having an end-to-end *encryption* on the service layer. Allowing friendly nodes is one of the requirements to a PN architecture but from a security point of view one should carefully reconsider this requirement [41].

5.2.3   PN Service Layer Functionality

Each PN cluster should be able to elect one *service management node* (SMN), where all services are registered. Wherever possible, the cluster SMN's will form one overlay network, where information about the available services can be exchanged. A PN service layer view is shown in Figure 5.14.

The PN service layer is responsible for the service access control based on the node ID. In a PN with only PN nodes and no foreign nodes, there is no reason why services should be denied for any PN node.

In an architecture with foreign nodes, however, several problems arise. First of all, one should be able to restrict access to specific services from the foreign node. The second, more important, problem is that in this case one requires end-to-end encryption. This will need to be addressed at the service layer.

5.2.4   PN Context Management Functionality

All context management functionalities take place through the *secure context management framework* (SCMF). The SCMF was discussed in Chapter 4 of this book. Further details can be found in [45]. The framework and its components allow the service of providing context and user profile information to any type of software client such as applications, services or other networking components to gain easy access to various types of information throughout the PN and PN- F environment. For example, the context information can be used when setting up PN federations

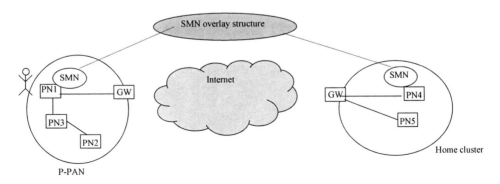

Figure 5.14   PN service layer view [41].

(PN-F) or selecting an appropriate gateway for the interaction with other clusters or the Internet. On the PN level, the context management functionalities are carried out through the set of interconnected *context management nodes* (CMNs) in each cluster (similar to the SMN) through provided interfaces. These interactions are shown in Figure 5.15.

In terms of security needs, at the PN level, the user should be allowed at all times to access his own personal and context information, since the security infrastructures on lower layers protect him from attacks. In the rare case of infrastructure failure or absence, the *context aware security manager* (CASM) integrated within the SCMF can alert the user for the low security level and disallow for sending any sensitive information over unsafe communication channels.

A more challenging case from a security point of view, is the PN-F. A PN-F overview architecture is shown in Figure 5.16.

In the PN-F overview of Figure 5.16, the components do not correspond to individual nodes, but they gather functionalities that can be coexisting on a single node or distributed among different nodes. In a PN-F, there would be external context and personal information requests that must be handled and this requires the availability of access control to this information. The CASM is responsible for managing and enforcing the context and personal information related policies, and, therefore, performing the access control and privacy filtering required. Security can be handled by security primitives and security policies. A security policy enforcement proposed in [41] for support of PN interactions is shown in Figure 5.17.

5.2.5   Networking Aspects of the PN-F

A PN-F (see Figure 5.16) is a co-operation on the service level between PNs. One of the PNs will have the role of the PN-F *owner* or *creator*. The *creator* is responsible to manage the PN-F and controls the access of new PNs in the federation. To support federations, a new entity, called the *federation manager* (FM) was introduced in [4]. The FM is responsible for the secure access within a federation.

It is the responsibility of the creator to manage the PN-F. To start a PN-F, the creator activates the FM service to create a new federation. The FM requests the security manager to define a new public/private key-pair and to create the root certificate for the PN-F. In addition, the user, via the FM, defines the rules of operation

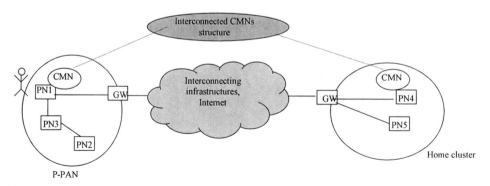

Figure 5.15   Context management interactions on a PN level [41].

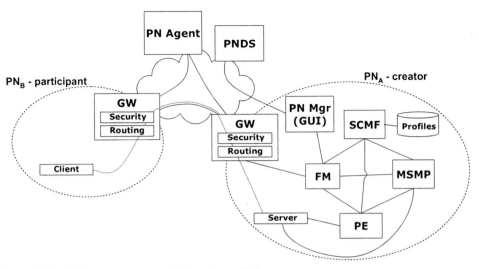

Figure 5.16 PN-F components and interactions [46].

for the PN, in the so-called *PN-F profile*. This PN-F profile is stored in the PN-F database.

The PN-F profile is used in the advertisement of the federation to potential members. In an infrastructure-supported federation, the PN-F profile can be advertized by a *PN-F broker*. This broker can also store profiles of potential members,

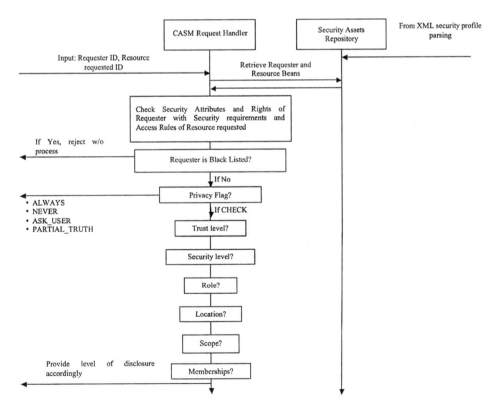

Figure 5.17 Security policy enforcement algorithm [41].

showing their interest for a specific type of federation. In an ad hoc mode, the profiles can be advertized via broadcasting over one of the radio interfaces. The result in any of these cases is that the FMs of the creator and the potential member can contact each other. The PN-F management functionality is shown in Figure 5.18.

After the discovery of an interesting federation, the potential PN-F member needs to define, under which identity to participate in the federation. For this, the member creates a VID.

To avoid any relation with the PN, a new key-pair must be created for participation in the federation.

The participation of a PN in a PN federation starts with the authentication of the PN member by the PN creator and visa versa. Knowing each other does not mean that access would be granted. Granting access or authorization has to do with the *trust* that the person behind the PN has in the other party. The authentication method will be defined by the creator in the PN-F profile.

In an infrastructure mode, the *trust level* of a user can be bootstrapped by having the PN's VID authenticated and registered by a *trusted third party* (TTP) certificate. Whether or not a potential member decides to join a PN-F is dependent on the users' knowledge of the reputation of the creator, PN-F or TTP and in the registration procedure.

In an ad hoc mode, the authentication of the other party can be done by the users based on the existence of an auxiliary *proximity-authenticated channel* (PAC), in a similar way, as the first stage of the CPFP imprinting protocol. Whether or not the access or participation with the PN-F is granted depends on the users' knowledge of the reputation of the other party.

In both cases, when the access is granted the member receives a *token* to access the PN. This can be either a *group key* or a *PN-F certificate*. The group key is used in combination with a TTP certificate.

After becoming a PN-F member, there will be a point when the users would want to utilize the federation (formation) in order to allow for services access between the PN's. To initiate the formation, a secure connection needs to be created between the federation managers of two PN's. The location (IP address) of the gateway to set-up this initial contact can be retrieved from the PN-F database at the PN creator.

At the connection set-up between the gateways of two PN's, the authentication is done based on their TTP or PN-F certificate. The group key or the PN-F certificate is used to validate the access rights to the federation. Outcome of the authentication process is a *pair-wise master key*, which is unique for one pair of

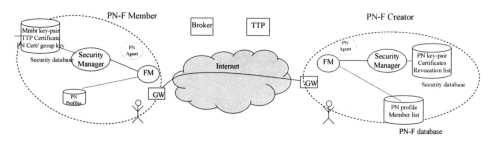

Figure 5.18 PN-F management functionality [41].

PN's. This key-pair is used to derive session keys for the secure connection between the two gateways.

Ad hoc PN-F discovery is obtained by using beacons at the neighboring cluster gateways. In this way, the roaming PN clusters can discover the federating PNs. A secure connection between their gateways can be established using the group key or PN-F certificate.

The formation within a federation might be done proactively or reactively [43].

The *PN manager* can be implemented as a graphical application used to provide an easy and intuitive GUI to the user where the most relevant information concerning the user PN is displayed. From the registration phase of the new PN to the complete PN view where all her nodes are interconnected, all the steps can be followed by the use of this graphical interface. An implementation performed in [3], is shown in Figure 5.19 with the screenshot of the registration procedure.

The user identity is initially bound to her mobile number within the PN domain server, which upon a successful registration returns the user password via an SMS. The mobile phone is considered one of personal nodes of the user that is carrying any time anywhere, so in this sense it is considered the optimal place as the password retriever. This registration phase only has to be fulfilled once in a whole PN life, as the next times the user logs in, only the password would be requested.

5.2.6   PN Interactions with External Service Frameworks

The PN concept extends the use of a portable device or a client to a larger network; the entire PN can be seen as a big UT that can be contacted by an external network that transparently perceives it as a single node [46].

An interesting aspect of this concept is to enable the access to an external service framework, from a node inside the PN. Examples of such an access could be Web surfing, Internet banking, remote login, and so forth. An external to the PN backend server is contacted by an internal node within the PN, through a gateway node and other foreseen entities, such as *Network Address Translating* (NAT) boxes and firewalls. There are also cases that an internal PN node should be discovered and contacted from an external node, to provide a shared service to the outside world, or to receive and take an external call.

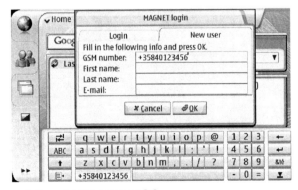

Figure 5.19 Implementation of a PN manager [4].

5.2.6.1   PN Interaction with SIP-Based Services

SIP sessions are established by the involvement of a number of proxies for finding the destination and redirecting the calls initiated by *user agent clients* (UACs). [42] examined how in bound and outbound PN calls are handled by via service proxies and redirect servers. Some scenarios involve PN agents, while others rely on service gateway nodes or gateways to handle calls and control sessions.

For supporting SIP inbound calls, further enhancements need to be designed and implemented. The issues described here are related to the extension of the PN entity set with an enhanced SIP component for the following purpose:

1. Retrieving among the entire set of SIP-enabled PN devices the one that has to receive the incoming call. The MSMP could be used for getting the list of descriptions of all the active SIP *user agents* (UAs) of the PN user that have been previously registered as active services. Afterwards, the context information retrieved from the SCMF should allow the selection of the most suitable SIP-enabled PN device for receiving the incoming call.
2. Redirecting the inbound calls to the PN user terminal retrieved during the previous step.

Two solutions are envisioned depending on whether the PN enhanced SIP entity is collocated with the PN agent, or whether it is installed in the PN cluster gateway/ SMN nodes. These are shown in Figure 5.20 and Figure 5.21, respectively.

Figure 5.21 shows the scenario where an enhanced SIP proxy is chosen to be collocated with the PN agent. The following message sequence occurs.

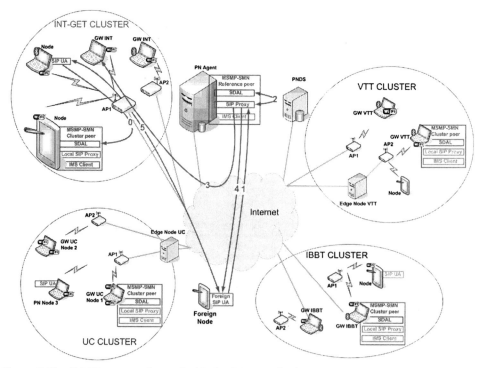

Figure 5.20   PN SIP proxy collocated with the PN agent [46].

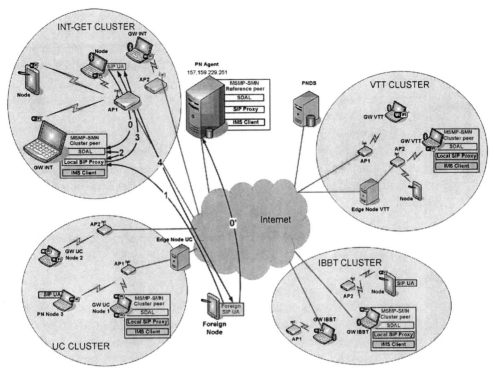

Figure 5.21   Local SIP proxies in the PN cluster gateway/SMN nodes [46].

Any active *user agent* (UA) of a PN user (i.e., a SIP enabled terminal), registers its information in its cluster SMN. This information is stored as a name (e.g., for service descriptions), and describes the UA in terms of user name, public SIP IP address, and additional location information as for example:

[PNId=alice_PN][clustereId=PPAN][userName=alice][publicSIP@=sip:alice@openims.test] [IP@=157.159.229.252].

This corresponds to step 0 of Figure 5.20. Therefore, any active PN UA can be discovered from any PN SMN through the SMN SDAL (i.e., through service discovery function calls) and the PN service overlay.

A foreign SIP UA that calls a PN user first contacts the PN enhanced SIP proxy collocated with the PN agent. This SIP proxy will redirect the incoming call to the appropriate PN user SIP terminal (UA). This corresponds to step 1.

When the PN SIP proxy receives foreign UA calls, it performs the following:

- Service discovery function call to its cluster SMN SDAL for discovering the PN user active SIP UA;
- Extracts the location/contact point of the PN active SIP UA from the description name received from its Cluster SMN during the discovery operation.

This corresponds to step 2.

After having retrieved the PN user SIP end point, the PN enhanced SIP proxy will ask the PN Cluster gateway, which will receive the incoming call to "open a gate" (i.e., to forward all the SIP messages coming from the foreign UA to the cluster SIP UA that should receive this SIP call). This corresponds to step 3. The PN SIP proxy redirects the call from the foreign SIP UA to the PN user SIP UA (i.e., the one discovered during step 2) and the SIP call is finally established. This corresponds to steps 4 and 5.

Figure 5.21 shows the scenario when the enhanced SIP proxies are collocated with the cluster gateway/SMN nodes.

The following message sequence occurs.

Any active UA of a PN user (i.e., a SIP-enabled terminal), registers its information in its cluster SMN. This information is stored as a name, and describes the UA in terms of user name, public SIP IP address, and additional location information as follows:

[PNId=alice_PN][clustereId=PPAN][userName=alice][publicSIP@=sip:alice@ openims.test] [IP@=157.159.229.252].

This corresponds to *step 0* of Figure 5.21.

A foreign SIP UA that wants to call a PN user first contacts the PN SIP proxy. This SIP proxy, collocated with the P-PAN gateway/SMN, will redirect the incoming call to the appropriate PN user SIP terminal (UA). This corresponds to *step 1*.

If the information of the PN SIP proxy is unknown, the foreign UA can retrieve this information (the SIP proxy location/contact point information) from the PN agent through dedicated function calls. This corresponds to *step 0*.

When the PN SIP proxy receives foreign UA calls, it first:

- Performs a service discovery function call to its cluster SMN SDAL for discovering the PN user active SIP UA;
- Extracts the location/contact point of the PN active SIP UA from the description name received from its cluster SMN during the discovery operation.

This corresponds to *step 2*.

The PN SIP proxy redirects the call from the foreign SIP UA to the active SIP UA of the PN user (i.e., the one discovered during step 2), and the SIP call is finally established. This corresponds to *step 3* and *step 4*.

Outbound SIP calls can be supported by the SMN and the gateway nodes. The only point is to find the external SIP proxy for finding the destination UA. This proxy is addressed by a public name and address. The name resolution in an external outbound call can be made by linking the PN naming system with an external DNS server. After resolving the name, the external proxy server is contacted via the personal UA and finally the signaling will be established between the personal UA and proxy. The other signaling will continue in a normal way of SIP session establishment.

5.2.6.2 PN/PN-F Outbound Internet Access

By using a network overlay solution [42], both PNs and the PN-Fs can be organized as *network overlays* with their own addressing space. Every node in a PN/PN-F has

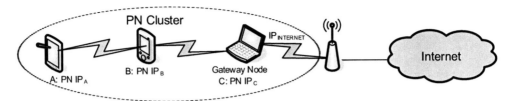

Figure 5.22   Outbound Internet access scenario for PNs [46].

its own unique PN/PN-F overlay IP address. By means of *neighbor discovery* and *ad hoc routing*, the neighboring nodes of the same PN/PN-F form clusters, resulting in a *secure ad hoc IP communication environment*, in which the neighboring PN/PN-F nodes can communicate using their private PN/PN-F IP addresses. The remote clusters are interconnected through the establishment of tunnels between their *gateway nodes* (i.e. the nodes that have access to the Internet, and the exchange of routing information over these tunnels).

By means of the *network overlay solution*, all nodes within the overlay are able to securely communicate with each other using their private IP addresses and this means that the gateway nodes are the only nodes that can communicate with the Internet. However, in many scenarios it would be interesting that also nongateway nodes in a PN or PN-F are also able to have access to the Internet. A possible outbound Internet access scenario is shown in Figure 5.22.

Nodes *A*, *B*, and *C* form a cluster, in which they can securely communicate using their overlay IP addresses. Only node *C* is able to access the Internet using the IP address *IPINTERNET*. The outbound Internet solution will enable nodes *A* and *B* to also have Internet access. In order to share the Internet access, node *C* must know it is a gateway node, and thus that it can access the Internet (i.e., context information).

5.2.6.3   PN Interaction with an IMS Platform

When the PN is viewed as a *virtual access network* by the IMS platforms, the available multiaccess paradigms can come into play to help integrate the PN in the general multiaccess context including other access networks such as WLAN, UTRAN, xDSL, cable. Typically, this interface between access, core networks, and IMS has been achieved via wireless access gateways [5, 45]. A tighter collaboration between the PN architecture and the IMS frameworks should be envisaged as a better way to achieve service bundling or composition and provisioning.

Similarly to the interactions with SIP-based services, one can extrapolate a number of basic and obvious requirements on the PN node and, especially, the nodes involved in the service sessions with the external service platforms. Some of the requirements for the PN are summarized in Table 5.1

With the advances in the area of P2P networks, an integration with IMS< PN, and PN-Fs can also be envisaged. This would require an additional study to operate in a P2P mode between nodes from the same or different PNs. The federation of PNs is of particular interest in this case with providers and IMS platforms playing mediating and mending roles for PN federations or simply for communications

Table 5.1 Requirements for PN Integration with an IMS Platform [46]

IMS Entity	PN-IMS interoperability requirements.
IMS Client	PN nodes must be provided with an IMS client and a SIP UA with a SIP URI for each node. Furthermore PN nodes must support Gm interface to communicate with the IMS service plane and especially the P-CSCF.
P-CSCF	This entity is responsible for service-based local policy control that enables the IMS operator to authorize and control the usage of bearer traffic based on SDP parameters negotiated at IMS session. Moreover P-CSCF is defined as the contact point to IMS, hence it must be located in IMS operator's domain.
I-CSCF	This entity is the IMS user's home network entity that interacts with HSS in order to find out the capabilities of available S-CSCFs and to select a suitable S-CSCF for IMS client. Hence we follow the standards and locate I-CSCF in the IMS operator's domain.
HSS	This entity contains the IMS user profile and credentials and must be located in IMS operator's domain.
S-CSCF	This entity is regarded as the brain of IMS and is responsible for session establishment and service invocation.
	A possibility is to install a local S-CSCF in the PN clusters in order to enable inter cluster session establishment, service invocation, service interoperability and service composition. In this case the local S-CSCF can be installed or deployed in the PN clusters as well as in edge nodes or routers belonging to trusted third party.
Application Server / Service Capability	Service capability is a modular and self-contained service building block that can be shared and reused by various application servers (such as a presence service capability that could be reused and shared by multimedia conferencing, chatting, and multiparty gaming). In order to ensure call or session forwarding, control and management, application servers and service capabilities can be designed in PN clusters to enable PN to use its local and personalized services and to manage the composition of service capabilities locally to introduce new integrated services. This proposal can be realized through a distributed control between providers and PN users on the basis of viable and acceptable agreements to these actors. However, PN must also support the access of PN nodes to application servers and service capabilities that are in IMS operator's domain.

between peers while ensuring privacy protection and overall security. Authentication and authorization support would be provided by a trusted third party.

5.3 Architectures for WSN

The first generation of WSN applications was characterized by a number of sensors connected to a single sink responsible for collection of data sensed on the field. The sink acted as a gateway bridging the WSN with the other types of networks. With a sudden increase in the number of nodes, however, the amount of information gathered at the sink might exceed the communication capability of the sink [47].

To overcome the scalability problem, one solution is to use multiple sinks, properly interconnected through the transport network that forwards the sensed data to the application servers and storage systems [48]. Opposite, to the multiple-sink approach, where the sinks are specific devices with relatively high deployment cost, an novel option, which can lower the cost of WSN deployment, is to use a new type of WSN gateways (not specifically deployed for the interaction with the sensor nodes), properly merging the WSNs with other existing wireless networks.

Sensor networks should be deployed and operated so that the following objectives are met [49]:

- the sensors cope with bandwidth limitations especially in the case of transfer of bandwidth-demanding information;
- they must use their limited energy resources judiciously and thus extend the network lifetime;
- they must monitor or measure and represent a physical process subject to some specified accuracy; and
- the network must be able to handle a certain amount of sensor failures by creating a fault-tolerant system.

The first goal is to identify analogies between the themes of source and channel coding and some problems that arise in the context of wireless sensor networks. It is planed to identify and exploit such similarities in flat sensor network architectures and hierarchical sensor network architectures.

In particular one have to adhere to translation of ideas from the areas of information theory, rate distortion theory, vector quantization and error correction coding in order to model and address fundamental problems that arise in the context of sensor networks. Then, already established results could be used to evaluate the performance of sensor networks with respect to some fundamental performance measures.

The next goal is to use ideas from other sciences in order to address issues that arise in sensor networks. In particular, we plan to elaborate on percolation theory and its applications in sensor network-related problems. Issues related to connectivity and transport capacity will be quantified by using this theory [49].

5.3.1   Hybrid Hierarchical Architecture

The *hybrid hierarchical architecture* (HHA) proposed by [47] (see also Chapter 4 of this book) is characterized by the coexistence of terminals with different computation capabilities and several transmission technologies, in which the WSNs populate only one layer in the hierarchy. To make it a viable concept, the HHA requires further research into the functionalities and structure, together with the identification of a number of most relevant application scenarios that can be accommodated by the HHA [47]. The HHA (see Figure 4.4) was based on the development of an optimal routing technique for WSN to realize the *publish-subscribe* paradigm within the HHA, and considering the reliability and energy efficiency issues.

For example, in an indoor scenario (e.g., office), the hierarchy would refer to the levels 1 (*mobile gateways → subscribers*) and 2 (*fixed sensor motes → publishers*) of the HHA. In this scenario, the interaction between a subscriber (mobile) and one or many publishers must be supported. Mobile gateways are represented both by the mobile phones or PDAs carried by people and, therefore, with a mobility limited to the pedestrian speed.

The interaction can be realized in a *distributed* way by multihop transmissions within the WSN (level 2) with the exploitation of a small number of nodes elected as sinks. Basically, once the mobile gateway transmits a subscription, the closest neighbors would elect themselves as sinks. They flood the network looking for the publishers. Upon the reception of the flooded packets, the publishers follow the reverse path to the sinks with the requested information. Because the subscribers are movable, when the information reaches the sinks, the subscriber has changed the position. An algorithm can be devised to track the subscriber, so that it can be reached through a multihop path from one of the sinks.

5.3.1.1 Optimal Placement Methods in Hierarchical Sensor Networks

The problem of the optimal placement of super-sensors in hierarchical sensor network architectures, is important in relation to the minimization of a generic cost function that captures several special cases and identifies the analogies to a vector quantization problem. The cost function may represent energy-related metrics either in single or in multihop communications from an ordinary sensor to a super-sensor [50, 51].

A well-known method for reporting the content of a source with infinite entropy by using only a finite number of bits, but introducing some distortion, is quantization. In other words, quantizer was used that accepts as input the source, and produces as output one of a finite number of quantization levels, chosen so that the distortion between input and output is minimal. It is well known that quantization is suboptimal (i.e., it cannot achieve the minimum distortion for a given rate constraint).

On the other hand, translated to WSNs, quantization can be performed in a distributed manner, by dividing the network in a number of clusters, each cluster corresponding to a different quantization level. Therefore, it offers a significant advantage. Event reporting algorithms motivated by quantization algorithms were developed and evaluated in [7] in terms of their efficiency.

It is worth investigating the hierarchical sensor coverage problem that involves deploying a set of sensors with sophisticated sensing capabilities over a grid of ordinary sensors in terms of channel coding strategies by defining appropriate analogies. The question is to identify the amount of redundancy that needs to be added to the system (in the sense of additional sensors) so as to achieve a certain level of performance, which in this case can be described by the probability of the error in the measurements. The fundamental connection between the two can be summarized as the need for the removal of the redundant information, and the efficient insertion of redundancy [52].

Another direction of investigations was related to WSNs and percolation theory n order to identify the fundamental tradeoffs and relations between the end-to-end throughput, transmission rate, connectivity and transmission range of WSNs [52].

5.3.2 Design of Middleware

In order to decouple the development of applications from the hardware features of the mote platform, it is necessary to design an efficient middleware layer that can perform various tasks, such as query requests, command dissemination, and wireless mote reprogramming. A suitable middleware layer can allow for the rapid deployment of new applications or for mechanisms for self-composing applications, and ensure that the WSN could serve a number of purposes and facilitate an evolution of services [53].

The interaction between any generic application and the API is one design requirement; how to parse the generic queries and requests from the applications, delivering them to the actual sensors. An efficient design, in order to optimize the energy consumption and latency is a challenge, especially in view of pervasive systems [49].

5.4   Protocol Architecture for Opportunistic Networks

Within a network of collaborative opportunistic terminals, the communication of sensing information or context, in general, in the course of distributed and collaborative sensing and decision-making must be suitable for both the link layer and the network layer transfers [13]. The link layer protocols should be fast and lightweight but suitable only for information dissemination in the direct vicinity (e.g., by one-hop broadcast such as in ad hoc scenarios). The network layer protocols can be designed for assured point-to-point and point to multipoint communications such as in multihop communications in ad hoc scenarios for a network-wide context data exchange.

The protocols must be extensible: it must be possible to add new types of sensor information to a stable protocol specification and to encode complex data structures such as context/situation descriptors, spectrum policies, or capability profiles for collaboration.

Additionally, it must be possible to assign dedicated communication channels providing different capacity to different categories of context information depending on their communication frequency (e.g., short but frequent information data sets according to sensor update rate), their relevance (e.g., with respect to the vulnerability of algorithms against delayed delivery, inaccuracy or loss of information), or their volume (e.g., for infrequent but large information data sets). If using opportunistic channels for signaling, this allows for aligning the statistical properties of the opportunities with dependability requirements of the signaling bearer. The reliability of signaling protocols then is not necessarily better but can be estimated more precisely compared to the multiplexing of different types of signaling to a single opportunistic channel.

The signaling in opportunistic networks is delay-sensitive with respect to both timing requirements and the requirements on synchronicity of context data transmission between the multiple terminals either in centralized or peer-to-peer schemes. Thus, there is already a strong demand for allocating a significant amount of bandwidth, exclusively to distributed sensing and collaboration signaling channels. The high capacity signaling channels, in turn, must be allocated from available spectral opportunities in competition with user communications of the local terminal and with other opportunistic terminals. The resulting demand for robustness against the rate variations, depending on the bandwidth availability and the channel capacity, as well as against the packet loss implies the following:

(a) Optimized opportunity allocation strategies that differ from those utilized for user data channels;

(b) Optimization of protocol design towards a stateless message-based interface;

(c) A PDU encoding and message size that can cope with the variation of the statistical properties of opportunities and primary user characteristics.

From a protocol architectural perspective, the opportunistic terminals must provide (at least) three different types of virtual communication channels each optimized for a dedicated objective. An opportunistic terminal thus is inherently multihomed, denoting that it utilizes connections to different virtual networks simultaneously via dedicated network interfaces. Each of these interfaces relies on a dedicated goal and strategy in accessing a single opportunistically used shared medium. The model in [13] was assumed to simplify the interfacing with the widely used IP suite of protocols that is in principle agnostic of the dedicated functionality and the configurable *service access points* (SAPs), namely with the following aspects:

- The user data channel/interface provides access to opportunistic user data, which is used by user applications and distinct management applications;
- The control and management channel/interface provides access to opportunistic peer-to-peer signaling, which is dedicated to sensing and collaboration and context/situation dissemination/ retrieval;
- The meta-signaling channel/interface provides access to a well-known or easy to detect low rate signaling channel. It is used to establish opportunistic control & management channel(s).

The effective availability of the different channels depends on the current operating environment, the terminal configuration capabilities, the primary user behavior, and the available (spectrum) resources. In a "fine-grained" configuration the bearers carrying the user data and signaling are one-to-one mapped onto opportunities characterized by suitable statistical properties. For example, periodically updated sensing information can be mapped to periodically available opportunities, resulting in less packet drops or less delay variation as if statistically multiplexed with other channels.

In realistic configurations, the opportunistic terminal should be able to allocate channels on demand (i.e., "fine-grained" configurations as well as "single channel" configurations should be selectable depending on the current terminal's situation). The most convenient configuration in terms of terminal hardware requirements will be to provide one user data channel and one opportunistic signaling channel or a cognitive pilot channel, respectively [13]. Figure 5.23 shows a simplified terminal protocol architecture for opportunistic communication networks.

Establishing signaling connections via an opportunistically shared medium requires a dedicated meta-signaling protocol, such as a cognitive pilot channel in centralized scenarios or a meta-signaling channel in ad hoc, P2P scenarios to negotiate the access policies between the peers prior to the exchange of control and collaboration messages via the signaling or user data channels.

The QoS requirements from the dedicated applications or application data streams may coincide with a dedicated set of opportunities that potentially satisfy these requirements better than others due to their statistical properties. For

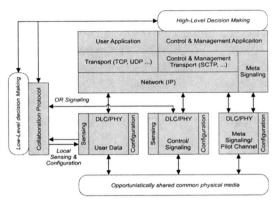

Figure 5.23   Generic opportunistic terminal protocol architecture [13].

example, a reliable signaling channel might be mapped to a dedicated frequency with low or well-predictable primary user activity. This channel or dedicated data stream can be mapped to a virtual network interface, potentially created dynamically upon request and backed by a dedicated DLC/PHY instance. This simplifies lower protocol layer management requirements by utilizing the multiplexing capabilities of the network layer protocols instead. Management functions thus are partly provided by decision-making [39, 54], which is unique for the opportunistic terminal architecture shown in Figure 5.23.

Sensing and collaboration data are communicated via a link layer protocol (e.g., for timing constraints and limited spatial relevance) or are communicated via dedicated network layer protocols (e.g., for end-to-end semantic and assured transport requirements). These shall rely on the same set of information elements (i.e., message payload) in order to allow for the flexible forwarding or routing. It is up to the decision making to select the most appropriate method to locate, disseminate, or collect the information. This approach implements a context-based cross-layer optimization of signaling as proposed by [13].

Although not explicitly shown in Figure 5.23, the control and management application should support the failover signaling streams, which is an inherent capability of the *Stream Control Transmission Protocol* (SCTP). This means that in case of a critically low QoS for the active signaling channel (i.e., the current network interface), the signaling stream should switch over to either a user data channel or to another signaling channel (i.e., another virtual network interface). Additionally, a fallback to the meta-signaling channel should be possible to prevent a potential degradation of the opportunistic network connectivity that may demand for a time-consuming reassociation of the terminals. The higher protocol layers and the higher layer decision making do not need to be involved if the problem of the channel QoS degradation can be resolved by the lower layer protocols or the decision making, transparently [13].

From a protocol perspective, providing multiple network interfaces each associated with a distinct set of opportunities instead of a multichannel MAC with layer-internal management of opportunities, simplifies the handling of application demands. Nevertheless, this approach requires a layer internal coordination function of comparable complexity. The benefit is in the fact that this partly can be resolved by high-level decision making, which supports a decision-making architecture that is no longer bound to the protocol layers but is cross-layered and bound to timing constraints instead. [39] addressed this issue by proposing to distribute the algorithmic and cognitive processes, accordingly. XML-based messages can be used to exchange complex context data and structured data such as policy rules, profiles, or complete ontologies between peers via the transport layer protocols [13].

5.5 Device Architecture for Enhanced User Experience

A composite access network environment can be based on diverse network technologies, including radio technologies in support of ad hoc communications (e.g., WLAN, WiMAX). Moreover, one *network domain* may belong to several operators. All these characteristics and the posed constraints by the composite environment were considered by [55] where a system architecture for B3G environments was proposed. The proposed service systems are based on service enablers. It is considered of interest here, because the network domain is composed of entities that draw data from the MTs, in order to form the content for provisioning of context-aware services. This content can be afterwards used by the candidate applications. Such an approach has the following three areas of applications:

- End-to-end-user experience enhancement as a tool to measure the radio coverage and network quality;
- Ubiquitous terminal assisted positioning (UTAP);
- Anonymous mobile community services (AMCS);

This approach can be very suitable in the context of composite network environments.

The *network domain* is a functional entity that processes, calculates, and provides the end-to-end user experience indices per access network, per device, and per service to different clients. Figure 5.24 shows the architecture of this function al entity. These clients could be the following [56]:

- Radio planning tools;
- SLA management software;
- Subscribers;
- Users and user terminals;
- Servers;
- External third party applications (terminal manufacturers, application developers).

The typical information flows between the functional modules of the integrated end-to-end user experience entity at the network side, as well as the flows between this entity and its peer entity at the terminal side are shown in Figure 5.25.

The support of the end-to-end user experience as proposed by [55] requires that the network domain would be enhanced with certain positioning capabilities.

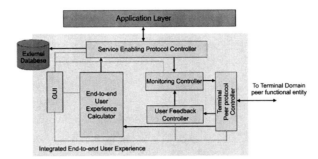

Figure 5.24   Integrated end-to-end user experience monitoring generic functional architecture [56].

A *ubiquitous terminal assisted positioning* (UTAP) entity was proposed in [56] as a method that increases the accuracy of positioning techniques. The functional architecture of the entity on the network side is shown in Figure 5.26.

The proposed architecture is in line with the architecture specified in the *secure user plane location architecture* (SUPL) architecture specification [57].

The *preprocessing* module is responsible for terminating the requests that derive from the UTAP-enabled terminals or the applications that require the position of a terminal.

The *user and privacy management* module has to ensure that the privacy of the terminals is respected.

The *controller* module controls the information flow within the UTAP network entity.

The *algorithms* module abstracts the components performing the actual location algorithms.

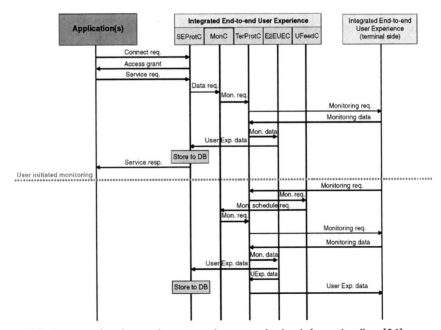

Figure 5.25   Integrated end-to-end user experience monitoring information flow [56].

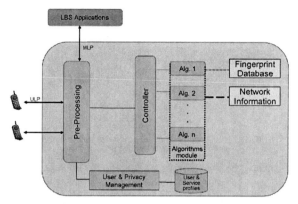

Figure 5.26 Ubiquitous terminal assisted positioning entity functional architecture.

The *GSM and 3G fingerprint component* (DCM) is based on the idea of the fingerprint methods and used to create a radio map of the area where the mobile terminals are to be located.

There are three different methods for locating WLAN users, namely: the strongest access point, triangulation, and location fingerprinting [56].

A typical high-level, nonerroneous, data flow through a supporting user entity is shown in Figure 5.27.

When the positioning request message arrives at this entity, the message is parsed in the *preprocessing* module. The *user and privacy management* module checks that the client has the rights to use the service. The *controller* module checks that the client has rights to use the service and it determines the client and session, in which the message belongs to, and it moves the session to appropriate state. Then, the *controller* determines the positioning algorithm to be used based on the *quality*

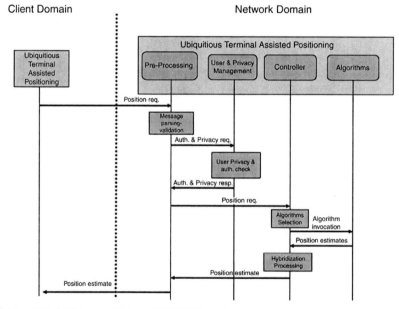

Figure 5.27 Typical information flow within UTAP functional entity.

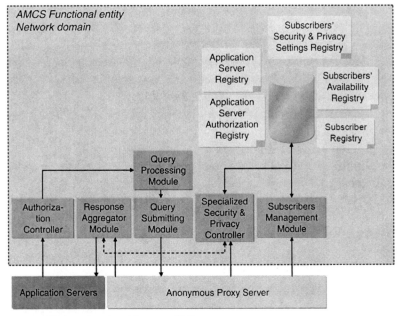

Figure 5.28   AMCS functional architecture [56].

of positioning (QoP) requirements and invokes the algorithm. The position estimate is encoded to a given protocol format and returned to the client.

To protect the privacy of the user, an *anonymous mobile community services* (AMCS) framework is introduced The AMCS functional entity is the heart of the

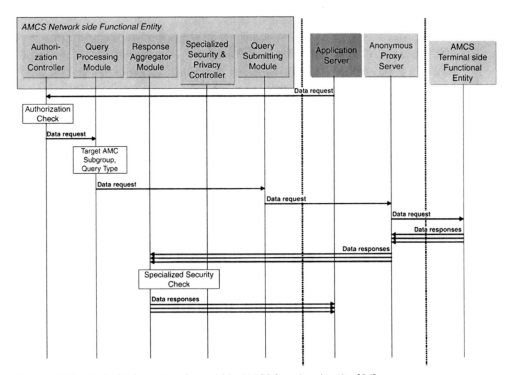

Figure 5.29   Typical information flow within AMCS functional entity [56].

AMC service provision system. It encompasses the basic functionality for the provision of advanced anonymous services. Figure 5.28 shows the functional architecture of the AMCS entity at the network side.

The *anonymous proxy server* fulfils the most central requirement of the AMCS (i.e., the anonymity). It complements the functionality of the AMCS entity in implementing the reference model *service enabling layer*.

A typical information flow within the AMCS functional entity is shown in Figure 5.29.

The requirements on the terminal architecture will regard the need for software development for the implementation of a source code for retrieval and processing of the collected data by the MTs.

5.6 Testbed Architectures

5.6.1 Body Sensor Networks

A *body sensor network* (BSN) test bed was developed in [58] consisting of a variety of different hardware and software components in order to demonstrate how the context information captured via WSN/BSNs, can be used by B3G/IMS services/applications in an end-to-end system implementation and by multiple B3G services/applications. Figure 5.30 shows an overview of the physical architecture of the BSN testbed developed by [7, 58].

The testbed in Figure 5.30 can be used to demonstrate key features of the personal application scenarios that require capturing the user's physiological and psychological data in order to provide security, comfort, or entertainment applications to a user [58]. The BSN consists of three sensor nodes with different sensor probes worn at the body of a person in order to capture the physiological data and a sensor node to capture the environmental information. A gateway device connects the BSN to a B3G network, which provides an IP multimedia subsystem core, *application servers* (AS), and the specifically developed by [7] service enabler. The AS can host

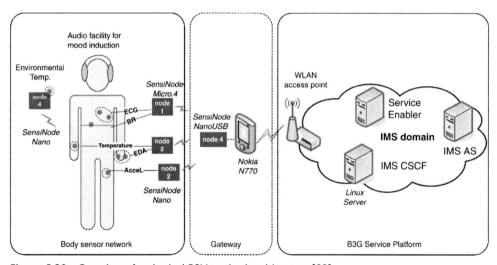

Figure 5.30   Overview of a physical BSN testbed architecture [58].

context-aware services and applications, which interact with the gateway of the BSN via the service enabler, in order to obtain context information.

5.6.1.1   Functional Requirements

The functional requirements can be summarized as follows [58]:

- The developed BSN has to be portable and wearable, which has to be reflected into the selection of the sensor probes as well as the sensor node platform.
- The BSN should be based on COTS components as much as possible, in order to minimize the design and development effort for custom hardware.
- The BSN needs to be able to capture a variety of different physiological signals of a person that are required for the inference of mood state and activity level of a person. This includes probes for breathing rate, ECG, EDA, skin, environmental temperature, and 3D accelerometers.
- The BSN platform shall allow the implementation of the WSN cross-layer protocol stack architecture (see Chapter 2 of this book) with the key WSN services.
- The BSN platform shall allow the evaluation of the developed protocol stack characteristics, such as memory footprint, processing, and communication delays, packet error rates, and so fourth.
- The BSN shall allow experimentation with a variety of different body sensor network type applications.

5.6.1.2   Communication Requirements (CR)

The communication requirements can be summarized as follows [58]:

- The BSN platform needs to support a peak data transfer rate of approximately 50 kpbs, according to the sampling rate requirements of physiological data and the amount of data generated.
- Within the BSN limited multihop forwarding of data (maximum of 2–3 hops) needs to be supported.

5.6.1.3   Processing and Memory Requirements

The processing and memory requirements can be summarized as follows [58]:

- The BSN platform needs to be able to provide sufficient processing and memory capabilities to execute the developed WSN protocol stack and one application.
- The BSN platform shall be able to support the application requirements of simple BSN applications. This includes the timings required for sampling (up to 1 kHz per channel) and packet processing (up to a rate of 50 kbps). In addition, simple data processing such as prefiltering of data (e.g., an average filter), need to be supported.

5.6.1.4   Hardware Requirements

The BSN testbed is comprised of several hardware components. This includes the actual sensor node platform, the gateway platform, and the B3G service platform.

Sensor Nodes
The body sensor network is comprised of four sensor nodes. Two sensor nodes are used to host the physiological sensors, a third one captures the environmental data with respect to a wearer of the BSN and the fourth one, is used for the communication interface to the gateway device.

Two different sensor node platforms were used for the realisation of the BSN Figure 5.30, as the hardware platform of the BSN was designed in two development stages.

Sensor Probes
The BSN provides six different sources of sensory information to extract a variety of different physiological parameters of the mobile user and information of its environment. The six different sensor probes attached to the sensor nodes of the BSN are the following:

- *ECG electrodes with snap leads*: used as a source to derive the heart rate and its variability of a person.
- *Piezo respiratory belt transducer*: used as a source to derive the breathing rate of a person.
- *Finger electrodes*: used as a source to measure the electrodermal response or activity of a person.
- *Skin temperature sensor*: used as a source to measure skin temperature.
- *Accelerometer sensor*: used to measure acceleration, to keep track motion pattern of a person.
- *Environmental temperature sensor*: used to measure the temperature of the environment a person is located in.

Signal Conditioning Units
Two custom signal-conditioning boards were developed for the different sensor probes. The first signal conditioning board hosts the necessary circuitry for the ECG and the breathing rate sensor, while for the ECG signal, an ASIC was specifically designed for the amplification and conditioning of biopotential signals [59].

The second signal conditioning board hosts a measurement bridge for the skin resistance of a person and an input for the skin temperature sensor. Both signal conditioning boards provide the conditioned signals as analogue output to the AD converters of the microcontroller on the sensor nodes. Adequate drivers that allow a configuration and interaction with the signal conditioning boards were developed.

The amplitude of the electrical signal delivered by the breathing strap sensor was found very low [58] and may require amplification before the proper breathing rate detection can be realized. In order, to avoid the additional hardware design, an off-the-shelf data acquisition board offering a *programmable system-on-chip* (PSoC) was used. In addition, specific PSoC software had to be developed to perform the desired amplification.

Figure 5.31   Outside and inside view of the BSN node including signal conditional units and sensor probes for ECG and respiration [58].

Figure 5.31 shows the completed physiological BSN node for the ECG/Respiration application. Besides the sensor node casing, the left side of Figure 5.31 shows the used ECG snap leads/electrodes and respiratory belt. The right side of Figure 5.31 shows the internals of the BSN node, including the sensor node board, signal conditioning unit, and batteries.

5.6.1.5   Sensor Node Platform

Both sensor node platforms, the Micro.2420 and the Nano.2430 make use of a real time operating system called FreeRTOS [60], in order to manage the execution of software components on top of the platform. The version used on the operating system provides a pre-emptive and cooperative scheduling support, coroutines, message queues, and semaphores. The protocol stack described in Chapter 2 of this book was implemented on top of the operating system as an instantiation of the WSN system architecture [40]. It is shown in Figure 5.32.

The implemented protocol stack module consists of four subsystems: the *connectivity* (CO), *middleware* (MI), *management* (MA), and *application* (AS) subsystems. Each subsystem is realized as an independent task running concurrently on the sensor node with different priority.

The CO subsystem provides 802.15.4 PHY and MAC protocol entities and provides IETF 6LowPAN support [61], including framing, mesh routing headers, and compressed IP and compressed UDP protocol entities. The MI subsystem is realized as a lightweight publish subscribe middleware [52] and provides forwarding techniques such as probabilistic forwarding [62] for data dissemination in the network. The MA subsystem offers a service discovery service, which allows for the sensor nodes to register their services with a local service directory in their cluster. Furthermore, a node discovery service makes it possible to discover the nodes in their neighbor hood as well as their respective roles. The AS subsystem can be configured according to the requirements of the experiment. Interaction between the subsystems follows the service primitives defined for the SAP in [52].

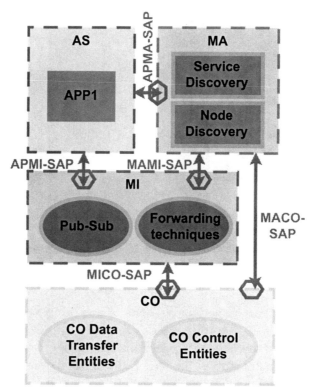

Figure 5.32 Protocol stack module [58].

5.6.1.6   Gateway Platform

The gateway implements all components of the protocol stack as described in the previous section. In addition the gateway protocol stack comprises a full TCP/IP protocol stack including SIP user agent support for the communication with IP based B3G networks. This includes the middleware interaction manager, the gateway manager, and the service promoter. The MI and MA subsystem run as a gateway daemon process named gated and communicate with the connectivity subsystem of the WSN via the nRouted process using a TCP socket. The IEEE 802.15.4 packets received by micro/nano USB module from local sensor network, simply convert into nRoute protocol format. nRoute protocol is used in communication between host Nokia 770 and a serial device to allow 770 access to the local sensor network. The nRouted process sends all received data via the USB serial interface to the application and forwards all incoming WSN data over the USB serial interface back to the gated process.

Figure 5.33 shows an overview of the signal processing chains used for the extraction of the desired features for the mood recognition algorithm.

The measured analog signals at the sensor probes are conditioned in the sensor processing boards at the sensor nodes, before being sampled at different rates. The ECG signal is sampled at 250 Hz, while the other remaining signals are sampled at 20 Hz. Digital pre-filtering takes place on the sensor node before the transmission to the Nokia N770. Further processing steps for the extraction of required features of the physiological signals take place in the Nokia N770.

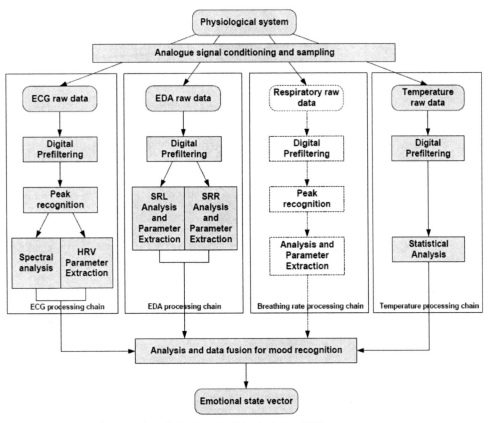

Figure 5.33   Signal processing chains for physiological data [58].

The most complex processing chain in terms of algorithmic and computational complexity is the ECG processing chain. A peak detection algorithm was implemented with an automatic threshold detection, in order to extract the RR peak distances of the heart beat. Additional mechanisms can increase the reliability of peak detection in the presence of noise artefacts in the signal or packet loss.

Several features are computed from the RR peak distances: the heart rate, RMSSD, and PNN-50. Furthermore, a spectral analysis can be performed to identify the high and low frequency components and their ratio. For the spectral analysis a 256 FFT was used.

The EDA processing chain considers both the short-term (ST) phasic components in the form of skin resistance response (SRR), and the long-term (LT) tonic components in form of the skin resistance level (SRL).

The skin temperature processing chain computes short- and long-term development of the skin temperature over time in the form of the mean and temperature rise.

All physiological data is continuously provided to the processing chains and buffered for further processing. The processing chain is invoked every 5 seconds on the currently buffered data set and necessary features are extracted and provided to the mood detection algorithm, which currently only detects the level of arousal. The interval of 5 seconds provided a good trade-off between required resolution accuracy and computational effort on the resource-limited system.

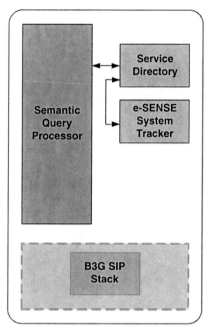

Figure 5.34 Implementation of the service enabler [58].

5.6.1.7   Service Platform

Several software components need to be deployed on physical devices ion order to emulate the commercial platform.

The service enabler processes the context information queries of the IMS applications via the semantic query processor. The knowledge repository used to map semantic queries into BSN services was hard coded in the semantic query processor [58]. In addition, the service enabler hosts a service directory that keeps track of the services available in a BSN system. The system tracker is realized by the home subscriber server (HSS) of the IMS platform. The implementation is shown in Figure 5.34.

5.6.1.8   Testing

The final system architecture of the BSN testbed is shown in Figure 5.35.

In order to verify the standard IMS presence supports (subscribe, publish, notify) from within the client application on the mobile user device (Nokia 770), subscribe, notify, and publish requests have been sent and received from and to the (Nokia 770) client via the IMS service platform to the IMS application server and the service enabler. When the client gateway receives a publish request from the sensor network, the client sends the publish request to the service enabler to keep the record of new services in the service directory. The client application (Nokia 770) sends a subscribe-mood request to the IMS application server and receives a notify indication from the application server. The application server then sends a subscribe request to the service enabler and receives a notify indication from the service enabler, which then forwards a notify indication to the IMS application gateway. The service enabler finds the corresponding sensor network gateway from the service directory and sends the subscribe request to the it. All these signaling processes be-

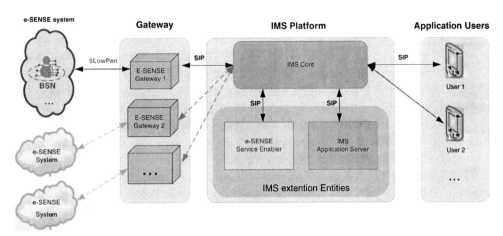

Figure 5.35   Complete implementation of a BSN testbed.

tween the gateways, the IMS application server, the service enabler, and application gateway have been completed with the IMS service platform and tested successfully in several stages. A Wireshark analyzer was used for the monitoring of the packet exchange between the IMS service platform and other entities.

5.6.1.9   Ambient Smart Signs Testbed

The *Ambient Smart Signs* test bed targets indoor office and public environments and aims at providing an environment- and person-aware system for guiding staff and visitors to their intended locations by means of a WSN [58].

The system itself consists of two functional parts; on one hand there is the intelligent and proactive *Smart Sign* application that deals with providing relevant information to each user at the right place and time and guiding them through the venue via ubiquitous displays, and on the other hand there is a meshing wireless sensor network called the Ambient Network which feeds the Smart Signs with environmental data in case of emergency which effectively making the system environment-aware. The wireless sensor network also provides location information of visitors and occupants to be used by the Smart Signs system to provide each individual or group of people with the right information only when they are in the neighbourhood.

The two functional parts (Smart Signs/Ambient Network) interact at two levels as follows [58]:

- SmartTags—worn by staff and visitors—interact with SmartSigns Server to provide location information which in turn is used for proximity detection;
- WSN gateway interacts with the Smart Sign server via a graphical WSN interaction tool developed by Ambient Systems (i.e., Ambient Studio) feeding it with a multitude of sensor readings (temperature, humidity, smoke, etc.) and events such as fire and location information of e-Sense sensor points.

The *Ambient Smart Sign system* as a whole is capable of guiding users, monitoring environmental readings, and informing persons in case of emergency situations. It effectively guides staff and visitors in case of emergencies out of hazardous areas.

In addition, it informs rescue staff of the current status of the building and wounded people or those who most urgently need help. The *Ambient Network* WSN essentially consists of three types of devices. A Gateway (triangle) is the "master" over the network and provides an interface to third party devices. MicroRouters (filled circles) provide a backbone infrastructure and maintain an active link (solid line) with the Gateway (when in range) and other MicroRouters, via multihop communication. Additionally, so-called EndPoints (white circles) can be added to the networks that do not have a permanent link with the network, but only communicates using short interactions (dashed line). In this way, hundreds of nodes can extend the MicroRouter infrastructure that is limited to a total of 32 nodes. Figure 5.36 and Figure 5.37 show the layout of ambient network and multihop behavior, respectively.

An office environment consists of several floors and side buildings. Most floors contain offices and conference rooms. Employees work in their offices and use the conference rooms on a regular basis. The company has close contact with its international partners and daily hundreds of people visit the office complex.

Employees all carry their personal WSN mobile tag, which identifies them uniquely. Visitors—upon arrival—are also given such a tag. At each key-point on each floor (e.g., near elevators and staircases) a Smart Sign exists that is aware of the persons in its vicinity (employees and visitors alike) and has knowledge of the person's target office or conference room. Users can request the location they would like to visit or leave messages for other people via several kiosks located at various places in the building such as the reception. A WSN is fitted throughout the office building and monitors a plethora of sensors such as temperature, humidity, smoke detectors, and so on. It also provides location information of the sensor tags (employees and visitors alike), which—combined with the proximity information from the Smart Signs—gives a clear indication of the position of a person.

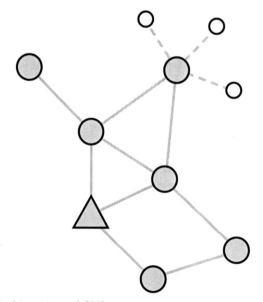

Figure 5.36   Layout Ambient Network [58].

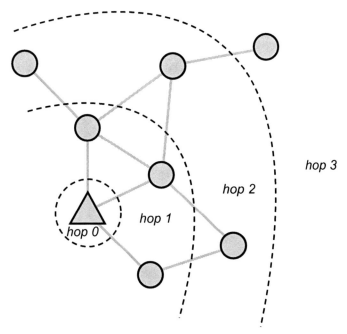

Figure 5.37 Multihop behavior Ambient Network [58].

5.6.2 Integration of Remote WSN Testbeds

The integration of remotely located existing testbeds can provide an operational and efficient way to make use of existing test beds, measurements, and experiences with different sensor platforms. In particular, the sharing of the testbeds might accomplish the implementing and testing of new protocols as well as the sharing of measurements to provide more realistic input data for further simulation studies.

In this respect, the availability of testbeds can enable some more concrete work for researchers whose current interests are mostly focused on solving challenging optimization problems that the sensor networking research field provides [63]. Several considerations regarding the integration of testbeds are related to the application area, network features, and node characteristics.

5.6.2.1 Possible Approaches for Integrations

Data Integration
A first, high-level approach to achieve the testbeds and perform distributed measurements integration could be accomplished with the fusion of high-level data collected in different testbeds and exchanged with the help of a metalanguage. As an example, the same experiment (e.g., control of the same device in domestic environment) could be repeated in two different testbeds and the results merged in a single database.

This kind of integration is based on the compatibility of application areas of the testbeds and the compatibility of the scenarios. The data generated in different testbeds should be consistent to each other [63].

Remote Web Integration

If a common user interface to different testbeds can be set up, a remote integration can be accomplished. Multiple experiments can be made in parallel on different testbeds, merging in real-time the state of the nodes and the information from the motes.

If the devices are remotely reprogrammable and the operative system is compatible, the same algorithms could be installed and verified in different testbeds. Another suggestion is to resort to the remote Java tool from DIKU, University of Copenhagen. It could be a very useful starting point for remote management and testing of sensor networks. It is being used with TinyOS and can also interface with other OSs [63].

Weak In Situ Integration

If there is the possibility of moving the testbeds, weak in situ integration means that the testbeds can be gathered in the same situation and linked using the gateway facilities. This level of integration is based on compatibility of gateway technologies, each gateway should be responsible for a different area of the field under monitoring, and results should be merged in a common database [63].

Strict In Situ Integration

If motes are compatible among each other, a single testbed could be made out of the nodes belonging to distinct testbed, to accomplish a larger testbed. As an example, study of scalability and performance of routing protocols, or authentication protocols may be more accurate if the testbed used is on a wide area. This level of integration requires compatible radio interfaces and compatible MAC [63].

5.6.2.2 Benefits of Integration

New and unexpected Internet-enabled applications and services make tremendous use of the emerging technologies and at the same time shape new requirements for future ones [64]. At a global level, the Internet is becoming more and more the backbone of the modern economy and society, to the point where new generations from industrialized countries cannot even conceive a world without the Internet. Such multiple interactions may have multifaceted and even unexpected consequences, at any technological, social, or economic level. Therefore, new proposals for Internet architectures, protocols, and services should not be limited only to paperwork, as they need early experimentation and testing in large-scale environments, even though some of these ideas might be implemented only in the long-term. The integration of test-bed sites is one strong step in this direction.

5.7 Conclusions

This chapter described the achievements of the FP6 IST projects in the area of networking and enabling technologies for ad hoc networks.

Enabling dynamic, ad hoc, and optimized resource allocation, control, and deployment, administration with accounting that ensures both a fair return-on-investment and expansion of usage, differentiated performance levels that can be accurately monitored, fault-tolerance and robustness associated with real-time trouble shooting capabilities are one's goals for the design of these architectures.

A network architecture for the support of ad hoc networks in an integrated environment should target self-organized and self-healing operations, cooperative network composition, service support, and seamless portability across multiple operator and business domains.

Migration paths and coexistence through overlay, federation, virtualization, and other techniques should be investigated to support several network and management architectures including legacy systems. Benchmarking capability of the proposed architecture(s) should be considered from the very beginning to help identify possible further improvements related to the successful deployment of any concept design. A clean slate or evolutionary approaches, or a mix of these, can be equally considered.

A conceptual framework helps to ensure that the architecture satisfies vital aspects of the future networks under investigation. The constituent concepts would then enforce the incorporation of solutions, which are in line with the societal behaviors, the current, and the emerging business models, and which also exploit new technology potentials.

The summarized notable innovations in the architectures described in this chapter are related to the seamless mobility support at different granularities, support for ubiquitous pervasiveness including dynamic service discovery and deployment, identity management, cross-domain issues of federations and integration of broadcast technologies in the system. Innovations can be detailed for each type of architecture depending on the purpose and scenario but in general, such innovations are related to how the subsystems and the specific components are arranged and how they interact in order to make a system.

References

[1] FP6 IST Projects at http://cordis.europa.eu/ist/ct/proclu/p/mob-wireless.htm.

[2] FP6 IST Projects in Broadband for All, at http://cordis.europa.eu/ist/ct/proclu/p/broadband. htm.

[3] FP6 IST Project Clusters Report, "Network and Communication Technologies," February 2008, at http://cordis.europa.eu/ist/ct/proclu/p/mob-wireless.htm.

[4] FP6 IST Projects MAGNET and MAGNET Beyond, at http://cordis.europa.eu/ist/ct/ proclu/p/mob-wireless.htm.

[5] FP6 IST Project DAIDALOS and DAIDALOS II, at http://cordis.europa.eu/ist/ct/proclu/p/ mob-wireless.htm.

[6] FP6 IST Project Ambient Networks, at http://cordis.europa.eu/ist/ct/proclu/p/mob-wireless. htm.

[7] FP6 IST Project E-SENSE, at http://cordis.europa.eu/ist/ct/proclu/p/mob-wireless.htm.

[8] FP6 IST Project MEMBRANE, at http://cordis.europa.eu/ist/ct/proclu/p/broadband.htm.

[9] FP6 IST Project FIREWORKS, at http://cordis.europa.eu/ist/ct/proclu/p/mob-wireless. htm.

[10] FP6 IST Project E2R and E2R II, at http://cordis.europa.eu/ist/ct/proclu/p/mob-wireless. htm.

[11]   FP6 IST Project WINNER and WINNER II, at http://cordis.europa.eu/ist/ct/proclu/p/mob-wireless.htm.

[12]   FP6 IST Project WINNER II, Deliverable D6.13.14, "WINNER II System Concept Description," November 2007, at http://cordis.europa.eu/ist/ct/proclu/p/mob-wireless.htm.

[13]   FP6 IST Project ORACLE, Deliverable 2.6, "Protocol Architecture and Sensing Parameters: Final Specifications," October 2008, http://cordis.europa.eu/ist/ct/proclu/p/mob-wireless.htm.

[14]   FP6 IST Project DAIDALOS II, Deliverable 122, "Global Architecture," November 2007, at http://cordis.europa.eu/ist/ct/proclu/p/mob-wireless.htm.

[15]   FP6 IST Project DAIDALOS II, Deliverable 421, "Architecture and Design: Pervasive Service Management," December 2006, http://cordis.europa.eu/ist/ct/proclu/p/mob-wireless.htm.

[16]   FP6 IST Project DAIDALOS II, Deliverable 251, "Architecture and Design: Broadcast Integration," December 2007, at http://cordis.europa.eu/ist/ct/proclu/p/mob-wireless.htm.

[17]   FP6 IST Project DAIDALOS II, Deliverable 212, "Report on Application of Key Concepts to Networks," December 2008, http://cordis.europa.eu/ist/ct/proclu/p/mob-wireless.htm.

[18]   3GPP TS 23.246, "MBMS architecture and functional description (Release 6)," March 2004, at www.3gpp.org.

[19]   FP6 IST Project WINNER II, Deliverable 3.5.1, "Relaying Concepts and supporting Actions in the Context of CGs," October 2006, at http://cordis.europa.eu/ist/ct/proclu/p/mob-wireless.htm.

[20]   FP6 IST Project, CRUISE D210.2, "Topology Control Framework and Algorithms," November 2007.

[21]   FP6 IST Project FIREWORKS, Deliverable 1D2, " Final System Requirements," September 2006, at http://cordis.europa.eu/ist/ct/proclu/p/mob-wireless.htm.

[22]   FP6 IST Project WINNER II, Deliverable 4.8.1, "WINNER II Intramode and Intermode Cooperation Schemes Definition, " June 2006, at http://cordis.europa.eu/ist/ct/proclu/p/mob-wireless.htm.

[23]   FP6 IST Project DAIDALOS II, Deliverable 124, "Final Global Architecture," December 2008, at http://cordis.europa.eu/ist/ct/proclu/p/mob-wireless.htm.

[24]   Gandham, S., R., et al., "Energy Efficient Schemes for Wireless Sensor Networks With Multiple Mobile Base Stations," in *Proc. of IEEE GLOBECOM*, 2003.

[25]   Luo, J., and J.,-P., Hubaux, "Joint Mobility and Routing for Lifetime Elongation in Wireless Sensor Networks," in Proc. Of IEEE INFOCOM, 2005.

[26]   Wang, Z., M., et al., "Exploiting Sink Mobility for Maximizing Sensor Networks Lifetime," in *Proc. Of 38th Hawaii International Conf. Sys. Sci.*, 2005.

[27]   Zheng, H., and M., Greis, "Ongoing Research on QoS Policy Control Schemes in Mobile Networks," in *Mobile Networks and Applications*, 2004, Vol. 9, pp. 235–241.

[28]   FP6 IST Project DAIDALOS II, Deliverable 311, "Concepts for Service Provisioning in Relation to Key Concepts," September 2007, at http://cordis.europa.eu/ist/ct/proclu/p/mob-wireless.htm.

[29]   Rosenberg, J., et al., "SIP: Session Initiation Protocol," RFC 3261, IETF, June 2002, at www.ietf.org.

[30]   NSIS Protocol Suite, www.ietf.org. April 2007.

[31]   Uszok A., et al., "KAoS Semantic Policy and Domain Services: Toward a Description-Logic Approach to Policy Representation, Deconfliction, and Enforcement," In *Proc. of IEEE 4th International Workshop on Policies for Distributed Systems and Networks (POLICY 2003)*, Lake Como, Italy, 2003.

[32]   Kagal, L., Finin, T., and A., Joshi, "A Policy Language for A Pervasive Computing Environment," In *Proc. Of IEEE 4th International Workshop on Policies for Distributed Systems and Networks*, June 2003.

[33]   Damianou, N., et al., "The Ponder Policy Specification Language," In *Proc. of IEEE International Workshop on Policies for Distributed Systems and Networks* (POLICY 2001), Bristol, the United Kingdom, Springer-Verlag, January 2001.

[34]   Patel-Schneider, P., F., "OWL Web Ontology Language Semantics and Abstract Syntax," W3C Recommendation 10, February 2004.

[35]   DMTF, CIM Core Model White Paper v2.4, DSP0111, August 2000.

[36]   Moore, B., et al., RFC 3060—Policy Core Information Model—Version 1 Specification, February 2001.

[37]   Moore, B., Ed., *Policy Core Information Model (PCIM) Extensions*, January 2003.

[38]   TeleManagement Forum specification, Shared Information/Data (SID) Model, GB922, Release 6.0, November 2005.

[39]   FP6 IST Project ORACLE, Deliverable 4.2, "Definition of Context Filtering Mechanisms and Policy Framework," December 2007, at http://cordis.europa.eu/ist/ct/proclu/p/mob-wireless.htm.

[40]   FP6 IST Project E-SENSE, Deliverable 7.3, "Final Activity Report," December 2007, at http://cordis.europa.eu/ist/ct/proclu/p/mob-wireless.htm.

[41]   FP6 IST Project MAGNET Beyond, Deliverable 4.1.3, "Final Secure Extended Architecture, " June 2008, at http://cordis.europa.eu/ist/ct/proclu/p/mob-wireless.htm.

[42]   FP6 IST Project, MAGNET D2.1.1, "Conceptual Secure PN Architecture," January 2005, http://cordis.europa.eu/ist/ct/proclu/p/mob-wireless.htm.

[43]   FP6 IST Project MAGNET Beyond, Deliverable 2.3.1, "Specification of PN Networking and Security Components," December 2006, at http://cordis.europa.eu/ist/ct/proclu/p/mob-wireless.htm.

[44]   Sanchez, L., et al., "Enabling Secure Communications over Heterogeneous Air Interfaces: Building Private Personal Area Networks," in *Proc. of Wireless Personal Multimedia Communications*, Aalborg, Denmark, September 2005, pp. 1963–1967.

[45]   Prasad, R., and A., Mihovska, *New Horizons in Mobile Communications: Networks, Services and Applications*, Vol. 2, Norwood, MA: Artech House 2009.

[46]   FP6 IST Project MAGNET Beyond, Deliverable 2.2.1, " PN Specifications and Interworking," February 2008, at http://cordis.europa.eu/ist/ct/proclu/p/mob-wireless.htm.

[47]   FP6 IST Project, CRUISE D010.4, "CRUISE Final Activity Report," February 2007.

[48]   FP6 IST Project, CRUISE D210.1, "Sensor Network Architecture Concept," November 2006.

[49]   FP6 IST Project, CRUISE M210.2, "Main Research Challenges Related to Topology Control Framework and Algorithms," August 2006.

[50]   FP6 IST Project, E-SENSE D4.1.1, "Distributed Services Concept," December 2006.

[51]   FP6 IST Project, e-SENSE Deliverable D1.1.2.

[52]   FP6 IST Project, e-SENSE Deliverable D1.2.1.

[53]   FP6 IST Project, e-SENSE Deliverable D1.1.1.

[54]   FP6 IST Project ORACLE, Deliverable 4.3, "Deliverable: OR Decision Making Engine Definition, " http://cordis.europa.eu/ist/ct/proclu/p/mob-wireless.htm.

[55]   FP6 IST Project MOTIVE, at http://cordis.europa.eu/ist/ct/proclu/p/mob-wireless.htm.

[56]   FP6 IST Project, MOTIVE D2.2, "MOTIVE System Architecture," August 2006, at http://cordis.europa.eu/ist/ct/proclu/p/mob-wireless.htm.

[57]   Secure User Plane Location Architecture, Draft Version 2.0 – 07 Mar 2006, Open Mobile Alliance, OMA-AD-SUPL-V2_0-20060307-D.

[58]   FP6 IST Project, E-SENSE D 5.2.2, "Test Bed-Upgrades," December 2007, http://cordis.europa.eu/ist/ct/proclu/p/mob-wireless.htm.

[59]   Yazicioglu, R., F., et al., "A 60μW 60 nV/$\sqrt{\text{Hz}}$ Readout Front-End for Portable Biopotential Acquisition Systems," in *Proc. Of IEEE International Solid-State Circuit Conference*, February 2006, San Francisco, CA.

[60] FreeRTOS, "A portable open source mini Real Time Kernel," http://www.freertos.org/, last accessed August 2007.

[61] "IPv6 over Low power WPAN (6LowPAN) working group," www.ietf.org/html.charters/6lowpan-charter.html, last access on 31.8.2007.

[62] Wu, J., et al., "A New Reliable Routing Method Based on Probabilistic Forwarding in Wireless Sensor Network," in *Proc. of the Fifth International Conference on Computer and Information Technology*, Washington, DC, September 2005.

[63] FP6 IST Project, CRUISE D 122.1, "Report on Existing Test Beds and Platforms," September 2006, http://cordis.europa.eu/ist/ct/proclu/p/mob-wireless.htm.

[64] European Commission, DG INFSO, "An Overview of the European FIRE Initiative and Its Projects," September 2008, at www.europa.eu.

CHAPTER 6

Feasibility Studies for Novel Network Concepts

Business models and standardization compatibility and interoperability are two important aspects that can ensure that any designed concept can be successfully deployed. This chapter describes business models and standardization efforts in the area of ad hoc network concepts developed in support of the designs proposed by the Framework Program Six (FP6) Information Society and Technologies (IST) projects [1, 2].

This chapter is organized as follows. Section 6.1 gives a short introduction and describes a business feasibility analysis for the concept of ambient networks developed within the frames of the FP6 IST project Ambient Networks [3]. Section 6.2 describes a feasibility study for relay-based networks, which was a contribution of the FP6 IST project WINNER and WINNER II [4]. Section 6.3 analyses the business aspects of a successful wireless sensor network (WSN) deployment performed within the scope of the FP6 IST project E-SENSE [5]. Section 6.4 describes the business analysis for the proposed within the FP6 IST project MAGNET Beyond concept of the personal networks (PNs) and PN-Federations (PN-Fs) [6].

6.1 Introduction

The main issues in ad hoc networks important for the successful deployment are how to ensure personalization of the networks and services [7]. Business issues can relate to the following:

- Networks, applications and services;
- Top-down versus bottom-up approach;
- Substitution and complementation;
- Users and buyers.

In a business consideration, an important differentiation should be made between the services and network solutions, which are operator centric and solutions, which are end-user centric. Users can have access to personalized applications and services on a basis, where they are serviced by the network and the service operators, or the users can set up the individualized applications themselves.

Customization means receiving a tailored product or a service; *personalization* means receiving a tailored experience; *individualization* means that a person builds his/her identity through choices or through comparing a personalized experience with an outside one. The market potentials for the different kinds of market players will heavily depend on the configuration of the systems used for either customized, personalized, or individualized applications or services.

In this connection, it is important to emphasize the complex character of the value networks or organizations involved in setting up personalized ad hoc networks and services. The market potentials for these different players will very much depend on how these value networks are constructed (i.e., the business models developed).

The market potentials of the different scenario cases would differ greatly. [7] showed that the market potentials for the network operators and service providers are far greater in one case than in another case. The reason is that the data rates necessary for most applications and services in the second case are relatively low, the data rates in the first one case can be high. Therefore, the network and service operators see better business cases in ad hoc network and services for high data rates than for low data rate scenarios. The opposite, however, is the case for equipment manufacturers, especially end-user device manufacturers. The simple reason is the sheer number of users in the two cases. For example, while an individual personal network (PN) for high data rate will generate much traffic, the number of users can stay relatively limited. The number of users for a network supporting a medical scenario, on the other hand, would be relatively large and, unfortunately, rising. There will, therefore, be a higher demand for a great number of devices for e-health users.

6.1.1   Business Feasibility Analysis

[8] analyzed the business feasibility of an ambient network concept. For the analysis, different types of viabilities as well as the perspectives of different types of actors were used. A two-dimensional analysis approach with both business opportunities and the perspectives of different actors in specific business situations would include the following "viabilities" [8]:

- Technical viability and management viability;
- The demand/supply aspects of the market model viability;
- The cost and price aspects of the financial viability.

The value network aspects of the market model should be described for selected cases with a focus on strategy, business drivers, business roles, and value nets. In the selected here case for ambient networks, the main concept is the role of the network composition and the designed functional entities.

6.1.1.1   Analysis Perspectives

Provided that the AN technology [3] will work, that the AN concepts are technically feasible, and that providers are willing to cooperate, a number of business opportunities can be identified as follows:

- The ability to "get access" to a larger number of potential customers;
- The ability to allow "own customers" to get access to better service quality, coverage, and capacity;

- To deploy and operate high capacity and high data rate networks more cost efficiently;
- To reduce the entry barriers due to the lower need to invest in the own networks;
- To extend the current business (e.g., the possibility for local actors to form or join more regional or national level alliances for access and value added services);
- To use the company competence to take on new business roles:
  - For a credit card company to be an ID and payment provider;
  - For an operator to be an ID an payment provider or access aggregator;
  - For a "yellow-page" company to be a broker;
  - For a facility manager to be a local operator.
- To re-use an established business relation (e.g., for an operator to act as an:
  - ID and payment provider;
  - Access aggregator;
  - Network manager.
- To use the AN to exploit the company assets, service offers, or valuable locations.

From the *end user* perspective, usability should be discussed in terms of the need for user interaction and the complexity in the establishment and the management of business agreements. This deals with the requirements of low complexity for the user, the usability, and the ready to use services as per the *market model viability demand side* [7]. From the *provider* perspective, the consumption of the network resources, the impact of the additional signaling load, and the consequences of a potentially large number of business relations and agreements should be considered.

End Users
Usability in terms of the need for user interaction and the complexity in the establishment and management of business agreements are critical elements in the viability of ambient networks.

The flexibility of the ambient network technology is a vehicle to manage the inherent spread within a mass market when it comes to handle the user interface complexity. The composition functionality can offer anything from automatic transparent multiple accesses to a complete freedom of the dynamic individual choice of services. Traditional product/service packaging methods, such as differentiation and bundling, are applicable to mitigate the complexity. New facilities such as *automated software agents* add to these means of assistance. Many users may like to have a single point of contact regarding their subscription and customer relation handling and not be forced to discuss with several providers. This will demand a kind of a broker functionality to abstract the user's relation from the connectivity and service provisioning if the user is using several operator networks and access services from several service providers [9].

The evaluation of user interfaces and the appraisal of presentations of the new mobile services are generally quite challenging [8]. The speed of penetration of new services should not remain well below the operator expectations, while sales of advanced mobile terminals develop well. There is no indication that the complexity of ambient networks should be more complex than the usage of traditional mobile terminals and services. The main challenge, however, is to create a mass market.

The user interface is more or less controlled by the *terminal* suppliers. It is not only that the supply of terminals remains a problem for all new technologies. Critical usability dimensions should be carefully considered from the point of you of the individual user, yet remain marketable.

In the case of ANs the built in flexibility of the designed technology makes it possible to address the multiplicity of the user competence to handle the complexity. The promises for low cost, low price wireless broadband should compensate for any difficulties in the adoption compared to high price or complex offerings [8].

Providers

The providers face challenges such as the consumption of networks resources, the impact of additional signaling load, and the consequences of a potentially large number of business relations and agreements.

[10] points at indications that the mobile voice traffic is quite concentrated on indoor locations. Typically around 60% of all mobile voice traffic seems to be initiated or terminated in stationary locations at the home or at work, around 15%—in public indoor locations, and only 25% is truly a wide area mobile usage. This spatial distribution of the traffic puts special demands on the voice-oriented mobile networks, because the indoor coverage is important. The future spatial distribution of mobile broadband traffic may be similar or maybe with less emphasis on the home/work where fixed-line broadband access may take a larger share than fixed-line voice. Indoor coverage will however remain an issue.

Sources [11–13] conclude that the network costs could be reduced significantly by complementing wide area macro architectures with already a moderate number of complimentary base stations. One critical observation is that the indoor picocells or the IEEE802.11a significantly relieves the load on wide area macro cells. This result highlights the potential in collaboration between the wide area operators and the local area access providers. The complexity of the traditional network and other management issues in the combined operation of wide area and local area networks is demanding. A collaboration based on the AN technology is a competitive alternative [3].

The signaling load is another potential cost driver of communication networks. The simulations performed in [8] for AN show that the resulting signaling load is very small compared to the amount of data transferred, between 0.1 and 0.7% depending on the number of networks. Hence, it can be concluded that the additional signaling load introduced by business-related signaling (even in extreme cases) is very small to the application session data.

One key concept in ANs is the network composition. Composition demands a *dynamic and uniform* framework that allows the heterogeneous networks to cooperate automatically. This cooperation includes both business and technical aspects and each relation can be described by a composition agreement between the networks and /or business entities. Many different types of cooperation are supported (e.g., network attachment of user devices, load sharing, and joint control of large networks, roaming into "visited" networks and also configuration of devices into *personal area networks* (PANs). In addition, the dynamic roaming feature supports situations where the user or the home operator does not have any previous agreement or relation with the operator of the visited network and, therefore, require that an agreement is established on the fly before the user is able to connect. Such open composition makes for a large number of potential business relations and agreements [9].

The maintenance of a wide cooperation network would bring its own implications. Governance issues including business agreements and information model as well as compensation aspects have to be considered. Management of risk, ability to measure business processes and money flow between business interfaces are thus also included.

The value net structure with more or less open composition introduces new challenges and complexity as well as opportunities compared to present mobile markets. Monitoring and control of service delivery is decentralized as well as quality of service (QoS) commitments to the end user. New types of business information have to be produced, exchanged and maintained concerning advertisements (offers) from providers to end users, agreements between providers and end users, advertisements (offers) between providers and agreements between providers. Charging in the heterogeneous network is a challenging task to be solved from information, control, and monitoring point of view. As AN provides the ability to act in real-time negotiations and cooperation the actors also need real-time information about their context in order to take the best decisions. The governance aspects can be expressed as transaction costs issues (i.e., search and information costs, bargaining costs, and policing and enforcement costs that are involved when activities are organized outside the organizational boundaries of one firm and therefore involve exchanges between firms). In the case, when the transaction costs are high, the activities can be organized in a hierarchy. This is similar to the market situation, in which the network operators and device manufacturers have dominant positions over the other players in the net. In the case, when they are low, the transactions will be organized outside the firm, and the control is not any more in one hand. This means that the organizations form more loose types of contracts in which they state what they will and will not do [16].

The *willingness to pay* can be expressed as a budget restraint on the user's total communications service portfolio. The fixed-line circuit-switched connection, the fixed-line broadband connection, mobile circuit-switched voice, mobile broadband as well as a number of value added services (VAS) are to some extent complementary and occupy the same user budget [8].

Significant changes of spending patterns are most often related to a large price cuts, new service with much lower price or a new service with significant inherent advantages. Such changes are largely independent of inferior quality of the new service in demand. Mobile circuit-switched voice has obviously inherent advantages over fixed-line circuit switched voice as it gains markets share of the total voice volume quite rapidly in spite of higher price and lower quality. Pricing of international voice over IP also obviously compensates well for lower quality.

Mobile broadband competes with fixed-line broadband as well as complements it. Inherent advantage of mobility in nonvoice services remains to be proven. One indicator may however be the fact that the clear majority of computers sold are laptops already today. Users obviously perceive at least nomadic usage away from the fixed-line connection. Terminal user interface developments like the I-phone may add to demand for mobility [1].

Contemporary pricing of mobile broadband seems to have initiated change, in particular offerings of unlimited usage nights, weekends, and holidays for small amounts or even free. Declining the prices of mobile voice will make room for mobile broadband in the users' communication budgets.

It was concluded in [8] that the successful provision of VAS, affordable prices, and quality differentiation possibilities depend on the significant increase of the network capacity, while keeping the price per unit amount virtually unchanged.

The literature on cost for wireless distribution networks concludes that affordable mass-market high throughput wireless broadband is not possible with a traditional macrocellular architecture [8]. The end user connection cost is almost directly proportional to the bandwidth provided. However, systems providing only a partial coverage show a much more favorable cost structure. In order to achieve a reasonable transmission cost for the wireless multimedia services, hot spot wideband schemes seem to be the only way to go [14]. For some hierarchical architectures, with wideband hot spot service supplemented by wide area, low data rate coverage, there exist some scenarios where amore favorable cost structure more suited for the wideband multimedia applications can be achieved. The implication of the results is to use multimedia terminals and services tolerant to variable data rates and communication quality not only in a transient introduction phase, but also in the long-term perspective.

The literature on network cost issues is to a large extent about the cost reductions from an increased capacity utilization. A general observation is that wireless network traffic show very strong spatial variations in traffic density [15].

If the spatial variation is low and the average traffic density exceeds the total area capacity of the first macro carrier, either a second macro carrier or additional macro base stations can be deployed. When the traffic is high and strongly varying with many relatively small peaks, then adding a carrier to the macrocellular layer or deploying macro base stations more densely is typically most efficient. Deploying a microcell in each of the many local traffic density peaks would require a large amount of microsites, and hence be costly, since each micro base station would have excess capacity and be poorly utilized. However, if the peaks in traffic density are very strong and rather few, so that only a few microcells need to be deployed then the cost for this is lower than that of extending the macro-layer.

A heterogeneous traffic density alone is thus not sufficient for motivating micro and picocell, or multiaccess solutions from an infrastructure cost perspective. There are also requirements on (1) a high overall traffic density, (2) strong variations in traffic density, and (3) special spatial correlation properties. Several studies show that the wireless network traffic exhibits strong spatial variations. That is not only among the general categories such as rural, suburban, and urban areas. Variations are also very strong at the local level with special concentrations indoors, in homes, working-places, and public premises. [11] showed that the network costs could be reduced significantly through the implementation of heterogeneous networks. [11] Network architectures targeting lower distribution costs through better matching of the deployment to the spatial traffic demand and fully exploiting the available infrastructure as well as the radio spectrum were studied in [11]. Their methodology is applied to different combinations of cellular and wireless local area systems. Combinations including either picocells or IEEE802.11a yield the lowest costs, some 50% below the costs for macro/micro architectures. Another example showed that the maximum feasible guaranteed data rates in *high speed packet access* (HSPA) systems of 1.5 Mbps (downlink) and 0.5 Mbps (uplink) can be offered with almost a full-area coverage already at a moderate number of complimentary BSs.

[15] showed savings in infrastructure capacity resulting from the AN-enabled use of cooperative radio accesses both within and across the business boundaries. It was shown that in vertically integrated value chain business models, high degrees of competition, AN-enabled multi-RAT cooperation) there are large potential cost savings for all operators in suburban and urban areas, but not in rural areas. For another scenario with multiradio and multioperator cooperation there are clear potential cost savings in rural areas. Finally, in a third, disintegrated market AN scenario, for a case with a large mobile network operator renting capacity in high-user density areas from a local access provider, there are clear cost savings advantages from the operator's point of view. These cost savings arise when assuming costs associated with the AN functionality for cooperation.

The increased capacity utilization may be used to increase the capacity of a given network as well as to make it possible to build less costly network architecture. More access points offered to the end users can increase the network availability. Higher capacity utilization may, however, deteriorate the quality of the connection service in terms of throughput, bit errors, transfer delay, and so forth [8].

Collaboration Between Different Players
Collaboration between different types of players facilitates the more efficient use of other critical industry resources than the wireless network.

The cost advantages presuppose the collaboration between players in the markets for the wireless services. They can be important potential drivers of change towards the collaboration within an industry with ever declining prices of the core voice service. Continuing price erosion will represent a serious threat unless faltering voice revenues can be replaced with affordable new services for a mass market. Potentially attractive services generally require high bandwidth. Very high bandwidth may be necessary to stand up against the indirect competition with fixed line broadband services. It was shown in [14] that an important requirement for wireless wideband services is that they could not be substantially more expensive than the voice services offered. The connection revenues cannot be expected to be proportional to the bandwidth provided. The average revenue per user cannot be expected to increase other than slowly with a mass market perspective. Mobile network operators need to cut costs significantly when revenues are fairly stable and they are required to grow the network capacity significantly. They must in fact revise the network architecture in a situation with increasing the enterprise risk. The core revenues decline and new services are not safe bets. New capacity requires capital investments.

It was evaluated in [8] that the revenue pressure from declining the voice prices would stimulate a change towards a collaborative usage of the shared networks as well as specialization towards a separate network provider and service provider market players.

A collaboration opens other possibilities that may change the industry towards a more efficient market behavior. Collaboration facilitates specialization of functions that are internal to vertically integrated organizations (e.g., mobile network operators). Economics of scale and flexibility should benefit the independent providers of functions such as compensation clearing, access brokerage, and service aggregation. The markets of collaborating partners should ease barriers of entry for new players to join the existing collaborations or form their own [8].

Regulators should continue to stimulate the development of mobile markets towards increasing openness and enhanced competition. Small steps are taken continuously, such as price caps on international roaming, turning spectrum licensing technology neutral, and imposing licensing conditions, which abolish terminal restrictions and choking of competitive services. Even if the present market situation does not seem to favor immediate introduction of Ambient Networks technology, we evaluate that there is slow but steady change in that direction. Regulatory authorities have started to stimulate increased competition. Margins on mobile voice continue to decline and incumbents emphasize more and more the need to cut costs.

Walled garden approaches to value added services have not been very successful and incumbents show signs of aligning with independent providers of VAS. Mobile broadband offers are increasingly supplied with WLAN access both to an operator internal access points as well as external local operator networks [8].

6.1.2 Feasibility Conditions Specific for Ambient Networks

To make a business feasibility analysis of the AN concept the inherent properties of the concept can be taken as the basic building blocks [8]. These properties describe the main strategic goal of the concept and can be physically realized mainly by the appropriate architecture. The strategic goals are used as guidelines and/or requirements for the physical implementation. If these strategic goals are actually viable in a real market environment may however be discussed. There are several arguments stating that something needs to be done with the existing way of using communication resources. Due to an increasing need for high band width operators need to build or use their networks differently. This can be argued since the cost of building out capacity is proportional to the capacity need. New ways of using the networks and taking an advantage of different network characteristics, cooperating with other operators to share cost etc may be plausible ways to go to meet the falling margins. The AN concept can move the communication industry in this direction. There are both external and business internal aspects that can impact the actual realization of AN. This depends on issues such as the actual demand for the functionality, a potential business benefit in implementing the concept in their networks, whether the architecture fits into a smooth migration path and so forth [8].

For a successful network roll-out, the customer must be supported with VAS to create a demand for the concept. [3] Expert interviews were carried out in [3] and showed that a driver for the customer to acquire an AN capable terminal with multiradio functionality are the VAS.

For example, mobility management and dynamic roaming provide the customer with more flexibility, choice, and probably better quality. All these services are limited to the connectivity level. On top of that VAS that takes an advantage of the functionality such as availability of content should be offered for an extra revenue. Many of the strategic goals of a next generation network design are linked to the demand and that the users sees the benefit of becoming an AN customer. Directly or indirectly these properties will create added value for users.

New entrants could focus on providing ready to use or the most advanced VAS that offer new possibilities.

For small operators it might be difficult to provide VAS ready for use and those can only provide only the connection to these services. Niche players that focus on, for example, *personal* VAS could use context-awareness to enhance their specialty.

Security and privacy are two other important aspects in the design of every network. WLANs have undergone several developments for security enhancements. One of the most important issues seen from the users point of view is that they can feel confident that the bill is correct and that their communication is not abused by others. With the increased e-commerce applications (e-banking, e-trading, etc.) high security levels are a must-have feature for all communications. This is important for the wholesale communication with an agreement handling as for the retail communication to end users. The differentiation of QoS support levels is an important mechanism to segment the user market and create various types of subscriptions. By this all user groups may find a subscription that is optimal for their specific usage [8].

Trust is built in cases where something goes wrong but is solved. With many ad hoc cooperating actors it will be difficult for the end users to contact the right actor. So someone needs to take care of that but also for a business actor it can be difficult to assess what went wrong and whose fault it is.

Leading operators and their followers have an experience with the end user complaints and how to handle them. They have also build a brand and many customers trust as a well-known brand. Small actors have not such brands but can boost their advantage that they can react quicker and be closer to the customer in certain situations. New entrants might bring a brand from another sector of the industry (credit card company) but usually need to earn the trust of the customer.

The theory of dominant design was introduced in [16]. This theory argues that many components need to be in place before a technology will diffuse to the next hierarchical level and at the end be widely used and be regarded as the dominant design. At the component level it is regulation, standardization, and availability of technical components that need to be in place. For economy of scale, the technology needs to be highly standardized and both the end-user terminals and the network equipment needs to be available.

In the business context the manufacturers of handsets should ensure the evolution of mobile devices. Without a mass-market of the terminal availability new network concepts will not succeed. The terminals must be available, be capable of accessing various network types, have a long lasting battery life-time and have the ability to support the novel functionalities.

The needed network equipment must be available and the evolution costs from legacy to next generation technology must be within the acceptable limits for the various actors and strategic groups. Migration packages make the steps from legacy to next generation networks smaller and easier to take [8]. Standardization is an important prerequisite for the availability of components and is taken up in several important standardization bodies. The price of the components, the migration packages or a full system is difficult to assess and can be seen as an uncertain factor [17].

The access to and usage of heterogeneous networks belonging to different business entities result in a higher degree of complexity in terms of control of multiradio access, mobility support, and network management.

Many different types of cooperation would be supported by next generation networks: network attachment of user devices, load sharing, and joint control of large networks, roaming into "visited" networks, and also configuration of devices into PANs. In addition, the dynamic roaming feature supports situations where the user or the home operator do not have any previous agreement or relation with the operator of the visited network and therefore require that an agreement is established "on the fly" before the user is able to connect.

[3] developed a composition framework that provides a set of "tools" for communication and cooperation between networks and between user devices and networks. Networks and operators can advertise service and access offers to end users. The user devices have a functionality for the discovery of offers, evaluation of offers, and decision making for the selection of network and access. Before a session can start the user devices (1) need to be attached to the selected network, which includes establishment of basic connectivity and security, and (2) terms and conditions for the specific session need to be agreed which is done by negotiation of a *composition agreement* (CA) as shown in Figure 6.1.

From the end user perspective the composition framework enables the distribution and the evaluation of a multitude of offers with different terms and conditions. The network attachment enables a basic (secure) connectivity to be established for information exchange and negotiation before a user session is actually initiated.

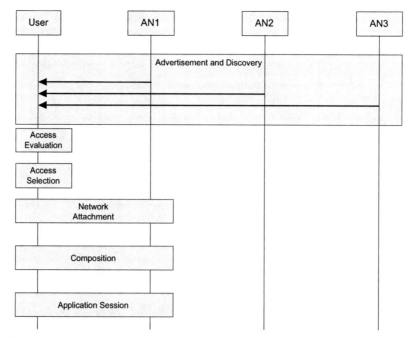

Figure 6.1 Composition steps before a user session can start connected to a provider AN.

When a negotiation is finalized, the CA is stored and used for monitoring of the ongoing session. After the session is ended, the used CA is stored and can be retrieved and reactivated. This feature is of large importance because users without a subscription can be more easily identified and trusted and hence the attachment procedure is facilitated.

It is important to note that the composition for operators supports the automatic handling of roaming (access to networks) and business relations with other (unknown) providers. For the end-users the composition process makes sure that the network attachment, the ID and the trust management and provisioning and establishment of application sessions are handled automatically without any need for a user interaction. The composition framework also includes a payment support called compensation. The AN supports both direct compensation schemes between parties as well indirect schemes where a compensation helper supports the other parties [8].

6.1.2.1 Economic and Financial Model Viability

The economic and financial viability is concerned with the simple concept of whether providers are able to obtain a sustainable profit to remain viable. This relies a lot on the cost and supply of services from the providers, and the demand of the services from the users. Large actors with large networks would have a larger overall costs compared with smaller actors. However, larger networks are capable of serving more users, and users that require mobility. This economy of scale can translate to a lower cost per user if enough demand is captured.

The pricing strategy used dictates the amount of demand experienced. The lower the price, the higher the demand tends to be, depending on the pricing elasticity of demand. The pricing set by the providers hence needs to find the optimum strategy through the correct balance between charging high prices to increase margins, but not so high as to reduce the overall demand. The pricing strategy of the providers can therefore range from charging very low prices but getting a high number of users to charging very high prices for low numbers of users.

The user's willingness to pay for a service obviously influences both the level of demand and the profitability of the service. The user's willingness to pay, however, may not be reflected in the actual cost of providing the service. For example, SMS messaging is a very profitable service in terms of $ charged per bit as the amount of network resources it uses is negligible. Conversely, applications such as high resolution video streaming is much more costly in terms of network resources, but if users do not perceive much value in it, the low willingness to pay may not make it a profitable service to provide.

The user density has a highly significant impact on feasibility. Areas with very low user densities, from a technological point of view, may be more suited for wide-area networks, which are typically deployed by large actors. Higher user densities on the other hand, are areas where demand is large enough to provide business feasibility.

6.1.2.2 Selected Business Cases

A number of selected cases and value networks were analyzed in [8], such as a large operator that strengthens the leading position, a market entrant who wants to use the existing networks, a middleman, and third parties enabling an access to any network, companies exploiting established customer relations, small players exploiting local company assets, value added service provider using connectivity.

A simulation model was used to examine the above cases involving ANs. The cases were modeled and described, along with the parameters used. Two cases were simulated: the "home operator" case where network operators are able to use third party access providers to supply the network access to their subscribers, and the "free user" case, where users do not have fixed subscriptions, and are able to select their access providers freely. These cases were chosen because they demonstrate business models that are fundamentally different from current approaches, and include, in the "free user" case, an advanced implementation of an AN functionality. At the same time, however, the cases also reflect some of the more disruptive trends that can be observed in the mobile communications market such as:

- The opening up of networks by operators where the walled garden approach is being replaced by unrestricted access to the Internet, leading to wireless network operators playing a role similar to wireline ISPs.
- Increasing competition amongst operators and the commoditization of network access.
- The focus of network operators away from access provision (e.g. through the use of RAN sharing agreements and wireless tower companies) and onto the provision of services and applications.

The simulation results obtained by [8] provided indications of the effects of the business models, such that the consequence on the feasibility can be gauged. The simulations concentrated on the effect of the pricing and the revenue, as these are key parameters involved in the feasibility of the business model. In addition to the simulation model, a calculation model called business feasibility model, was created within the project. The purpose with the business feasibility model was to create a tool to theoretically simulate and correctly calculate, if the overhead in signaling and the initial costs that AN creates can be compensated with an expected higher revenue. The outcome of the business feasibility model is the amount of revenue each actor involved in AN can expect per year and how big the signaling is in relation to the transferred data.

Home Operator Case
One of the cases possible with AN is for operators to allow its users to roam to other operators under certain conditions. The simulations were performed with the home operator using a pool of multiple access providers to supplement its network capacity. The simulation was set up with one home operator and three access providers. The 1 sq. km simulated area has a total throughput requested by all users in that area shown in Figure 6.2, with the total throughput requested at about 15 Mbps. The home operator has a wide area network that has a coverage over the whole area, while the access providers have a coverage in the more densely populated areas.

The home operator's network is deliberately set up such that the total amount of capacity it is able to offer to the users is 5 Mbps, and uses the access providers

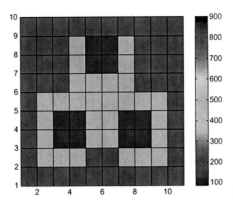

Figure 6.2   User densities showing three hotspots in a 1km × 1km area. The color bar on the right shows the colors representing the different user densities in users per 10,000 m² (i.e. a 100m × 100m area) [8].

to provide the capacity that it is not able to provide with its own network. Nearly all of the users served by the home operator's network are those located outside the hotspot areas, and the users within the hotspot areas are served mostly by the access providers' networks.

The effect of this is shown in Figure 6.3.

The home operator places requests for prices from the access provider every time a user cannot be supported by its own network. It is assumed that the service level agreements in place, between the home operator and access providers, have penalties in place such that the minimum QoS offered by the access provider to the user is guaranteed. Therefore, the criterion that the home operator uses to choose the successful access provider is the *lowest price*. Figure 6.4 shows the resulting pricing of the access providers over a period of six days.

The result shows a very fast reduction in the prices offered, due to the competition that exists between the access providers. The pricing eventually converges to $0, resulting in a low rate of revenue. This is shown in Figure 6.5.

The high rate of revenue collected in the early stages reflects the higher initial prices that are offered, but this dramatically falls off (represented by the decrease in the gradient of the curves) as the prices go down.

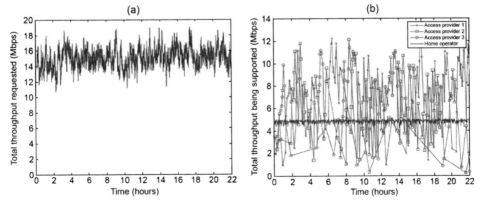

Figure 6.3   (a, b) Total throughput demand of all users in simulated area and user throughput supported by the networks of the home operator and access providers [8].

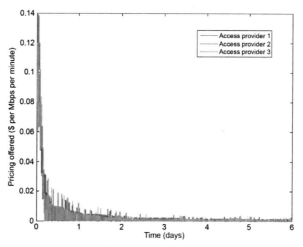

Figure 6.4 Pricing of access providers [8].

The access provider revenue numbers in Figure 6.5 are basically the amount that the home operator pays for the access providers to provide access to two thirds of the users in the simulated area. This cost is very low relative to the revenue collected by the home operator from the users themselves, as shown in Figure 6.6.

Free User Case
Another possible concerns the highly dynamic nature of ANs as one of the most distinguishing features that differentiates it from current networks. Aspects such as negotiations, agreements, compensation, and contract periods that normally tend to take a relatively long period of time can be performed in much smaller timescales using an AN. The composition procedure [16] also makes the desegregation of roles amongst multiple actors, and the interworking amongst these actors more straightforward. These factors, among many others, can have a significant impact on the business and economic aspects of the wireless communications market.

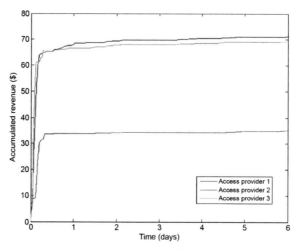

Figure 6.5 Accumulated revenue of access providers [8].

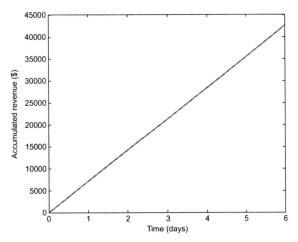

Figure 6.6   Accumulated revenue of home operator.

Consequently, the simulation case aims to examine these effects. The studied case includes a very dynamic and competitive environment, as this would pose a challenging environment, in which to determine the feasibility of an AN. The scenario is shown in Figure 6.7. The simulation case includes the following:

• Multiple access providers, each with their own deployed network;
• All access providers compete for users;
• Providers advertise their services to all users within range of its network;
• Users that do not have a long-term relationship with any one access provider;
• Users choose their provider on a per-session basis.

This case basically allows users to become "free agents" and select and change their operator on a much quicker timescale using an AN functionality. One of the

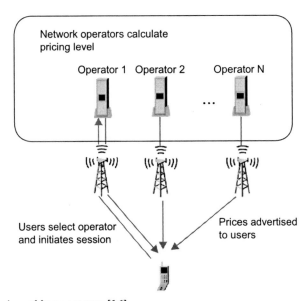

Figure 6.7   Overview of free user case [16].

main departures of this case with what is done currently is the user not having a long-term relationship with one operator. Currently, users are subscribed to a "home" operator, and are not allowed to roam freely onto other operators. Any roaming that occurs is also controlled by the home operator, and any business transactions concerned in roaming is between the operators and does not involve the users. Another difference due to the dynamic nature of ANs, is that the usage pricing is employed, with no monthly line rental and a flat-rate charging. This setup can be seen as amore long-term view of AN, compared with the home operator setup in the previous section. These network operators can be actors that are from various strategic groups, from leaders and followers such as established operators with national coverage to new entrants and small, specialized actors such as hotspot operators. The high-level sequence of messages and actions involved in setting up a connection between the user and network provider is shown in Figure 6.8.

Simulation for Competition Analysis
Competition arises when more than one operator is present in the market. Figure 6.9 shows the resulting prices when 2 and 4 operators are competing in the same case, respectively.

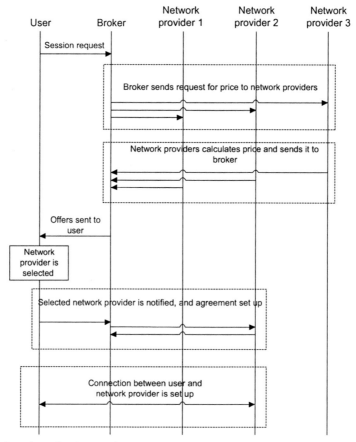

Figure 6.8   Overview of actions and messages to set up agreement to set up connection [16].

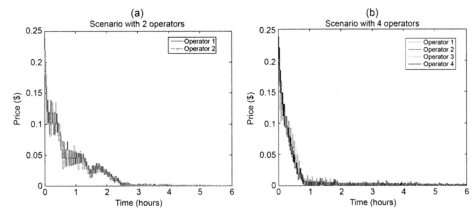

Figure 6.9   Pricing with (a) 2 and (b) 4 operators [16].

The most apparent effect that can be observed is the resulting reduction in prices, even when only 2 operators are competing. Initially, the prices charged by the operators went down quickly to match the user's willingness to pay level. However, the prices continue to fall beyond this level, before converging close to $0.

Figure 6.10 shows the accumulated revenue over time for the simulation runs.

These show how the rate of the revenue collected by all operators is decreasing as the prices start to converge, eventually settling down to a steady, but low rate. The difference in the amount of collected revenue for different operators is the result of the initial randomness in the pricing algorithms prior that the stable pricing is achieved. It is caused simply by the competition experienced by the operators, where a lower pricing results in higher revenues. The same effect is experienced for the other services offered by the operators. The implication of this is that the competition amongst operators in the dynamic environment of the AN can cause significant reductions in prices. This effect of the margin erosion is also prevalent in existing networks, although the effect is not as pronounced as the case simulated. The result can be said to be expected as the resulting case when the supply is much higher than the demand.

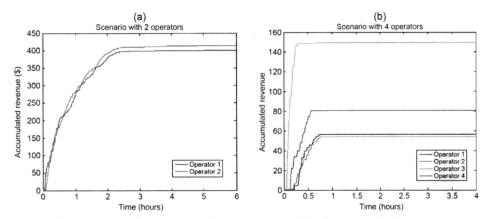

Figure 6.10   Accumulated revenue with (a) 2 operators and (b) 4 operators [16].

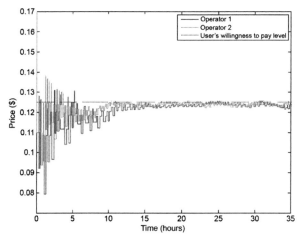

Figure 6.11   Pricing with 2 operators with network congestion [16].

The operator's networks are given some limits of the number of simultaneous sessions they can support. Therefore, it would be interesting to determine how competition is affected by the network congestion. [16] simulated the users in the 1 km^2 area, which is equivalent to approximately 89 Erlangs of voice calls, to emulate a situation with congestion. The networks serving in that area were set such that the total voice traffic capacity was lower than 89 Erlangs. This is a very simple method of configuring the networks for congestion, but serves to create the desired effect for examination. Figure 6.11 shows the resulting prices for two operators, each with 40 Erlangs of network capacity in the simulated area, giving a total of 80 Erlangs with the two networks combined.

The results clearly show that once the capacity is reached, the operators will revert to the same pricing level shown in situations where competition is absent and will begin converging on the user's willingness to pay level.

To summarize, the simulation results show that the introduction of a dynamic competitive environment can lead to relatively quick changes in pricing. The level of pricing is sensitive to the level of demand, and the amount of competition. When there is enough demand and competition exists, the access providers can experience significant margin erosion. At this point, the costs of the access provider must be lower or comparable to its competitors through the careful provisioning of network capacity, avoiding excessive over provisioning.

While the importance of lower costs is obvious, its effect is more apparent in such an environment. On the other hand, in situations where demand is high and competition is low, there is an opposite effect on the pricing, the prices can rise dramatically.

6.2   Cost Analysis for Relay Deployment

The main motivation for a relay-based deployment is to decrease the overall network cost, while still meeting the target performance (e.g., in terms of capacity density and coverage area) that has been specified for a given area. One of the main *operational expenditure* (OPEX) drivers in mobile networks is the requirement to connect the different radio access points deployed in a certain scenario to the network, and so the use of simple *relay nodes* (RNs) controlled by base stations (BSs), allows a minimum installation cost. In short, the economic benefits of RNs based on lower *capital expenditures* (CAPEX) and OPEX than BSs (wireless backhaul, lower site acquisition costs, less costly antennas, lower cost and complexity, and faster deployment) are the main motivation for the inclusion of these new nodes in multihop cellular communication systems [18]. Finally, from an operational point of view, in the case when a relay stops work properly, the users served by this relay could obtain the service through its BS, probably with less throughput, but avoiding a drop of the communication link. Some important aspects concerning the cost analysis for relay-based deployments, are the extension of the cost methodology based on iso-performance curves to the uplink direction, and the results of different kinds of simulations performed for evaluating the relay based deployment for different scenarios [18].

6.2.1   Relating the Uplink to the Downlink in the Cost Methodology

A cost methodology based on the *indifference* or the *iso-performance* curves was introduced in [19] enabling a trade-off between the number of RNs and the number of BSs in a network. Thus, the least-cost network configuration could be determined. The methodology was developed to enable comparisons between three types of access points: for example, BSs, RNs, and microcells. In [20], it was shown how alternative deployment options and spatial processing influence the shape of the indifference curve.

The methodology includes both the downlink and the uplink. Multiple services (e.g., voice and data) flowing in the same direction could also be compared in the same way [20].

Figure 6.12 shows an indifference map containing the indifference curves for the uplink and the downlink of a multihop cellular wireless network. Although the service requirements for any point on an indifference curve must be the same, they need not be the same for each curve on the map. However, in this example, the service

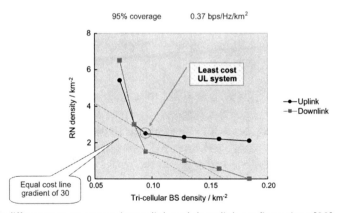

Figure 6.12   Indifference map comparing uplink and downlink configurations [20].

requirements are the same for the uplink and the downlink: a capacity density of 0.37 bps/Hz/km^2 and 95% area coverage.

The uplink behaves differently to the downlink, and the same configuration of BSs and relays only meets the requirements for both directions where the two indifference curves cross. However, this point is not necessarily the least cost configuration.

To determine the least cost configuration, which meets the requirements of both the uplink and the downlink, two equal-cost lines are drawn. For the example in Figure 6.12, a BS to relay the cost ratio of thirty is assumed, which determines the gradient of the equal-cost lines. The equal-cost line for the downlink is tangent to the corresponding indifference curve closer to the origin, indicating that the least-cost configuration for the downlink has a lower cost than for the least-cost uplink configuration. The least-cost downlink configuration does not meet the service requirements for the uplink, which requires a higher density of relays for this density of base stations.

Assuming that the same relays will serve the uplink and the downlink, the least-cost configuration for the system as a whole can be determined by whichever of the points where the equal-cost line is tangent to an indifference curve is furthest from the origin. In Figure 6.12, the limiting direction is the uplink.

In a wide area evaluation, an iso-performance curve to evaluate the different multiaccess configurations consisting of macro BSs and relays can be created. The core idea with the approach is that different deployment alternatives can provide the same service to a network area. In this section the different deployment alternatives are different mixes of macro BSs and RNs. One alternative is that only BSs are deployed to serve the traffic in the network area. Other feasible deployments, i.e. other points on the iso-performance curve, consist of less BSs and more RNs. A sequential decrease in the number of BSs will result in a sequential increase in the number of RNs. It should also be pointed out that there is a lower limit to the number of BSs that needs to be deployed in the network (i.e., the deployment cannot consist exclusively of RNs).

The starting point is a network area with uniform or non-uniform traffic density, and the next step is to sequentially deploy RAPs (e.g., macro BSs and RNs) until full coverage is reached. To evaluate whether all users are satisfied, a snapshot calculation of the resources in the network is performed. The following steps comprise the procedure:

The deployment procedure in bullets 1-5 is performed for 10 different randomly generated non-uniform traffic maps. The plotted result is a mean value of these deployments. The procedure below is described for RNs as the complementary RAP, but it is also valid for micro BSs.

1. A traffic map with size 5×5 km is created.
2. (a) An initial deployment consisting of 10 macro BSs is performed. These are unable to serve all the users in the network (only DL is considered).
   (b) RNs are deployed until the users are satisfied. The result is the leftmost point in the curve, [i.e., 10 BSs and around 42 RNs (the value is a mean)].
3. (a) Two additional BSs are added. Evaluate whether this deployment can serve the users by calculating whether the resources in the DL are sufficient; this is done for one snapshot.

(b) RNs are deployed until the users are satisfied. The result is a combined deployment of 12 BSs and 37 RNs.
4. Same as 3 (a) and (b). Another point in the ISO-curve is generated.
5. Continue until the users are satisfied with only BSs. In this scenario, this is on average achieved by 28 BSs.

For every macro BS added to the initial deployment, the number of RNs or micro BSs required to achieve full coverage is naturally lowered. The deployment continues until the network is covered by macro BSs only. The outcome is shown in Figure 6.13.

The result is a line with a constant slope for both the RNs and the micro BSs. This result means that the economical trade-off between adding RNs or micro BSs compared to macro BSs, is independent of the existing macro BS density. In this case, the slope is around 3.3 for the RNs, thus implying that 3.3 RNs can be traded for 1 BS, still achieving equal performance. It also implies that if the cost of 3.3 RNs is lower than the cost of 1 BS, it is economically advantageous to deploy RNs, and this is independent of the existing BS density in the network. The same reasoning can be applied for the micro BSs case, where the slope is 1.55, indicating that 1.55 micro BSs can be traded for 1 BS.

Both complementing RAPs (RNs and micro BSs) perform well compared to the macro BS. One reason for this is the heterogeneity due to the traffic distribution and fading model, making it advantageous to deploy many smaller RAPs compared to a lower number of macro BSs. Furthermore, the urban propagation has a significant attenuation factor for NLOS transmissions. The difference in the number of relays is about 50% higher than the number of micro BSs (leftmost points corresponding to 28 micro BSs and 42 RNs). This implies that the RNs are economically beneficial if the micro BSs are 50% more expensive than the RNs.

6.2.2   Simulations in Real Dense-Urban Scenarios

The purpose of these simulations is to validate the relaying concept from a cost perspective in a particular dense urban scenario (e.g., real cartography of Madrid), comparing different kind of deployments, with and without RNs, but with the same

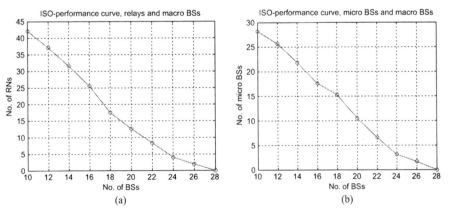

Figure 6.13   Indifference curve for RNs (a) and micro BSs (b) [20].

or similar performance from a capacity density and coverage percentage points of view. The simulations only contemplate the downlink direction. These are system level class III simulations (static or quasi-static behavior of the system) based on a 3D ray-tracing model for the estimation of the SINR over a real cartography of Madrid city. The preliminary results [19] had indicated that the total cell capacity decreased with the inclusion of RNs, but then the service area enlarged. The reason for the reduction of the capacity was the inefficient resources partitioning used, and so it was decided to implement a strategic more efficient in order to reduced the wasted resources.

These simulations are focused in the comparison of traditional and relay-based deployments from a cost analysis viewpoint, and protocol aspects were not considered, assuming a correct operation of all protocol functions, concentrating the analysis in the performance of the deployment exclusively from a radio propagation perspective.

A target user throughput of 2 Mbps in the whole of the deployment was assumed (i.e., equivalent to an *equal throughput scheduling*, which provides to each served user the same throughput). The users with better spectral efficiency (single for sectors or mixed for relays) are served until the allocated resources of the RAP are consumed.

Table 6.1 Simulation Parameters for Feasibility of Dense Urban RN Deployment [18]

Parameter	Value	Comments
Duplexing scheme and asymmetry	TDD (1:1)	Only DL
Carrier central frequency (DL)	3925 MHz and 3975 MHz	Frequency reuse of 2
Channel bandwidth	50 MHz per sector	
BS location and height	Below rooftop at 10m from the street floor	
Maximum transmit power per sector	37 dBm (5.012W)	
Number of antennas per sector and type	1 antenna K733337XD	Similar radiation pattern to the proposed in WINNER baseline assumptions
RN location and height	Below rooftop at 10m from the street floor	
Intersite distance of BSs in the same street	Around 600m in vertical streets and 700m in horizontal streets	In order to avoid either the crossroads or streets corners
Number of sectors per BS	2	
UT height	1.5m	
Elevation antenna gain for UT	0 dBi	
Receiver noise figure for UT	7 dB	
RN location and height	Below rooftop at 10m from the street floor	
Maximum transmit power per RN	30 dBm (1W)	
Number of antennas per RN and type	1 antenna with omnidirectional pattern	
Elevation antenna gain for UT	7 dBi	
Receiver noise figure for RN	5 dB	
Distance between the sector of a BS and its associated RN	Around 300m	In order to avoid either the crossroads or streets corners

A flexible resource-partitioning scheme in the RECs was used based on a fair load balance in terms of the users with best spectral efficiency (single versus mixed). An iterative process for each REC beginning with a certain partitioning (7/5 and 3/5 for single and multihop communications respectively in two consecutive MAC frames), and searching the optimum partitioning from a capacity density perspective. This process is repeated for BS-RN and RN-UT links in order to minimize the wasted resources in the first hop.

In this occasion, an active user density was assumed. The value of this parameter used in the simulations is 1600 users/km^2 (corresponding to a hot spot in a microcellular dense urban area). In order to reach this user density, the interdistance of the common grid used in the simulations as reference points, was of 25m. It is assumed that the users are fixed in the points of this grid. Table 6.1 summarizes the simulation parameters for the dense urban scenario.

Figure 6.14 shows the distribution of frequency bands and the resource partitioning used in the current simulations.

The amount of resources dedicated to the BS-UT, BS-RN and RN-UT links, respectively, in the relay-based deployment, is adjusted to the intermediate results of the simulations, so that the allocated bandwidth to the BS-RN link in the first MAC frame depends on the spectral efficiency obtained in each site (sector or RN), and of course in the BS-RN link.

The RAPs included in the simulations correspond to the micro BS and the RN with output power of 37 and 30 dBm, respectively.

From a cost-analysis perspective, it is important to note that the micro BSs used in the simulations are composed by two sectors, each one with its antenna and equipment but sharing of course the site acquisition and transmission line costs.

Table 6.2 shows that the simulated deployments, BSs only and relay-based, tried to achieve the same performance and for at least five different configurations

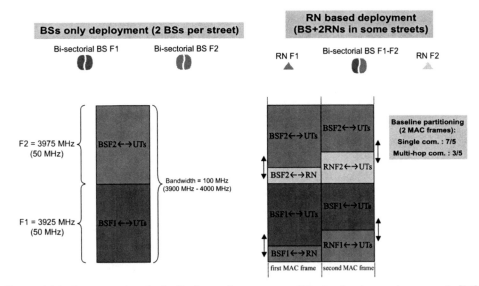

Figure 6.14   Frequency bands distribution and resources partitioning for dense urban scenario [18].

Table 6.2 Different Configurations Used in the Simulations for Relaying Concept Validation

Configuration	Number of BSs	BSs Density (km²)	Number of Sectors	Sectors Density (km²)	Number of RNs	RNs Density (km²)
1 (BSs only)	32	18.93	64	37.87	-	-
2 (relay based)	30	17.75	60	35.50	2	1.18
3 (relay based)	30	17.75	60	35.50	4	2.37
4 (relay based)	28	16.57	56	33.14	8	4.73
5 (relay based)	31	18.34	62	36.69	4	2.37

(one for BSs only and four for relay-based in order to apply the indifference curve methodology), and so to get the least cost deployment in terms of cost ratio between BS and RN. Due to the limitations of the particular scenario (area of 1300m per 1300m with irregular streets and size of blocks), the number of RAPs as well as the location were restricted to certain values so that it was impossible to obtain enough points (BSs and RNs configurations) with the same performance for delineating one iso-performance curve.

There were two deployments with the same performance, one using only BSs and the other one replacing two sectors by four relays. For the same reason, and in order to avoid the location of RAPs near to the corners, the separation between adjacent BSs was 600m in the horizontal streets and 700m in the vertical streets. For the RECs the distance between the BS and its associated RNs was around 300m.

An optimal deployment was found to be the one locating two micro bisectorial BSs per street. For example, a BSs-only deployment with three BSs per street

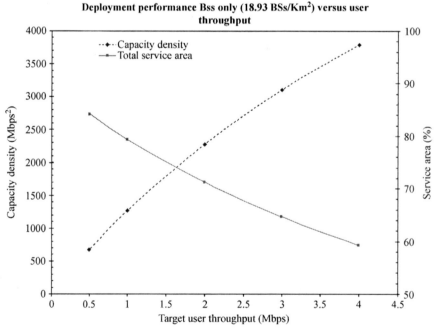

Figure 6.15 Variation of capacity density and total service area (outdoors and indoors) versus the common target user throughput for the deployment with BSs only (scenario area: 1.69 km², 8 horizontal streets per 8 vertical streets, 2 bisectorial micro BSs per street) [18].

Table **6.3**   Performance in Terms of Density Capacity and Service Area [18]

Configuration	Total number of Active Users @ 2 Mbps	Density Capacity (Mbps/km²)	Total Coverage (%)	Outdoors Coverage (%)	Indoors Coverage (%)
1 (64 sectors)	1925	2278.11	71.19	93.12	60.47
2	1850	2189.94	68.44	89.49	58.15
3	1866	2208.88	69.03	90.26	58.64
4	1824	2159.17	67.47	88.23	57.35
5 (62 sectors + 4 RNs)	1930	2284.62	71.39	93.36	60.62

obtained only an increase of 5% in the total coverage but, practically, with the same outdoors coverage, wasting besides a part of the frequency band (50 MHz) allocated to each sector.

In order to characterize the baseline deployment using only BSs, the behavior of the performance for different user throughputs was analyzed. Figure 6.15 shows this analysis, assuming a user density of 1600 users/km² (slightly upper of the typical for a hot spot in dense urban scenario), showing for a given user throughput the capacity density and service area that it is possible to obtain for such deployment.

Table 6.3 summarizes the results of the simulations highlighting the two configurations; a BSs-only and a relay-based one, which showed the same capacity density and coverage. The indoor coverage detected is very poor in all the cases due to the high frequencies used by the RAPs, whereas the coverage along the streets is above than 93% for the best deployments. The spectral efficiency of the users was approximately in the range from 0.2 to 4 bps/Hz, being lower than 1 bps/Hz for 40% of the served users (average of all RAPs included in the deployment).

Table 6.4 shows some examples of CAPEX and OPEX for different RAPs (micro base station and relay node). The relay of this table has a transmit power of 33 dBm.

The OPEX costs are represented by their net present value (assuming a lifetime of ten years and a discount rate of 6%), and in this way the CAPEX and OPEX

Table **6.4**   CAPEX and OPEX Cost Elements Example for Different RAPs

Cost Element	Unitary Cost for CAPEX or Net Present Value for OPEX (K€)	Cost Type / Comments
Micro BS Equipment	5	CAPEX
Micro BS Site Acquisition and Deployment	6	CAPEX/Small footprint
Micro Fixed Line Connection	0.05	CAPEX/Connection to mass-market ADSL line
Micro BS Site Rent, Maintenance, and Power	23.4	OPEX/no back-up batteries
Micro BS Fixed Line Connection Rent	6.24	OPEX
RN Equipment	7	CAPEX/Small footprint and not backhaul (max. transmit power of 33 dBm)
RN Site Acquisition and Deployment	4	CAPEX
RN Site Rent, Maintenance, and Power	15.6	OPEX/no back-up batteries

can be combined, for comparison purposes of different RAPs. According to this approach the total costs of a micro BS and a RN (output power of 33 dBm) are 40.79 and 26.6 K€ respectively, yielding a BS/RN cost ratio around 1.5.

It can be concluded that the inclusion of RNs in the simulated scenario would be beneficial from a network costs point of view whenever the total costs of the relay used in the simulations be lower than the half costs of a sector included in the micro BS. All kind of costs (CAPEX and OPEX) should be included and the peculiarities of the scenario analyzed (e.g., availability of transmission lines and prices, in order to make a proper evaluation for deciding the best economical option).

6.3 Business Models for WSN Technologies and Applications

This Section analyzes the environment surrounding the mobile operators when WSN are integrated with the normal cellular infrastructure [21].

Figure 6.16 shows a business model at the center of the business organization, strategy, and *information communications technology* (ICT). All are influenced by the social, legal, competitive, customer, and technological surroundings.

To establish the basis of the business model the following essential technological issues related to WSNs are highlighted below:

- Energy consumption of wireless sensor nodes. The constraint of the battery consumption of the nodes will drive the maintenance cost of the end user network.
- The existence of alternative technologies and the capacity to assure service level will drive the use of different technologies for different business segments like business or residential use.
- The size of the network around a 3G gateway will drive the profitability of the service.

Although mobile services have become part of every day life [21], the possible disadvantages of WSN technologies and applications include the reliability of the

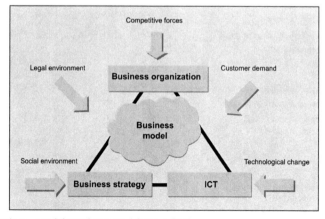

Figure 6.16 Business models and external factors [21].

technology and the privacy. RFID, personal body networks, and home networks add to the sensitive information that could be captured by hackers. This is a genuine fear for many potential customers. Other citizens object to the access of this information by governments and corporations.

The sensitive information that can be carried by WSN technology and applications implies privacy laws to protect the confidential information of users. The implications of the protection of personal data to the development of WSN services and markets was analyzed in [21].

One of the largest influences on the mobile communications sector is liberalization. The telecommunications industries of the world were privatized during the 1980s and 1990s following the lead of the United States and later the United Kingdom. This legal situation is what permits the competition between operators in America, Europe, and Asia. Deregulation is considered more conservative in most of Europe compared to the United States. A type of a law that affects the telecommunications sector is the typical regulation restricting the amount an operator with existing infrastructure may charge an operator with none to use the infrastructure.

WSN technology may be deployed in healthcare applications. A very high standard of quality will be required to ensure the safety of the public. Governments will doubtlessly regulate products and service related to wireless healthcare applications.

The sensitive information that can be carried by WSNs may cause regulation because of privacy concerns. Telecommunications companies are already demanded by law to protect the private information of citizens.

Some of the slowing forces behind a competitive market are the bargaining power of customers, the bargaining power of suppliers, the threat of new entrants, and the threat of substitutes; influence the fifth factor, competitive rivalry. This is shown in Figure 6.17.

Mobile services for WSNs will have a wide range of customers. The market segmentation section will separate buyers into thirteen groups: homeowners, athletes, sports

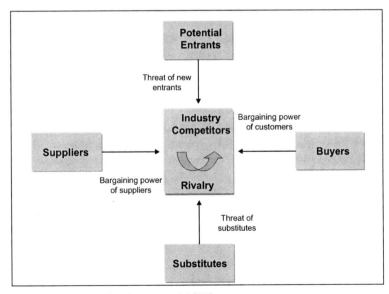

Figure 6.17 Michael Porter's five forces model [21].

fans, consumer healthcare, weight watchers, health-conscious, retailers, entrepreneurs, small business shipping companies, customers, industry, agriculture, healthcare industry, and environment. A number of factors influence the bargaining power of customers. In the case of a segment with high power, the customer will have greater influence over the prices and profitability of the services provided to them.

The number of consumers in each segment affects that group's bargaining power. For groups with a low number of customers, operators as suppliers are more dependent on each customer and customers in that segment have higher power.

The consumer segments—homeowners, athletes, sports fans, consumer healthcare patients, weight watchers, and the health-conscious—are characterized by a high number of buyers. No individual customer has significant influence. These groups are discussed in detail in the market segmentation section.

The commercial, industrial, and public groups are sufficiently large so that no one customer has a special sway over operators. The most notable exception may be retailers. Retailers in European countries are few in number but very widespread. Some, like Walmart and Carrefour operate in multiple countries. These retailers have high economic power.

High volumes also lend economic power to customers. Customers in consumer segments are likely to use e-Sense technologies and applications on a regular basis.

Retailers and hospitals will generate high amounts of traffic from daily use. The industrial segment transmits a large amount of information because of the number of nodes required to monitor large factories. Surveys of targeted manufacturers indicate a preferred number of wireless sensor nodes anywhere between 10 and 500.

Backward integration is a type of vertical integration whereby a customer acquires its supplier. The option of backward integration often serves to increase customer buying power. However, operators face a very low probability of combining with their customers.

In cases where there is a large number of small customers, backward integration is unlikely. Operators' multibillion-euro value makes the company too large for most customers to consider buying and therefore backward integration unlikely. Diversifying into unrelated industries is unpopular among modern finance specialists. The market segments described in this report have little, if any, involvement with telecommunications. It is unlikely that an operator's customers would acquire the company ignoring a conflict of core competencies and an absence of synergies.

Depending on the national market, there are a number of substitutes for each mobile operator. Prohibitive factors for nonwireless networks include the amount of cable needed and the process of installation. The absence of substitutes for WSNs reduces the bargaining power of customers while the number of operators raises it.

Differentiation of Telco operators' products will make their customers less likely to switch to a competitor. If the SBS of all companies offer the same utility and consumers perceive no difference between products, they will not hesitate to switch to a different operator in favor of lower costs. Brand equity differentiates the company in consumers' minds. The range of SBS that an operator offers will also serve to differentiate its bundle of services. A thorough and consumer-oriented segmentation will differentiate operators' services by adding value.

Switching costs are those costs that a customer incurs to switch to a different provider. They are a deterrent for changing suppliers, reducing the bargaining power of the customers. A common switching cost is the time spent searching for and evaluating new options. Mobile customers have lower switching costs than fixed-line customers because there is no permanent infrastructure for the connection to the operator.

Customers choosing a new mobile operator may need to adapt themselves and their acquaintances to a new telephone number. They may also have to adapt to the new interface of the phone and its software. The user may suffer wait times while new equipment is installed and be required to complete paperwork. Any feature of the old operator, for example a SBS, not provided by the new one is a cost to be considered.

Users of WSNs technologies and applications will also face switching costs. Similar to those of mobile customers, they may include search costs; wait times, the loss of valuable features, and new fixed costs for hardware. It may take time to learn the software to operate the network. Although switching costs exist, they do not preclude customers from switching operators. This is a source of power for customers.

Operators face some difficulty because of the bargaining power of customers. Because of the number of substitutes, customers are presented with choice: a market willing to cater to their needs. However, with a good market segmentation and a customer orientation, operators can turn the power of customers into an advantage.

The bargaining power of suppliers is relatively low. However, the threat of new entrants is high, raising the degree of competitive rivalry in the marketplace. To improve its place in the competitive environment, there are a number of strategies that telco operators could pursue.

Operators might choose to compete on nonprice issues. This would avoid driving down margins for the industry and being at risk of being beaten by low-cost new entrants. One nonprice issue is customer service. The quality and novelty of applications provided will allow operators to differentiate themselves from their competition. The companies can use their brand equity as leverage if it is strong.

Although brand equity can be an asset, it can also cause problems. Operators need to avoid being complacent because of a strong position. They need to maintain a commitment to customer service and innovation; otherwise they will eventually lose their place in the market to inventive new entrants.

Even though price competition does not seem like an appropriate strategy, prices, for services derived from wireless sensor networks especially those for segments characterized by high competitive rivalry like the retail industry, should be designed to achieve competitive parity. High brand equity may exist, but does not justify high prices. Expensive services will probably drive away customers considering the high level of competitive rivalry.

To conclude, in order to deal with the elevated amount of competition that exists, operators should avoid price competition in favour of high quality services, innovation, superior partners, and consumer-oriented segmentation.

6.3.1   The Value Network

The value network for WSNs is the group of actors that adds value to the process of conveying information from sensor networks to end users [21]. It is essentially the channel of distribution for sensor-based mobile services.

Value networks are becoming more complex over time. The original value network (which could be considered a linear value chain) for mobile services consisted simply of the end users and an operator who supplied all of the services. The entrance of new types of service providers has complicated the model [22].

A model for WSN value network is shown in Figure 6.18.

The shape and components of the value network will vary for each market segment. The model shown in Figure 6.18 encompasses all possibilities for the purpose of discussion.

The end-to-end value chain is composed by the following groups of players:

- Products providers:
  - WSN nodes manufacturers;
  - Hardware integrators;
  - Software providers;
- Services providers:
  - Network transmission provider (UMTS operator);
  - Content provider;
  - End service provider.

Figure 6.18 represents also the operational flows for the generic services. The interrelated nature of the members is what makes the model more of a network than a chain.

6.3.2 WSNs and the User

The sensor nodes, whose type depends on the intended use, transmit information to a special node called the gateway.

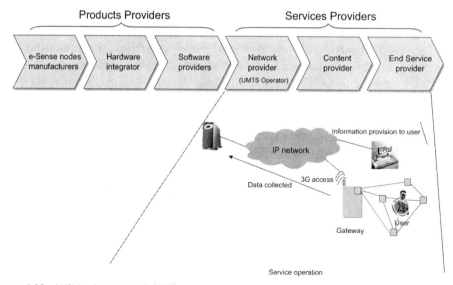

Figure 6.18   WSN value network [21].

The gateway may have 3G or other features to transmit the data collected to a server application that process the information and provided this information to final user via mobile.

Users, or consumers, are the customers who use the services created by the value network. Users are subdivided into the market segmentations developed in earlier sections. Each member of the network strives to create value for this group. All of the activities of the other parties must be designed to meet the needs of the users. They are connected to the value network by the operator.

The operator occupies a central role in the value network. It provides the infrastructure that makes mobile services possible. Content providers, application providers, and end users are all connected to the operator. For customers, operators are the face of the e-Sense value network. They will collect money from the end users and distribute it to content providers, application providers, and manufacturers.

Traditionally, the role of the operator or network provider ends with the capabilities to transmit information. Nowadays, the operators wants to provide services and not only work as a carrier, so their role could be extend to the content provider because, in fact these kinds of services could be machine to machine services, so the network operator has competitive advantages to provide advanced services using their M2M platforms.

In this case, the mobile operator will take the carrier role and the end service provider role, so for customers, operators are the face of the WSN value network. They will collect money from the end users and distribute it to content providers, application providers, and manufacturers.

Application providers supply three main classes of software to help in the service operation:

- The network management system that will help the service provider to monitor, manage and assure the network.
- The application software that is an essential part to manage and transform the information to the user.
- The network provision to help the provision of the sensor network and to help the operators to restore the sensor network when incidents occur.

This software provides interfaces with these devices that enable users to read, request, send, and use data from a WSN [23]. According to mobile application providers are mainly small, rather unknown startups. A business model for WSN services may distinguish the concept of a WSN operator. This addition to the value network would act as an *application service provider* (ASP) making an API available for the application providers and content providers to design services and to facilitate communication between technologies of different standards.

To conclude, WSNs are a resource of great importance. The technology is possible to imitate, which means that in the long run WSN technologies and applications will not be rare or provide a sustained competitive advantage. However, the services provide value for customers and will create competitive parity if not sustained competitive advantage. There are associated resources that will provide sustained competitive advantage, such as brand equity and partnerships.

6.4   Feasibility of WPANs

In a personalization framework, attributes and attribute values link the users and the content together and form the user interface [7]. Attributes of the content are matched up with attributes of the users. The user specific attribute values are paired with content information in order to determine which content to display and how to present it at any given time.

Some key trends concerning the working life and professional individuals' environments have to be identified in order to analyze the impact of WPAN technologies on future customer needs. These trends should be evolved with respect to a derived use cases and scenarios to evince the close relevance to the WPAN concept.

6.4.1   Use Case: Service Portfolio of a Journalist

An identification of the service portfolio that can be extracted for a journalist as a user is shown in Figure 6.19. These services can be targeted from the customer domain up to the mobile business domain suitably addressing in detail the PN service business model. The corresponding value chain model is shown in Figure 6.20.

One important actor in the journalist case is the final content consumer. The identification of all the different types of journalism users could possibly unveil interesting new sources of income in the value chain. One could say that the provisioning content type may determine the target user but for the sake of our analysis,

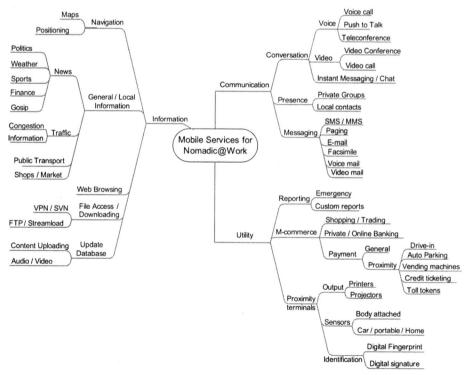

Figure 6.19   Service portfolio of a journalist [7].

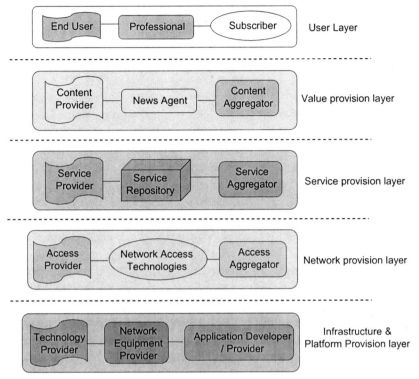

Figure 6.20   Value chain for a journalist service portfolio.

let us consider the inverted case. Therefore, following the analysis of the innovation journalism, the users can be categorized as follows:

Inventors are the people who work on new technologies, inventions, and innovations. They are interested in finding ways to the market, or stay updated in the experience and the knowledge of others. They might also try to find innovations close to their own area of interest and detailed technical information. Inventors might also be interested in finding a market for their products, and turn to media in different forms to find this market.

Business development people develop new lines of business or change the way a company is working. They constantly look for new or existing products and tools to drive that change effectively, choosing the best solution with the less possible cost.

Marketing departments include people who work with marketing of new products may be interested in other people's work and experiences in marketing new products. They will also be interested in other companies or organizations with products focusing on the same market niche.

PR agencies are public relations people who look for new companies to have as customers, and want to understand how the market is changing for their current customers. Magazines, and newspapers are also a market for the efforts of PR agencies.

Headhunters are people who are constantly trying to find talented people for specific jobs or tasks. They are interested in reading about the people behind the projects and the people with the new and groundbreaking ideas. Also human resource departments are interested in knowing who is doing what in their market.

Actors on the financial market are people connected to the financial markets are probably interested in acquiring information about innovations and their way to market.

People interested in new technology and their use, without being professionally involved in innovation or even being consumers of new technological products.

The aforementioned user types expect from mobile technology to find ways for delivering to them the content on the right time. In other words, they may be willing to pay for personalized content and services in order to perform their job in the best way.

Innovative journalists and, consequently, content/service providers or aggregators should be in position to cooperate effectively in order to produce, gather, process and deliver personalized content according to the use case, at the reasonable tariffs.

The simplest example of innovation journalism is the Weblogs or blogs. These let anyone with a web browser publish any individual content (text, picture, video, etc.) simply by uploading it to a Web server from anywhere as long as he has Internet access at the point of his presence. In this way, knowledge and information sharing is performed at once with the least cost. However, no one can guarantee the reliability of the content provided. This may not be a good journalism example case but still remains highly innovative.

A business model is described as the architecture for the product, service and information flows, including a description of the various business actors and their roles and a description of the potential benefits for the various business actors and a description of the sources of revenues [24]. The impact of ad hoc networking on the business models for different actors is analyzed in [25].

The PN devices are the central point of the PN value network. In the selected here journalist case, the local content provision via the usage of intelligent PN devices is the central point of the value chain [7].

The journalists will pay the PN *device manufacturer* (DM) for using the devices and the respective PN enabling functions. They will also pay the *WLAN site owner* (WISP) for the provision local connectivity and the ISP, mobile operator or the satellite operator for their global connectivity and the content providers (CPs) and ASPs for services demanded. There might be a local SP that offers billing services for CPs.

The infrastructure vendor wants his share of the total revenue. Part of the share comes from the DM and part of the share comes from the usage of intelligent PN devices. The DM also has to pay his subcontractors like operating systems providers for their patents and services.

The above are the set-off points for making an estimation how the mobile business model would affect the adopted use case scenario.

Figures 6.21 and 6.22 show the differences when the user represents the supply and the demand side, respectively. In Figure 6.21 the end-user is buying the content that the journalists have produced.

The local service production and consumption locks out the *mobile operator* (MO) from the value network, and its role decreases. For global communications,

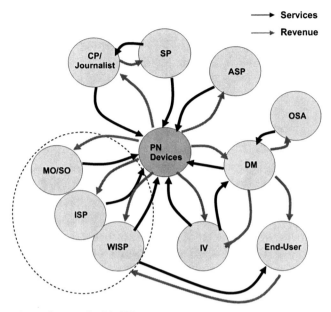

Figure 6.21   User is on the supply side [7].

however, the MOs will be important players if they could create sustainable partnerships with the ISPs, the infrastructure equipment vendors, handset manufacturers, and application developers.

The operators of big local sites (WISPs) will most likely support PN networking in their communication environment. They will probably also offer special portal services for a special event such as special offerings, programs, and result services.

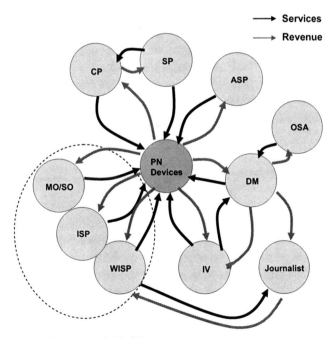

Figure 6.22   User is on the demand side [7].

In case of congestion, a *satellite operator* (SO) could be used to distribute generated content to a specially chosen ISP.

Infrastructure equipment vendors (IV) provide the mobile infrastructure for the site owner. Key activities include research and development, as they are fundamentally driving the innovation in the mobile industry. They also support set-up and operation of infrastructure. Equipment vendor skills might be valuable to assure a proper functionality and scalability for establishing the necessary PN networks, WLANs and PN Federations at a football stadium, for instance.

The DMs are influenced because the nodes of the PN and PN-Fs are differing from the existing devices, due to the need for gateways, routing facilities, and trust capabilities. Moreover, there will be new kinds of devices such as cameras, smart phones, PDAs, tablet PCs, and printers with P-PAN communication capabilities.

The PN and operating system architects (OSA) will be important resources in developing the site unique technology, and the device manufacturers will need their supporting expertise in order to create feasible solutions for the events taking place. If journalists will need special connectivity and QoS arrangements this could be arranged for.

PNs will not replace the existing mobile markets but could provide an important complement to present solutions. ASPs could discover new markets for infotainment and communication applications. Betting services could be one application being offered for the public visitors. Special information services could be arranged for the Journalists. Many services could be run without of reach for the traditional mobile operators and new pricing models are likely to emerge.

Key activities for the MO are the management of the customer interfaces (e.g., service and support as well as marketing and sales, and operation of wireless infrastructure). The MOs are fundamentally influenced by PN networks. There will be an increased demand for global communication from the arena while the need for operator carried local communication will decrease. In total the value network influence of the mobile operator is expected to decrease.

Finally the user team (e.g., journalists) could act as content providers (CP) in cooperation with WISPs and mobile operators. Other types of content providers valuable for the journalist work might be data base search engines. There might be special services provider (SP) services like enhanced security and content storage services offered for the user teams.

6.5 Conclusions

This chapter described some of the feasibility studies related to novel networking technologies proposed, researched and developed by the FP6 IST projects. The studies involved innovations from different aspects of various ad hoc networking technologies. It can be summarized that feasibility studies can provide a useful insight in the degree of acceptance of the new technologies by users and providers and it should be considered as an important add on to research advances for any technology.

A variety of factors would influence the future of any technology, therefore, it is important to identify before commencing with research, the user and usage scenarios and technical requirements. Feasibility studies, then can be conducted together with simulations investigating the performance of a given concept.

Without business possibilities, any technology might be only short-lived. Therefore, business models should be considered with any new concepts. The FP6 IST projects provided a valuable legacy of models based on thorough market analysis and technological trends, which can serve as a roadmap to the improvement and final successful deployment of the proposed concepts.

References

[1] FP6 IST Projects at http://cordis.europa.eu/ist/ct/proclu/p/mob-wireless.htm.

[2] FP6 IST Projects in Broadband for All, at http://cordis.europa.eu/ist/ct/proclu/p/broadband.htm.

[3] FP6 IST Project Ambient Networks (AN), at http://cordis.europa.eu/ist/ct/proclu/p/mob-wireless.htm.

[4] FP6 IST Project WINNER and WINNER II, at http://cordis.europa.eu/ist/ct/proclu/p/mob-wireless.htm.

[5] FP6 IST Project E-SENSE, http://cordis.europa.eu/ist/ct/proclu/p/mob-wireless.htm.

[6] FP6 IST Project MAGNET Beyond, http://cordis.europa.eu/ist/ct/proclu/p/broadband.htm.

[7] FP6 IST Project MAGNET Beyond, Deliverable 1.5.1, "Inclusion of Organizational, Customer, and Market Aspects of Business Requirements," June 2006, at http://cordis.europa.eu/ist/ct/proclu/p/mob-wireless.htm.

[8] FP6 IST Project AMBIENT NETWORKS Phase 2 D14-A.5, "Business Feasibility," 2007, http://www.ambient-networks.org/.

[9] Rietkerk, O. et al., "Business Roles Enabling Access for Anyone to Any Network and Service with Ambient Networks," Proc. Helsinki Mobility Roundtable, June 2006.

[10] Blomgren, M. et al., "Novel Access Provisioning—Final Report," http://www.wireless.kth.se/projects/NAP/publications.php, Stockholm, January 2007.

[11] Johansson., K. et al, "Modelling the Cost of Heterogeneous Wireless Access Networks," International J. Mobile Network Design and Innovation, Vol. 2, No. 1, 2007.

[12] Johansson, K. et al., "Cost Efficient Deployment of Heterogeneous Wireless Access Networks," Proceedings IEEE VTC Spring 2007.

[13] Johansson, K. et al., "Capacity Extension for Non-uniform Spatial Traffic Distributions," Proceedings IEEE PIMRC 2007.

[14] Zander, J., "On the Cost Structure of Future Wireless Networks," Proceedings IEEE VTC 1997.

[15] Prytz, M., "Infrastructure Cost benefits of Ambient Networks Multi-Radio Access," 63rd IEEE Conference in Vehicular Technology, VTC Spring 2006, Australia, May 2006.

[16] FP6 IST Project AMBIENT NETWORKS Phase 2 D8.A-3, "Business Role Models," 2006, http://www.ambient-networks.org/

[17] Cedervall, et al., "Initial Findings on Business Roles, Initial findings on business roles, relations and cost savings enabled by multi-radio access architecture in Ambient networks, Wireless World Research Forum.

[18] FP6 IST project WINNER II, Deliverable 3.5.3, "Final Assessment of Relay-Based Deployments for the WINNER System," September 2007, at http://cordis.europa.eu/ist/ct/proclu/p/mob-wireless.htm.

[19] FP6 IST project WINNER II, Deliverable 3.5.1, "Relaying Concepts and Supporting Actions in the Context of CGs," October 2006, http://cordis.europa.eu/ist/ct/proclu/p/mob-wireless.htm.

[20] FP6 IST WINNER II Deliverable 3.5.2, "Assessment of Relay Based Deployment Concepts and Detailed Description of Multi-hop Capable RAN Protocols as Input for the Concept Group work," June 2007, http://cordis.europa.eu/ist/ct/proclu/p/mob-wireless.htm.

[21] FP6 IST Project E-SENSE, Deliverable 6.4.1, "Impact on Business Models," November 2007, at http://cordis.europa.eu/ist/ct/proclu/p/mob-wireless.htm.

[22] "Telefonica Buys Cesky Telecom," Light Reading. April 2005–July 2006, available: http://www.lightreading.com/document.asp?doc_id=71900.

[23] Traptec, "Traptec," PowerPoint presentation, August 2006, available: http://www.traptec.org/traptec_demo.ppt.

[24] Timmers, P., "Business Models for Electronic Markets,"in Journal on Electronic Markets, 1998, Vol. 8, No. 2.

[25] Stanoevska-Slabeva, K., and M., Heitmann, "Impact of Mobile Ad Hoc Networks on the Mobile Value System," MCM institute University St. Gallen, 2004.

CHAPTER 7
Conclusion and Future Vision

An ad hoc network is a self-configuring network of wireless links connecting mobile nodes. These nodes may be routers and/or hosts. The mobile nodes communicate directly with each other and without the aid of access points, and therefore have no fixed infrastructure. They form an arbitrary topology, where the routers are free to move randomly and arrange themselves as required.

Each node or mobile device is equipped with a transmitter and receiver. Nodes can be purpose-specific, autonomous, and dynamic. This compares greatly with fixed wireless networks, as there is no master slave relationship that exists in a mobile ad hoc network. Nodes rely on each other to established communication, thus each node acts as a router. Therefore, in a mobile ad hoc network, a packet can travel from a source to a destination either directly, or through some set of intermediate packet forwarding nodes.

In a wireless world, dominated by Wi-Fi, architectures which mix mesh networking and ad hoc connections are the beginning of a technology revolution based on their simplicity.

This book captured important innovation in the area of ad hoc networks proposed by projects funded by the European Union in the period 2004–2008. All the chapters are based on the main contributions realized by the FP6 EU-funded projects in the area of Information Society Technology (IST) and under the themes *Mobile and Wireless Systems Beyond 3G* and *Broadband for All*. Some of the described concepts are stand-alone innovations, others were made in the context of larger system concepts targeting an integrated environment.

Chapter 1 explained the basic concepts of ad hoc networks and types on the background of FP6 EU related work. This includes, mobile ad hoc networks (MANETs), wireless sensor networks (WSNs) and wireless mesh networks. With respect to MANETs, the self-organization, mobility, security, privacy, and trust issues are very crucial and these have been covered. The requirements, protocols, and architectures and also security, privacy, and trust aspects of WSNs were presented. The design of wireless mesh backhaul networks has been explained along with the multihop concepts in next generation wide area systems. Pervasive communications and systems are the future trend in communication networks. Paradigms based on decentralized, bottom-up approaches (like bioinspired ones) are of particular interest for overcoming the limitations of more conventional top-down design approaches. Dynamic networks, networks that are mobile and can be created in an ad hoc manner, are core to the pervasive computing paradigm.

Chapter 2 described protocols and algorithms for ad hoc network and models and algorithms for routing and forwarding. It also covered multicast routing protocols, routing and forwarding in infrastructure based multihop networks including gain estimation, hybrid routing schemes, multiconstrained QoS routing,

449

centralized, and distributed routing. Protocols and algorithms for self-organizing and opportunistic networks have been deeply investigated. Multihop communications is one way to increase the coverage but it brings about the challenge of designing appropriate routing, QoS, and mobility protocols. This chapter described the role of MAC protocols and link layer techniques for sensor networks to achieve energy efficiency and the benefits of combining the design with novel enabling technologies, such as smart antennas. Various routing schemes such as hierarchical protocols, data centric networking, location aware protocols, and cross layer routing were also described. It is important to identify the limitations that come from the architectural specifics of each type of network while applying protocols. Interoperability between solutions proposed for the support of inherently different scenarios create another challenge for the optimal design of protocols and increases the complexity of algorithms. Information management and data processing is a huge challenge for wireless sensor networks. The decision making, topology construction, and self-recovering mechanisms for self-organization and topology control are essential building blocks for the proper performance of wireless sensor networks. Cross-layer optimization strategies could be extremely useful. The elements of uncertainty and autonomicity are extremely important to consider, while designing protocols and algorithms for future communication systems. The main FP6 IST project contributors to this chapter are DAIDALOS, DAIDALOS II, ADHOCSYS, WINNER, CRUISE, and E-SENSE.

Chapter 3 described issues related to the channel network capacity and moves onto explain concepts like diversity, scheduling, resource allocation, power control, and cross-layer techniques with emphasis on routing. The focus was on technologies for enhancing the performance of physical networks for different types of ad hoc networks. It was studied that in wireless networks, the network capacity enhancement can be achieved as a tradeoff with the transmission delay. Methods to enhance the performance of relay systems have been explained, which include distributed coding strategies, coded bidirectional relaying, cooperative cyclic diversity, fixed relay assisted user cooperation, joint routing, and resource partitioning. The introduction of the SDMA approach had huge impact on the PHY and MAC layer architectures, which requires a modification of the basic strategies developed for traditional (e.g., cellular) wireless networks. It was seen that scheduling can be conveniently combined with linear precoding in multiantenna systems and in this context can be studied as a specific problem of power allocation or matrix modulation design. Algorithms for enhancing the performance of wireless mesh networks were presented such as scheduling algorithms for wireless backhaul mesh networks and cross layer framework for wireless mesh networks. Capacity enhancements in wireless personal area networks were investigated which also covers the parent child communication model along with the parent child scheduling solution. The capacity was analysed though a simulation model. The use of cognitive radio and related techniques can enhance the intelligence of the nodes. The main FP6 IST project contributors to this chapter are WINNER, WINNER II, FIREWORKS, MEMBRANE, ORACLE, CRUISE, E-SENSE, PULSERS, PULSERS II, MAGNET, and MAGNET Beyond.

Chapter 4 focused on achieved innovations for context aware and secure personalized communications and described optimized and intelligent ad hoc communications, such as self-organization, autonomous decisions, and distributed protocols and algorithms along with issues related to service, context management, security, trust, control, and privacy. The main FP6 IST project contributors to this chapter are CRUISE, E-SENSE, MAGNET, MAGNET Beyond, ORACLE and RESOLUTION. Data aggregation and fusion in the context of modern ad hoc networks are essential as well as their interaction with topology control and various networking and security paradigms. Context aware service discovery frameworks include distributed service discovery for wireless sensor networks and the directory clusters. The service adaptation can be understood as a source of information for the personalization and adaptation to user behavioral patterns. Novel architectures for wireless sensor networks were presented. Lightweight security aspects had been proposed for MANETs, PNs, and WSNs as the basis for support of pervasive systems characterized by a huge number of possible scenarios. It was found that an adaptive security framework can help in dynamically managing security, privacy, and trust to always be able to apply the correct mechanisms for the current situation, and in addition, could significantly save on resources, such as battery power. The concept of virtual identity management was also introduced together with the studies on the role of the identity management and anonymity infrastructure and how it can be supported through different security frameworks. Hence, some of the key enabling techniques for providing the future user with reliable and fulfilling communication and services are self-monitoring, self-adaptation, and self-management.

Chapter 5 described FP6 innovations in the area of ad hoc network architectures, including enabling architectures for QoS, security, service discovery, and devices. A conceptual framework helps to ensure that the architecture satisfies vital aspects of the future networks under investigation. The constituent concepts would then enforce the incorporation of solutions, which are in line with the societal behaviors, the current, and the emerging business models, and which also exploit new technology potentials. Ad hoc networks are highly vulnerable to security attacks and dealing with this is one of the main challenges of developers of these networks.

An integrated target architecture in the scope of future networks should support personalized rich media networking, machine-to-machine communication, wireless sensor networks, ad hoc connectivity networks, as well as personal and body area networks. It should also be wireless-friendly, natively support mobility, be spectrum- and energy-efficient, and support future very-high-data-rate all-optical connections as well as heterogeneous wired/wireless access domains. Routing and location-independent addressing or naming, dynamic peering, signaling, resource virtualisation, and end-to-end content delivery techniques are related research issues.

Chapter 6 described the approach towards the successful adoption and deployment of a novel technology. Most of the FP6 projects performed user, business and feasibility studies in support of the scientific technological concepts. The trend of sharing content is present in the market, too. Tools for content production and sharing can make it quite easy for an average customer to become a content or even a service provider. These are aspects that need to be considered when developing new technologies.

Proliferation of information, content, products, and services provided, and owned, by individual users introduces new opportunities and challenges to operators.

Protecting the vast amount of personal information from nonauthorized access in a multioperator/provider environment proves to be a challenge and an opportunity for future operators.

Users want hassle-free connection to the Internet: a user will normally prefer a limited number of trusted relationships with selected operators and providers. They need a transparent interaction with the telecommunications infrastructure. Customers are willing to pay for hassle-free infrastructure primarily, which is capable to provide easy access. Operators must be able to make sophisticated use of information stored in the infrastructure to cater to individual user needs.

Mobile ad hoc networks are the future of wireless networks because of their practicality, versatility, simple concept, and relatively low cost.

Spectrum scarcity limits the growth and the penetration of mobile services. Ad hoc communications (e.g., opportunistic radio) is a widely discussed paradigm for mobile and wireless communications to overcome the spectrum scarcity by an optimized exploitation of the relevant dimensions space, time, and frequency.

The future will see a massive increase in the complexity and heterogeneity of the network infrastructure characterized by convergence (fixed and mobile), and support for sensors, mobility, and a variety of new highly dynamic services.

To handle this change it is necessary to design architectures and concepts for uncertainties and autonomicity. This underpins a fundamental shift in the design focus, from performance-oriented design to design for robustness and resilience. To cope with the dynamism, intrinsically present in such systems will require evolveability, defined as the ability to dynamically adapt and evolve in an unsupervised manner. New bottom-up paradigms (including bioinspired approaches) will be required to address the issues of scalability, reliability/resilience, interoperability, security, and limitations of power, mobility, and spectrum.

About the Editors

Ramjee Prasad was born in Babhnaur (Gaya), India, on July 1, 1946. He is now a Dutch citizen. He received his B.Sc. in engineering from the Bihar Institute of Technology, Sindri, India, in 1968, and his M.Sc. and Ph.D. from Birla Institute of Technology (BIT), Ranchi, India, in 1970 and 1979, respectively.

Prasad has a long path of achievement and rich experience in the academic managerial, research, and business spheres of the mobile communication areas.

He joined BIT as a senior research fellow in 1970 and became an associate professor in 1980. While with BIT, he supervised a number of research projects in the area of microwave and plasma engineering. From 1983 to 1988, he was with the University of Dar es Salaam (UDSM), Tanzania, where he became a professor of telecommunications in the Department of Electrical Engineering in 1986. At UDSM, he was responsible for the collaborative project Satellite Communications for Rural Zones with Eindhoven University of Technology, the Netherlands. From February 1988 through May 1999, he was with the Telecommunications and Traffic Control Systems Group at Delft University of Technology (DUT), where he was actively involved in the area of wireless personal and multimedia communications (WPMC). He was the founding head and program director of the Center for Wireless and Personal Communications (CWPC) of International Research Center for Telecommunications—Transmission and Radar (IRCTR).

Since June 1999, Prasad has held the chair of Wireless Information and Multimedia Communications at Aalborg University, Denmark (AAU). He was also the codirector of AAU's Center for Person Kommunikation until January 2004, when he became the founding director of the Center for TeleInfrastruktur (CTIF), established as a large multiarea research center on the premises of Aalborg University.

Prasad is a worldwide established scientist, which is evident from his many international academic, industrial, and governmental awards and distinctions, his more-than-25 published books, his numerous journal and conference publications, a sizeable amount of graduated Ph.D. students and even larger amount of graduated M.Sc. students. Under his initiative, international M.Sc. programs were started with the Birla Institute of Technology in India, the Insititute of Technology Bandung in Indonesia. Recently, cooperation was established with the Athens Information Technology (AIT) in Greece.

Under Prasad's successful leadership and extraordinary vision, CTIF currently has more than 150 scientists from different parts of the world and three CTIF branches in other countries: CTIF-Italy (inaugurated in 2006 in Rome), CTIF-India (inaugurated on December 7, 2007 in Kolkata), and CTIF-Japan (inaugurated on October 3, 2008).

Prasad was a business delegate in the Official Business Delegation led by Her Majesty The Queen of Denmark Margarethe II to South Korea in October 2007. He is a Fellow of the IEE, a Fellow of IETE, a senior member of the IEEE, and a member of NERG. He was the recipient of the Telenor Nordic Research Award (2005), the Samsung Electronics Advisor Award (2005), the Yearly Aalborg-European Achievements Award (2004), and the IEEE Communication Society Award for Achievements in the area of Personal, Wireless, and Mobile Systems and Networks (2003). Ramjee Prasad is a member of the steering, advisory, and program committees of many IEEE international conferences.

Prasad is the founding chairman of the European Centre of Excellence in Telecommunications, known as HERMES, and now is an honorary chair of HERMES. HERMES currently has ten member organizations from Europe. He is the founding cochair of The International Symposium on Wireless Personal Multimedia Communications (WPMC), which has taken place annually since 1999.

Prasad has been strongly involved in European research programs. He was involved in the FP4-ACTS project FRAMES (Future Radio Wideband Multiple Access Systems), which set up the UMTS standard, as a DUT project leader. He was a project coordinator of EU projects during FP5 (CELLO, PRODEMIS), and FP6 (MAGNET and MAGNET Beyond), and is currently involved in FP7.

He was the project leader for several international industrially funded projects with NOKIA, SAMSUNG, Ericsson Telebit, and SIEMENS, to name a few. Prasad is a technical advisor to many industrial international companies.

Prasad is the founder of the IEEE Symposium on Communications and Vehicular Technoliógy (SCVT) in Benelux. He was the chairman of SCVT in 1993.

He is the founding editor-in-chief of the *Springer International Journal on Wireless Personal Communications*. He is a member of the editorial board of other international journals and is the series editor of the Artech House Universal Personal Communications Series.

Albena Mihovska was born in Sofia, Bulgaria. She completed her B.Sc. in engineering at the Technical University of Sofia, Bulgaria, in 1990, followed by her M.Sc. in engineering at the Technical University of Delft, the Netherlands (1999). Since 1999, Mihovska has been with Aalborg University, Denmark, where she is currently an associate professor at the Center for TeleInfrastruktur (CTIF).

During her years of employment at Aalborg University, Mihovska has gained extensive experience in the administrative and technical management of EU-funded research projects. Further, she gained experience initializing industrial research cooperation as well as research cooperation funded by the EU.

She joined Aalborg University as a research engineer in July 1999 and was appointed to the European Union–funded technical management team within the FP4 ACTS project ASAP until its successful completion in 2001.

From September 2001 until April 2005, she was the project coordinator of the European Union–funded FP5 IST project PRODEMIS as a special support action instrument until its successful completion. The project was a main supporting project of the EU IST projects within the mobile and satellite area.

The outcome of the project was published as two books by Artech House in 2005, as well as in a number of technical research publications in peer-reviewed journals and conferences, an e-conference on mobile communications, a joint workshop, and a technology roadmap for the future development of mobile communications.

From January 2004 until December 2005, Mihovska was the research coordinator of the research team within the EU-funded IST FP6 project WINNER, which continued from January 2006 to December 2007 as WINNER II. The main objective of the project was the design of a new air interface that could be a competitive candidate for next-generation systems, in the scope of standardization activities within the IMT-Advanced ITU group. The project was a part of the WWI initiative and, as such, had close and required cross-issue collaboration with the rest of the WWI projects. Mihovska was part of the research teams working toward the identification of the system requirements and the design of interworking mechanisms between the newly designed system and other systems. Within the project, she proposed a concept for cooperation between different systems based on an autonomous decision framework. Based on this research idea, the theoretical approach was put forward as a development activity in the second stage of the project and was successfully demonstrated at a number of international events, including the Wireless Radio Communication (WRC) '07 Conference held in Geneva from October through November 2007. The experimental setup is now being considered for use in other projects, such as the CELTIC project WINNER+ and the FP7 project FUTON, working toward an architecture design for converging heterogeneous systems and service provisioning in which AAU is a consortium member. Mihovska was part of the research group within the project consortium who developed the final system concept requirements for the air interface.

From September 2006 to March 2008, Mihovska was the deputy technical manager of MAGNET Beyond. Therein, she contributed to the overall technical work progress and to the finalization of the MAGNET Beyond Platform system requirements. Further, she was involved in AAU-related research activities in the area of security for personal networks (PNs).

Since April 2008, Mihovska has been involved in research activities within the CELTIC project WINNER+, working toward advanced radio system technologies for IMT-Advanced systems. She is conducting research activities within the area of advanced radio resource management, cross-layer optimization, and spectrum aggregation.

The work proposed in pursuit of her Ph.D. degree from Aalborg University is a novel concept for interworking between radio resource management entities in the context of next-generation mobile communication systems. It is based on research activities commenced prior to and continued within the frames of the WINNER project. The concepts proposed within her Ph.D. thesis have been successfully implemented in the overall WINNER concept and have resulted in a number of peer-reviewed journal and conference publications, including the demonstration activities mentioned above. Further, she has a number of project-related publications, presentations, and various international and EU events.

Mihovska is a reviewer for *IEEE Communication Letters* and *The Springer Journal of Telecommunication Systems*. She has been part of the organizing and TPC committees of a number of international conferences, such as WPMC 2002, WCNC 2007, VTC 2008 Spring, the IST Mobile Summits 2002-2007, ATSMA-NAEC 2009, IEEE Mobile WiMax 2009 Symposium, and several workshops.

Index

The Artech House Universal Personal Communications Series

Ramjee Prasad, Series Editor

Towards the Wireless Information Society: Heterogeneous Networks,
Ramjee Prasad, editor

Towards the Wireless Information Society: Systems, Services, and Applications,
Ramjee Prasad, editor

Universal Wireless Personal Communications, Ramjee Prasad

WCDMA: Towards IP Mobility and Mobile Internet, Tero Ojanperä and Ramjee Prasad,
editors

Wideband CDMA for Third Generation Mobile Communications,
Tero Ojanperä and Ramjee Prasad, editors

Wireless Communications Security, Hideki Imai, Mohammad Ghulam Rahman and
Kazukuni Kobara

Wireless IP and Building the Mobile Internet, Sudhir Dixit and Ramjee Prasad, editors

WLAN Systems and Wireless IP for Next Generation Communications, Neeli Prasad and
Anand Prasad, editors

WLANs and WPANs towards 4G Wireless, Ramjee Prasad and Luis Muñoz

For further information on these and other Artech House titles, including previously
considered out-of-print books now available through our In-Print-Forever® (IPF®)
program, contact:

Artech House	Artech House
685 Canton Street	16 Sussex Street
Norwood, MA 02062	London SW1V 4RW UK
Phone: 781-769-9750	Phone: +44 (0)20 7596-8750
Fax: 781-769-6334	Fax: +44 (0)20 7630-0166
e-mail: artech@artechhouse.com	e-mail: artech-uk@artechhouse.com

Find us on the World Wide Web at: www.artechhouse.com